T0135780

Wolfram Schenck

Adaptive Internal Models for Motor Control and Visual Prediction

Logos Verlag Berlin

λογος

Bibliografische Information der Deutschen Nationalbibliothek

Die Deutsche Nationalbibliothek verzeichnet diese Publikation in der Deutschen Nationalbibliografie; detaillierte bibliografische Daten sind im Internet über http://dnb.d-nb.de abrufbar.

Dissertation, Universität Bielefeld, Technische Fakultät

ISBN 978-3-8325-1899-8

Logos Verlag Berlin GmbH
Comeniushof, Gubener Str. 47,
10243 Berlin
Tel.: +49 030 42 85 10 90
Fax: +49 030 42 85 10 92
INTERNET: http://www.logos-verlag.de

To Tanja
and to my mother

Acknowledgements

First of all, I would like to thank my supervisor Ralf Möller, who was always encouraging and supportive of my work. Our discussions were very helpful in shaping new ideas and in getting precise pointers to relevant background knowledge. At the same time, he gave me the freedom and the resources to pursue my own research goals. I also sincerely appreciate that Holk Cruse and Ralf Möller agreed to review this thesis. Furthermore, I am indebted to Jens Stoye and Sven Wachsmuth for joining my thesis committee.

I received support from my fellow doctoral students in Munich and Bielefeld as well, especially from Heiko Hoffmann. During many informal discussions in our shared office in Munich he introduced me to advanced neural network techniques and provided me with physical and mathematical insights on what happens in very high-dimensional spaces. Furthermore, the model on grasping to extrafoveal targets is a continuation of previous cooperative work with Heiko. He kindly agreed to reprint Figs. 6.7 and 6.8. In Bielefeld, I received help from Frank Röben in maintaining our server and backup services, from Tim Köhler in soldering cables for my robot setup, and from Olaf Gerstung in supervising the robot arm during the collection of learning examples for the study on extrafoveal grasping. Furthermore, I appreciate the many joined kilometers together with Birthe Babies and Lorenz Gerstmayr on our way through the long hallways of Bielefeld University from our department to the famous "cafete".

I also would like to thank my former diploma students Dennis Sinder and Silvio Große. The study on block-pushing is based on cooperative work with Dennis during his diploma thesis, the software implementation of the model is mainly his work. Christian Dax, Janos Kovats, Henryk Milewski, Jan Moringen, Josch Pauling, and Helmut Radrich contributed to the software development. Matthias Behnisch helped to rebuild the robot arm setup. Further technical assistance in setting up the robot hardware was provided by Fiorello Banci and Hans Brinkmann. Moreover, I am indebted to Angelika Deister for her organizational help.

Finally, I would like to express my gratitude to my family for their ongoing support. To my mother who endorsed and backed me throughout my life. To Eduard and Ilse Schenck for their intellectual encouragement and interest. To my wife Tanja for her active support and for her patience with someone who was glued to his desk. To all other family members, including my family in-law, for keeping their fingers crossed for me.

This work was supported by the Max Planck Insitute for Human Cognitive and Brain Sciences in Munich and by the Faculty of Technology at Bielefeld University.

Contents

Chapter 1

Introduction

1.1 Motivation

Sensorimotor coordination is at the basis of any advanced behavior. Sensory data provides an agent — a biological organism or an artificial agent like a robot — with the necessary information about the state of the environment and the agent's body, and motor commands are generated by the agent to move its actuators in the right way to fulfill its goals.[1] Although this looks like a straight input-output relationship at the first glance — sensory data in, motor data out — both theoretical reasoning and experimental evidence in the cognitive sciences have shown in recent decades that sensorimotor processing happens in an interconnected and reciprocal way. While the straight input-output relationship between sensory and motor data is still an adequate way to treat many motor control problems, other areas like perception and cognition require a different approach to sensorimotor processing, in which the clear distinction between perception and action is replaced by an integrated and dynamic view.

A study by Held and Hein (1963) illustrates the need for an integrated approach: In their experiment, pairs of kittens were harnessed to a carousel. The active member of each pair was able to walk and to move the carousel around, while the passive member was suspended to the air without contact to the ground. In this way, the passive member was subject to the same movements as the active member, but without initiating them, and both kittens received the same visual stimulation over a prolonged time period (outside the carousel the kittens grew up in complete darkness). Nevertheless, subsequent tests showed that the active member developed normal vision, while the passive member suffered from severe deficits in depth perception and paw-eye coordination. This result strongly suggests that active self-movement is necessary to develop sen-

[1] One may argue that only complex cognitive systems like humans can have goals as explicit representations of desired states; however, also lower biological organisms have at least implicit goals like survival and reproduction, which are expressed by their behavior.

sory skills. It is not sufficient for an organism to passively retrieve sensory data, instead it has to generate motor commands, to move, and to observe the resulting sensory effects. In this way, the learning of sensorimotor contingencies gives rise to perception and to an understanding of what one perceives (Noë, 2005).

The idea of integrated sensorimotor processing is closely linked to other theoretical approaches in cognitive science. "Situativity theory" (Clark, 1997; Greeno and Moore, 1993; Suchman, 1987) emphasizes the mutual interdependence between an agent and its environment in contrast to systems in classical artificial intelligence that are defined without any relation to a concrete environment. "Embodied cognition" (e.g., Brooks, 1991a; Varela et al., 1991) supposes that cognitive functions can only emerge in agents which are equipped with a real-world body, since this body enables and conditions the specific way in which the agents can perceive and understand the world. "New artificial intelligence" (e.g., Pfeifer and Scheier, 1999) incorporates these principles into areas like computer science and robotics. Psychology provides support via the "ecological approach" to visual perception (Gibson, 1979) and the "common coding approach" (Prinz, 1997), to name influential research directions. Furthermore, also the "motor theory of perception" (Berthoz, 2000; Jeannerod, 1997), which originates from the field of cognitive neuroscience, lays great emphasis on perception as an activity.

Everyday experience shows that sensorimotor coordination has to be acquired to a large extent through learning. For example, motor learning is a lifelong necessity: An infant has to learn even the most basic skills (from an adult perspective) like grasping and walking, an elderly person has to adapt to its shrinking body proportions and weakening muscles. In addition, the above-mentioned study by Held and Hein (1963) on the development of vision in kittens shows exemplary that perceptual competences are subject to learning and adaptation as well.

Considering the importance of integrated adaptive sensorimotor coordination for our understanding of motor control, perception, and cognition, this thesis explores computational models of sensorimotor processing. These models relate to hypotheses on how sensorimotor processing in biological organism may be organized at an abstract level; furthermore, these models and their specific implementations offer solutions for technical problems in the domain of adaptive robotics. For this reason, this thesis addresses technical, psychological, as well as biological aspects. The latter concentrate on the central nervous system (CNS) of higher vertebrates, especially humans and other primates. However, since modeling takes place on a rather abstract level, the suggested computational principles can be applied to a large variety of agents[2] that have to process sensory and motor data.

On the one hand, this thesis focuses on the learning of so-called internal models: "forward models", which predict the sensory consequences of the agent's own actions, and "inverse models", which act like motor controllers and generate motor commands. In this area, new strategies and algorithms for learning are suggested and tested on both simulated and real-world robot setups. In this way, this thesis contributes to the understanding of the "building blocks" of integrated sensorimotor processing. On the other hand, this thesis suggests complex models of sensorimotor coordination: In a study on the grasping to extrafoveal targets with a robot arm, it is explored how forward and inverse models may interact, and a second study addresses the question how visual perception of space may arise from the learning of sensorimotor relationships. Especially the latter study aims on providing support for the embodied approach to perception.

The author started to work on this thesis at the Max-Planck-Institute for Psychological Research (Munich) in the Cognitive Robotics Group. The main part of the research took place later on at the Computer Engineering Group of the Faculty of Technology at Bielefeld University. This location change has contributed to the dual nature of this thesis between cognitive psychology and computer engineering.

1.2 Outline and Contributions

This chapter (**Chapt. 1**) continues with a closer view on sensorimotor processing. The cognitivist approach and the embodied approach to sensorimotor processing are contrasted with each other, providing evidence from psychological and neurophysiological studies in favor of the latter. It is outlined how the application of robots fits into the embodied approach as research method which is used extensively throughout this thesis. Furthermore, internal models are defined in a formal way, and an overview of their role in models of perception and cognition is provided, with a special emphasis on anticipation and predictive forward models.

Chapter 2 presents a thorough overview of internal models in motor control. It is intended as theoretical introduction to our own studies on motor learning. Although we deal only with kinematic problems in our own work, some motor learning strategies have been developed originally for dynamical

[2] Troughout the thesis, the term "agent" is used in a very general meaning to denote both biological organisms and artificial agents like robots.

problems. For this reason, both kinematics and dynamics are covered by this overview. Moreover, trajectory planning and state estimation are addressed. As an important part of this chapter, a new learning strategy for kinematic control problems is presented which has been developed by the author ("learning by averaging"; Sect. 2.2.7). Learning by averaging has already been applied to saccade learning in the following studies:

- Schenck, W., Hoffmann, H., and Möller, R. Learning internal models for eye-hand coordination in reaching and grasping. In *Proceedings of the European Cognitive Science Conference*, pages 289–294. Erlbaum, Mahwah, NJ, 2003.

- Schenck, W. and Möller, R. Staged learning of saccadic eye movements with a robot camera head. In Bowman, H. and Labiouse, C., editors, *Connectionist Models of Cognition and Perception II*, pages 82–91, London, NJ, 2004. World Scientific.

- Hoffmann, H., Schenck, W., and Möller, R. Learning visuomotor transformations for gaze-control and grasping. *Biological Cybernetics*, 93 (2): 119–130, 2005.

- Schenck, W. and Möller, R. Learning strategies for saccade control. *Künstliche Intelligenz*, Iss. 3/06: 19–22, 2006b

Chapter 3 describes the computational methods that are used throughout the thesis. These methods comprise algorithms for the adaptation of artificial neural networks and for optimization. The presentation in this chapter concentrates on the core parts of each method without any derivations. The intention is to provide the reader with all the information that is necessary for an exact re-implementation of all algorithms.

In **Chapter 4**, a detailed comparison study of various motor learning strategies for kinematic problems is presented. To the best knowledge of the author, it is the first of its kind which directly compares the performance of "feedback error learning" (Kawato et al., 1987), "distal supervised learning" (Jordan and Rumelhart, 1992), and "direct inverse modeling" (e.g., Kuperstein, 1987) on several learning tasks from the domain of eye and arm control. Moreover, an improved version of direct inverse modeling on the basis of abstract recurrent networks (Hoffmann and Möller, 2003; Möller and Hoffmann, 2004) and learning by averaging are included in the comparison. A small subset of the comparisons in this study has already been published by Schenck and Möller (2006) (see the list of published studies for Chapt. 2).

Chapter 5 is dedicated to the learning of a visual forward model for a robot camera head. This forward model predicts the visual consequences of camera movements for all pixels of the camera image. The presented learning algorithm overcomes fundamental problems of adaptive visual prediction. The core idea of this algorithm does not only extend to a feasible technical solution, but offers also a plausible starting point for biological modeling. The first version of this algorithm (as presented in Sect. 5.1) has been described in

- Schenck, W. and Möller, R. Training and application of a visual forward model for a robot camera head. In Butz, M. V., Sigaud, O., Pezzulo, G., and Baldassarre, G., editors, *Anticipatory Behavior in Adaptive Learning Systems: From Brains to Individual and Social Behavior*, number 4520 in Lecture Notes in Artificial Intelligence, pages 153–169. Springer, Berlin, Heidelberg, New York, 2007.

The permission for the republication of this paper in a slightly modified and updated form is kindly granted by Springer.

In **Chapter 6**, a model for grasping to extrafoveal targets is presented. It is based on the premotor theory of attention (Rizzolatti et al., 1994), but adds as specific hypothesis that this grasping task involves the internal application of a visual forward model for eye movements. In this model, several of the already presented learning methods are combined and implemented on a robot arm setup. Based on this model, several grasping modes are compared; the obtained results are qualitatively congruent with the performance that can be expected from human subjects. This model has been proposed by Schenck and Möller (2007) (see the list of published studies for Chapt. 5) as potential application of the visual forward model from that study. A full publication of the model for extrafoveal grasping has been recently submitted (Schenck et al., to appear).

The study in **Chapter 7** is based on the theory that visual perception of space and shape is based on an internal simulation process which relies on forward models (Möller, 1999). This theory is tested by synthetic modeling in the task domain of block pushing with a robot arm. First, a visuokinesthetic forward model is learned which predicts the sensory consequences of small movement steps in this domain. Afterwards, an optimization process is applied to determine the right motor commands for arm control to push a block to a specified goal position; during the optimization, many movement sequences are generated as candidate solutions. The visuokinesthetic forward model is required for an internal simulation of the sensory effects of these movement sequences within the optimization process. Finally, the capability to find the right movement sequence by internal simulation is reinterpreted as a way for the perception of space. The work on this study was carried out in cooperation with Dennis Sinder, who wrote his diploma thesis (Sinder, 2006) under the author's supervision. An abbreviated version of this chapter has been published by Schenck et al. (2008):

- Schenck, W., Sinder, D., and Möller, R. Combining neural networks and optimization techniques for visuokinesthetic prediction and motor planning. In *ESANN'2008 proceedings — European Symposium on Artificial Neural Networks*, Bruges (Belgium), 2008. d-side publications.

In **Chapter 8**, the results of all studies in this thesis are summarized and discussed in relation to each other. Furthermore, an outlook of future research is given.

Appendix A specifies the robot arm setup which was used throughout the studies in this thesis (see also Fig. 1.1), and **App. B** provides a detailed description of the geometry of the used robot camera head. **Appendix C** contains detailed tables with all parameter settings of the comparison studies on motor control in Chapt. 4, while figures with detailed results of these studies are presented in **App. D**. Finally, **App. E** lists the often used notations and symbols.

Software The software for all learning algorithms (including the underlying neural networks) was developed by the author from scratch in the progamming languages C++ and Tcl/Tk except for the NGPCA algorithm (Sect. 3.4) and for "differential evolution" (DE; Sect. 3.5). For NGPCA, an implementation by Ralf Möller was used, for DE the implementation by L. Godwin (Godwin, 1998). Furthermore, software development partly relied on the following freely available libraries: ColDet (3D Collision Detection), GLUT (OpenGL Utility Toolkit), GSL (Gnu Scientific Library), ImageMagick, LAM/MPI (Local Area Multicomputer Message Passing Interface), SSL (Simple Sockets Library), and the TNT (Template Numerical Toolkit) matrix library. The control software for the robot arm is based on the PowerCube library by Amtec robotics. For the control of the pan-tilt unit of the robot camera head, we used the PanTiltRA library by the University of Tübingen. Matlab and the Netlab Toolbox (Nabney, 2002) were applied for quick prototyping and for data evaluation and visualization.

1.3 Approaches to Sensorimotor Processing

The term "sensorimotor processing" denotes information processing which relies on sensory and motor data as inputs and outputs. This kind of processing

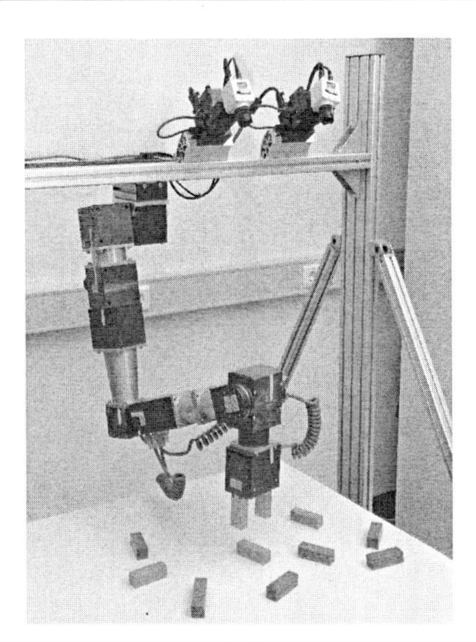

Figure 1.1 — The experimental setup which was used throughout the studies in this thesis, consisting of a robot arm with six rotatory degrees of freedom and gripper and a stereo-vision robot camera head (Fig. A.1 in App. A shows a slightly different version of this setup at a later point in time).

is required for motor control and perception, but it is also linked to cognition. Two opposing paradigms exist within cognitive science; they result in very different approaches to sensorimotor processing and to perception and cognition. The first paradigm, which we here call the "cognitivist view", relies on the assumption that perception and cognition are based on formally defined and implementation-independent processes in physical symbol systems. In the process of perception, symbolic sensory representations are created solely from the sensory input to the system. The second paradigm, called in the following "embodiment", contradicts this view. Embodiment emphasizes that perception and cognition emerge from integrated sensorimotor processing which involves "the dynamical interaction between agent and environment and their mutual specification during the course of evolution and the individual's lifetime" (Ziemke, 1999, p. 179).

In the following, we will describe both paradigms in more detail and argue that the cognitivist view has several severe shortcomings. To provide further support for the embodied view, we will present evidence for integrated sensorimotor processing from psychological and neurophysiological studies.

1.3.1 The cognitivist view

The core assumption of the cognitivist view is the "Physical Symbol System Hypothesis" (Newell and Simon, 1976).[3] This hypothesis states as necessary and sufficient condition that a physical system has to be a physical symbol system to exhibit intelligent action. In this way, cognition is defined as computation on symbolic representations (a symbol is an element that represents what it stands for; Varela et al., 1991). While the necessity for some kind of representation is widely accepted, the cognitivist view implies that the representations are actually realized in the form of a symbolic code in the brain.

Although the symbolic level is physically implemented, it is not reducible to the physical level. This distinction is quite obvious for the digital computer, which is definitely a physical symbol system: Looking closely at the hardware of such a computer would reveal transistors and capacitors at the physical level (among other electronic components), but no computer programs. In the brain, the physical level comprises the neurons and their connections (to put it simply). Furthermore, cognitivists propose a third level above the symbolic level, the semantic level (Pylyshyn, 1984) which concerns the purpose, the meaning, and the overall logic of the computation. In the formulation by Marr (1982), these three levels are the *implementational*, the *algorithmic*, and the *computational* level.

An important characteristic of the cognitivist view is the strong emphasis on representations. First of all, an agent has to build a sensory representation from the sensory data it receives from its sensors. For example, in Marr's theory of vision (Marr, 1982) the goal of vision is to create a representation of the three-dimensional world around the agent from the pattern of light detected by its visual sensors. In this process, important objects are recognized, and irrelevant information is sorted out. Afterwards, central processes use the internal representation of the world for planning or problem-solving. Finally, these computations generate motor commands which are sent to the actuators. Overall, the cognitivist approach follows a clear input-processing-output scheme in which these stages are clearly separable from each other and understandable each on their own without any reference to the agent's environment. With regard to sensorimotor processing, cognitivism in its strongest form would generally reject the idea of integrated sensorimotor processing below the symbolic level. Only

[3] The birth of cognitivism can be dated back much earlier to the year 1956, in which two important conferences took place (in Cambridge and in Dartmouth). At these conferences, the core ideas of cognitivism were formulated.

after the generation of the representation of the world, the perceived elements (e.g. a CUP) would be combined with actions (e.g., GRASP(CUP)) during the subsequent computation of action plans.

Well-known cognitive architectures from the symbol-processing approach are GPS ("general problem solver"; Newell and Simon, 1961) and STRIPS ("Stanford Research Institute Problem Solver"; Fikes and Nilsson, 1971). STRIPS generated plans for a robot, Shakey, who could move between a number of rooms, pushing boxes around and carrying out a small number of other actions. Such a symbolic planning system has to provide a formal structure for state description (e.g., the initial and the goal state of the world) and a set of applicable operators/actions. Beyond it, procedures must be available to produce a plan from the information about states and operators. For this purpose, STRIPS used the means-ends analysis, a planning technique from classical artificial intelligence. However, the operations of Shakey were restricted to its well-defined laboratory environment.

1.3.2 Limitations of the cognitivist approach

The cognitivist approach did not succeed in designing truly autonomous and "intelligent" agents[4] which can successfully act in the real physical world (and which could be used as model for intelligent behavior of biological organisms as well) (Pfeifer and Scheier, 1999). There are several reasons for this failure. To start with, symbol-processing systems are not well suited to cope with the stochastic and unpredictable nature of complex real world environments. Symbol processing requires that the sensory input from the world is reduced to a clearly defined set of different states in the sensory representation of the agent. The design of such a sensory representation can be possible for a reduced lab environment, but is infeasible for typical environments of human beings or biological organisms (these environments provide noisy, incomplete, and unreliable sensory data, they may contain unknown objects, and they may change rapidly in an unexpected way). Furthermore, even if the problem of the sensory representation was solved, action planning in the style of GPS or STRIPS becomes the more expensive (with regard to required computation time and power) the more complex the environment and its sensory representation are.

[4] Since it is far beyond the scope of this thesis, we do not want to engage in a discussion how intelligence should be defined. When we speak of an "intelligent agent", we have an agent in mind which can act autonomously in the real world in its specific ecological niche under a wide range of environmental conditions, and which has a non-trivial repertoire of actions to choose from. At the current state of research, robot engineers would be glad if their robots were as "intelligent" as mice or ants, to illustrate this point.

Thus, the symbol processing approach is not suitable for real-time action; its strength lies in the generation of "intelligent" behavior in restricted and well-defined task domains, e.g. for playing chess (as stated in Pfeifer and Scheier, 1999).

The problem of computational effort is closely linked to the frame problem (McCarthy and Hayes, 1990) which concerns the question how a model of the world can be kept in tune with a changing environment. An agent has to consider the side-effects of its own actions, but the number of potential side-effects is huge for an environment with realistic complexity. However, most of the side-effects are completely irrelevant, thus taking them into account is a pure waste of processing time. Unfortunately, determining which side-effects are relevant and which are not, costs computing power as well. To be fair, one has to state that the frame problem is intrinsic to any world-modeling approach, not only to the cognitivist one (Pfeifer and Scheier, 1999).

A further fundamental problem of the cognitivist view is the symbol grounding problem: How are the symbols connected to the real world, how do they get an intrinsic meaning? The semantics of a symbol-processing system are just constituted by the rules that define relations between symbols and govern symbol manipulation, but they are not linked to the outer world (Ziemke, 1999). Most likely, the designer of a symbol-processing system had certain meanings of the different symbols in mind, but this assignment of meaning is extrinsic to the system itself. Pfeifer and Scheier (1999) argue that the symbol grounding problem is a specific artifact of symbolic systems and disappears in other approaches.

Furthermore, Möller (1996, 1999) brings forward a threefold argument why the cognitivist approach is flawed especially with respect to perception. First, the form of the sensory representation should not be created by an externally imposed design process because this holds the strong risk of under- or overdesign. The designer could easily miss hidden features of the sensory data which are important for the motor tasks of the agent. On the other hand, he may impose his own (conscious) view of the world onto the agent although it has a very different morphology and sensor equipment.

Second, Möller (1999) emphasizes the process of self-organization which is supposed to take place in the neural networks of biological organisms. Self-organization relies on determining stable statistical interrelations between different data streams. Restricting this process to each sensory domain alone misses the interrelations between different modalities (e.g., between the visual and auditive impression of an approaching predator), and even worse, misses the strong interrelations between motor actions and the sensory consequences of these actions.

Third, the information processing approach suffers from the "homunculus problem". The rigid distinction between the sensory representation and the motor domain implies an "immaterial internal observer" who looks at the sensory representation and invokes the appropriate motor program. Thus, instead of explaining perception directly, it becomes necessary to explain the perceptual capabilities of the homunculus. Möller (1999) writes on p. 185: "An approach to perception should demonstrate that appropriate behavior is generated within the system." Otherwise, the designer runs into the danger of himself becoming the homunculus.

1.3.3 Embodiment

The problems of the cognitivist approach arise from the complete ignorance of the agent's body and of its environment, and moreover from the strict separation of sensory and motor processing. Several approaches in cognitive science address these issues, at the same time rejecting the claim of symbolic processing in the brain. First, "embodiment" emphasizes the need for a real-world body to enable the development of animal-like perception and cognition (Varela et al., 1991; Wilson, 2002). Second, closely related to embodiment, "situated action" stresses the mutual relationship between an agent and its environment, both from an evolutionary and from the individual's perspective (Clark, 1997; Greeno and Moore, 1993; Suchman, 1987).[5] Third, "enactment" elaborates on the influence of action on perception, even claiming that perception without action is not possible at all (Noë, 2005). These different approaches are generally compatible with each other and may be interpreted as different aspects of one and the same overarching theme, namely that perception and cognition can only emerge in active agents equipped with a real body in a specific environment. For simplicity, we refer to this research paradigm by the term "embodiment". In the following, we discuss the central concepts of embodiment for the design and the understanding of intelligent agents:

- *Physical incarnation:* Intelligent agents have to be physical agents. They exist within the physical environment and own a body with a certain morphology and certain sensory capabilities. This body is the basis of any motor action they are able to carry out, it defines their capabilities and their limitations. Higher-level cognition is grounded in the sensorimotor relationships this body experiences in the interaction with its environment.

[5] An important foundation for embodiment and situatedness is the field of "behavior-based robotics" set off by Brooks (1986, 1991a,b).

- *Situatedness:* Intelligent agents are situated. According to Pfeifer and Scheier (1999), this means that the agent "acquires information about its environment only through its sensors in interaction with the environment". Furthermore, the notion of situatedness points out that cognitive processes rely heavily on the agent's environment. Very often, intelligent agents use the environment as an external aid to find a solution for certain tasks or problems (e.g., speaking of humans, paper and pencil are used to take notes of intermediate steps during problem solving, or as an external memory extension). These actions modify the environment and change the state of the real world. This change may lead to new possibilities for the development of further ideas and actions. Clark (1997) designates this exploitation of external structure in the process of problem solving as "scaffolding".

- *Adaptivity:* Intelligent agents are learning systems. They adapt their behavior continuously to a changing and unpredictable environment. Most skills are not pre-wired, but are learned during the interaction of the agent's body with its surrounding. At the basis of this learning process is the acquisition of sensorimotor relationships.

Integrated sensorimotor processing in the embodied view does not necessarily imply the existence of sensorimotor representations, since the concept of representations is controversial in this field (e.g., Brooks, 1991b). In a more general account, integrated sensorimotor processing means sensorimotor coordination that takes place at a level far below any symbolic or object-like interpretation of sensory data. On the one hand, the sensory input is used for motor control, on the other hand — more important for the embodied approach — agents learn the sensory consequences of their own actions, in this way establishing an "understanding" of their sensory inflow in terms of their own action repertoire. Both aspects, motor control and sensory prediction, will be considered more thoroughly in Sect. 1.5 on internal models. In this section, we will also discuss how low-level sensorimotor coordination could give rise to perception and cognition.

In the embodied approach, the frame problem is not as prevalent as in cognitivism, since world modeling plays only a minor role, following the premise that the best model of the world is the world itself. A situated agent can always "look" at the world to obtain the most accurate and recent data about its state (Pfeifer and Scheier, 1999). Furthermore, a situated agent just requires very specific information about certain aspects of the world which are relevant for the task at hand. There is no need for an elaborated world model. This idea is related to the concept of "information pickup" in Gibson's theory of direct perception (Gibson, 1979) (see also Sect. 1.5.3).

Although the symbol grounding problem does not exist for embodied systems, Ziemke (1999) argues that the grounding problem nevertheless reappears in a different form. According to Ziemke (1999), the embodied approach faces the *body grounding* problem: Even if the control structures in an artifical agent have evolved to a large extent by self-organization, the human designer chose a certain robot body with specific sensors and actuators and a certain environment in advance. These choices are not intrinsically grounded in the interaction between robot and environment. In contrast, natural embodiment and situatedness of an animal are the result of a long history of evolution and individual development. Unfortunately, the body grounding problem cannot be overcome, it only can be relieved by reducing the amount of external design decisions as far as possible.

1.3.4 Evidence for integrated sensorimotor processing

In the following, some experimental findings from behavioral and neurophysiological studies are presented which provide evidence for the close interaction between sensory and motor data, even at the neural level. This section should be read as a small introductory "appetizer" without any claim for completeness. Further theoretical and experimental accounts to integrated sensorimotor processing are described in Sect. 1.5 and in Chapt. 2 on adaptive motor control.

1.3.4.1 Behavioral experiments

As described in Sect. 1.1, Held and Hein (1963) demonstrated the importance of active movement for the development of visually guided behavior in kittens. Related studies on visual adaptation in humans revealed that the speed and quality of adaptation depends on the amount and quality of active movements during the adaptation process. Held and Freedman (1963) showed this for a pointing task while the subjects had to wear wedge-prism goggles, Luria and Kinney (1970) for underwater pointing with diving goggles. Rossetti et al. (1998) were even able to shift the neglected region of left-hemispatial-neglect patients through a pointing task with prism goggles. All these studies support the claim that the development or change of perceptual skills depends heavily on the presence of agent-induced motor commands; passive movements do not have the same effect.

Further support arises from the phenomenon of ideomotor actions; this terms refers to body movements which are spontaneously evoked when observers watch other people performing certain actions (Knuf et al., 2001). They are an important observation within the framework of the common coding approach (Hommel et al., 2001; Prinz, 1997). Its main assumption is that perceived events and planned actions share a common representational domain. Prinz (1997) reviews various studies which support this claim, among them studies on the "Simon effect" (e.g. Hommel, 1993): In experimental studies on human subjects, it is usually observed that performance is clearly superior if stimulus and response position correspond with each other, as compared to non-correspondence. Within the common coding approach, a general explanation for this effect is that a corresponding stimulus already activates the (shared) code that is needed for the subsequent response. This effect is another example for a close sensorimotor coupling.

1.3.4.2 Neurophysiological evidence

In the field of neuroscience, Rizzolatti and colleagues have discovered several types of neurons in the rostroventral premotor cortex (area F5) of the macaque monkey which establish a link between sensory and motor information. The neurons in this area discharge during hand movements, but some of them also discharge on the observation of visual stimuli. "Canonical F5 neurons" (Murata et al., 1997; Rizzolatti et al., 1988; Rizzolatti and Fadiga, 1998) react to 3D objects whose shape and size corresponds to the prehension movement coded by the neuron, "mirror neurons" (Rizzolatti et al., 1996; Rizzolatti and Fadiga, 1998) react to hand actions carried out by other individuals which are similar to the coded movement. Rizzolatti and Fadiga (1998) proposed that mirror neurons are the neural basis for understanding the meaning of actions made by others. For the integrated sensorimotor approach, the canonical neurons are even more interesting. Their activation pattern suggests that sensory information is "understood" in terms of associated motor actions.

In the context of saccadic eye movement, there exists evidence for a "predictive remapping" of receptive fields. In various regions of the brain, neurons with visual receptive fields have been found which respond before an upcoming saccade to a stimulus that is not yet in their receptive field, but will appear there after the saccade. These neurons elicit a predictive sensory response to a planned but not yet executed eye movement, in this way demonstrating a close coupling between sensory and motor information on the neural level. Predictive neurons have been found in the superior colliculus (Walker et al., 1995), in the lateral intraparietal area (Duhamel et al., 1992), and in the frontal eye field (Umeno and Goldberg, 1997). In a recent behavioral study with human subjects, Melcher (2007) showed that the predictive remapping of retinal locations also extends to visual processing in an adaptation task.

1.4 Cognitive Robotics

The notion "cognitive robotics" describes a research method within the framework of embodied cognition. Its focus are models of sensorimotor processing which are implemented and tested on real-world robot platforms, following the design principle that intelligent agents need to be embodied and situated. Admittedly, there is a considerable gap between intelligent behavior on the one side and single sensorimotor processes on the other side. This gap has to be closed through ongoing progress in the field. At the moment, researchers try to understand how sensorimotor models can be learned, and how they can be used to guide behavior and to facilitate perception.

For example, the "Darwin" series of mobile robots (Almassy et al., 1998; Krichmar and Edelman, 2002) were developed to show by synthetic neural modeling how certain perceptual skills emerge by the interaction of the robot with its environment. On both Darwin V (Almassy et al., 1998) and Darwin VII (Krichmar and Edelman, 2002) a rather detailed model of the brain structures which are involved in visual processing and visomotor coordination was implemented. During the interaction with the environment, this synthetic brain developed through changes in the synaptic strengths. In this process, specialized neural structures emerged which are linked to perceptual skills like invariant object recognition or experience-dependent perceptual categorization. The authors claimed that by exhaustive analysis and manipulation of these artificial brain structures, this approach provides valuable heuristics for understanding the interactions in the real brain.

A second example for cognitive robotics research is the recent study by Bongard et al. (2006). They proposed an active self-modeling process for a four-legged robot. In this process, the internal sensorimotor self-model of the robot is first generated by model synthesis through directed exploration. Explorative movements are not generated at random, but chosen according to which movement would be the most informative for the identification of the self-model. Afterwards, the self-model is used for action generation until an unexpected sensorimotor pattern is detected. In this case, the self-modeling process starts again. The authors showed that the system is even able to cope with the loss of one leg. They concluded that their "work suggests that directed exploration for acquisition of predictive self-models may play a critical role in achieving higher levels of machine cognition" (Bongard et al., 2006, p. 1121). Further studies from the field of cognitive robotics are mentioned in Sect. 1.5.3.

Robot experiments and computer simulations belong both to the area of synthetic modeling, in contrast to analytical experiments on the behavioral or neurophysiological level. Especially, the robotics approach to cognitive science offers some genuine advantages for the process of model testing. Webb (2000) writes on p. 546 with regard to the related field of biorobotics: "Robots as models are a means by which hypotheses can be tested for adequacy and sufficiency to explain a set of data, and additional predictions from the hypotheses can be derived."

Moreover, Webb (2000) emphasizes that the use of robots enforces the researcher to characterize the problem thoroughly and to consider and understand the role of the environment. In a pure computer simulation, the simulated environment of the agent can lack important properties of the real physical world with regard to the tested model (see also Grasso, 2001). By using robots, this risk is reduced (although not completely eliminated; e.g., the sensor equipment of the robot may be insufficient, causing misleading results).

In addition, robot models enforce completeness: It is not possible to omit any part of the sensorimotor loop — the tested model works only in its full implementation, including the sensory and the motor part (and the integrated processing in between). This enhances the validity of the results, helps to identify missing or wrong parts of any model, and facilitates the generation of new hypotheses.

1.5 Internal Models

The transformations between motor commands and sensory states are determined by many factors, among them the properties of the environment and of the agent's body. There is ample evidence that the CNS of biological organisms represents these transformations internally (Schaal and Schweighofer, 2005; Shadmehr and Wise, 2005). Therefore, these representations are called "internal models". Kawato writes: "Internal models are neural mechanisms that can mimic the input/output characteristics, or their inverses, of the motor apparatus" (Kawato, 1999, p. 718). Moreover, internal models are used to describe sensorimotor processing, independent of their hypothesized existence in the CNS. The concept of internal models has its origin in control theory and robotics and plays an important role in the research on motor control.

There exist two main classes of internal models: inverse models (IMs), which generate motor commands to close the gap between the current and the desired (sensory) state, and forward models (FMs), which predict the (sensory) consequences of the agent's actions. Moreover, it is reasonable to suppose the existence of a more general category of FMs which predict external events (Miall and Wolpert, 1996; Schubotz, 2007), but this category is beyond the scope of this thesis. While IMs are mainly required for motor control (Chapt. 2), FMs (or more generally: anticipatory mechanisms) serve a multitude of purposes. Theoretical considerations and experimental findings suggest that FMs are required for sensory cancelation (Sect. 1.5.2), for state estimation (Sects. 2.1.1 and 2.1.4.5), and for context estimation (Sect. 2.1.5). Furthermore, FMs are a basic component of the motor learning strategy "distal supervised learning" (Sect. 2.2.4). The combined application of forward and inverse models is hypothesized for motor planning and for simulation theories of perception (Sect. 1.5.3) and cognition (Sect. 1.5.4).

Whenever internal models are accepted as building blocks of sensorimotor processing, the question arises how they are acquired. For biological organisms, internal models represent certain information processing capabilities of the CNS. Such capabilities can either be inherited or learned during the lifetime of the organism. But even inheritance involves learning by evolutionary forces in the process of phylogenesis (Ax, 1987). In robotics, a classical engineering approach would try to describe the "plant" (which comprises the robot and its environment) analytically as precise as necessary. Knowing the plant, one already knows the FM, and IMs (motor controllers) can be designed using the tools of control theory (Dorf and Bishop, 2004). But such an approach is only feasible when the plant is known analytically. This may hold for certain industrial applications, but not for autonomous robots in complex and changing environments (where the "body" of the robot might change as well due to wear and tear). Thus, not knowing the plant beforehand, the robot has to adapt its internal models to the environment in which it is moving and acting — it has to learn. In summary, the learning of internal models is an integral part of sensorimotor processing, both for biological and artificial agents. As consequence, an important focus of this thesis is on learning mechanisms for internal models.

In Sect. 1.5.1, we start with a formal description of internal models, based on state-space control theory. This description is intended as precursor for Chapt. 2 on adaptive motor control. Afterwards, the role of sensorimotor internal models for sensory cancelation, perception, and cognition is explored in detail.

1.5.1 Formal description

The following description of internal models relies on the formal approach of state-space control theory for dynamical systems (as textbook reference, see for example Dorf and Bishop, 2004). At a given moment, the agent and its environment are in a certain physical state. This state and the dynamic interaction

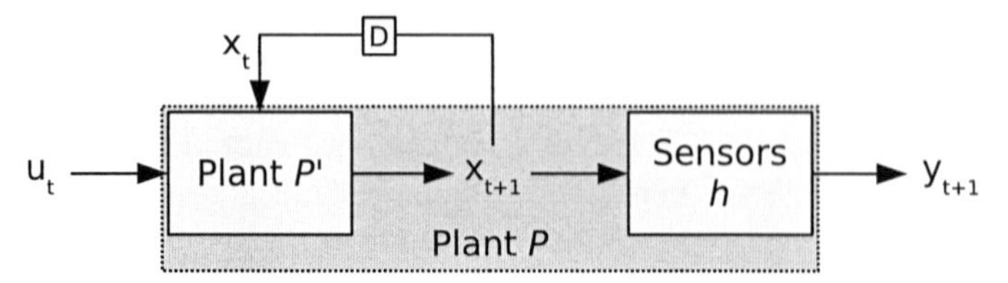

Figure 1.2 — The plant as abstract representation of the physical system. The box labeled *D* indicates a delay by one time step.

between the state variables determine how the system will develop in the future. All (non-redundant) state variables which are causally relevant for this development at time step t are collected in a vector $\mathbf{x}_t$.[6] In state-space models of physical systems, the number of state variables is usually equal to the number of energy-storing elements (Tao, 2003). The external input of the system in each time step is the vector $\mathbf{u}_t$. In motor control, it is the motor command which is generated and executed by the agent. The output of the system is denoted by the vector $\mathbf{y}_t$. In the domain of sensorimotor processing, $\mathbf{y}_t$ is sometimes identical with the sensor readings of the agent. Alternatively, $\mathbf{y}_t$ can also contain a more abstract representation which is derived from the sensor readings, for example the retinal coordinates of an object instead of the activation of thousands of fibers of the optical nerve. Generally, the output $\mathbf{y}_t$ is a function of the current state $\mathbf{x}_t$: $\mathbf{y}_t = h(\mathbf{x}_t)$. The important difference is that the output $\mathbf{y}_t$ is directly measurable by the agent, while the state $\mathbf{x}_t$ may be not. Depending on the context, we will sometimes refer to $\mathbf{y}_t$ as plant *output*, sometimes as sensory *input* of the agent. The first designation applies to a system-centered view, the second to an agent-centered view. Furthermore, $\mathbf{y}_t$ is sometimes also denoted as *sensory state* (whereas $\mathbf{x}_t$ is the *system state*, to which the abbreviated term "state" always refers).

The so-called "plant" serves as an abstract representation of the physical system. Figure 1.2 shows a block diagram of the important relationships. The system itself is represented by the plant P':

$$\mathbf{x}_{t+1} = P'(\mathbf{x}_t, \mathbf{u}_t)$$

The measurement process yields

$$\mathbf{y}_{t+1} = h(\mathbf{x}_{t+1}) \ .$$

[6] Throughout this thesis, we will mainly deal with time-discrete systems. Nevertheless, when it is more appropriate, we will sometimes use a time-continuous notation.

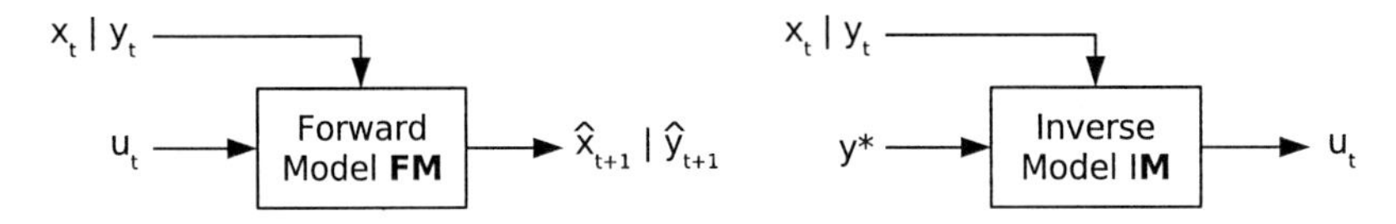

Figure 1.3 — Left: Forward model (the output is either a prediction of the state $\widehat{\mathbf{x}}_{t+1}$ or the sensory output $\widehat{\mathbf{y}}_{t+1}$ of the system in the next time step). Right: Inverse model.

For convenience, we define a plant P which combines the original plant P' and the measurement process:

$$\mathbf{y}_{t+1} = P(\mathbf{x}_t, \mathbf{u}_t) = h\left(P'\left(\mathbf{x}_t, \mathbf{u}_t\right)\right)$$

As already stated, there are basically two classes of internal models: forward models (FM) and inverse models (IM). FMs mimic the behavior of the plant and generate a prediction of $\mathbf{x}_{t+1}$ or $\mathbf{y}_{t+1}$ (see Fig. 1.3, left):

$$\widehat{\mathbf{x}}_{t+1} = \mathrm{FM}(\mathbf{x}_t, \mathbf{u}_t) \quad \text{or} \quad \widehat{\mathbf{y}}_{t+1} = \mathrm{FM}(\mathbf{x}_t, \mathbf{u}_t)$$

In addition, pure sensory FMs are also imaginable:

$$\widehat{\mathbf{y}}_{t+1} = \mathrm{FM}(\mathbf{y}_t, \mathbf{u}_t)$$

IMs represent the inverse relationship. While an FM is an approximation of the plant, an IM is an approximation of the inverse of the plant and acts as motor controller (see Fig. 1.3, right):

$$\widehat{\mathbf{u}}_t = \mathrm{IM}(\mathbf{x}_t, \mathbf{x}_{t+1}) \quad \text{or} \quad \widehat{\mathbf{u}}_t = \mathrm{IM}(\mathbf{x}_t, \mathbf{y}_{t+1}) \quad \text{or} \quad \widehat{\mathbf{u}}_t = \mathrm{IM}(\mathbf{y}_t, \mathbf{y}_{t+1})$$

$\widehat{\mathbf{u}}_t$ is an estimate of the motor command which would yield the respective plant's response in the next time step. Usually, one just writes $\mathbf{u}_t$ instead of $\widehat{\mathbf{u}}_t$. The IM with the inputs $\mathbf{x}_t$ and $\mathbf{y}_{t+1}$ is the most common one. Since $\mathbf{y}_{t+1}$ is not the real value in this context but the desired one, one writes $\mathbf{y}^*$ instead of $\mathbf{y}_{t+1}$. This finally yields

$$\mathbf{u}_t = \mathrm{IM}(\mathbf{x}_t, \mathbf{y}^*) \ .$$

This general introduction on internal models focusses on FMs and IMs for dynamical time-discrete systems. In Chapt. 2 on motor control, various types of IMs will be discussed whose input can differ from this introductory presentation. Furthermore, in addition to internal models for dynamical systems we will deal with internal models for kinematic relationships.

1.5.2 Sensory cancelation

FMs can be applied to differentiate between self-induced and externally induced sensory effects. The FM predicts the sensory state of the next time step, given the current motor command. If the FM is precise, the difference between the predicted sensory state and the real sensory state after execution of the motor command will be close to zero. However, if unexpected external processes cause additional changes of the environment, there will be a difference between the predicted and the real sensory state. In this way, the prediction by the FM can be used to cancel out self-induced sensory effects, and to detect externally induced sensory effects by the remaining difference to the predicted sensory state.

A classical example for this kind of reasoning is the reafference principle suggested by von Holst and Mittelstaedt (1950). They outline for several neuromuscular systems, among them eye movements and accommodation, how the efference copy cancels out the "reafference", i.e. the signal transmitted by the sensory receptors in response to the efferent motor command. For example, when subjects with temporarily paralyzed eye muscles elicit eye movement commands, they experience a shift of the visual surroundings in the intended movement direction (because the expected eye movement does not take place). Vice versa, when we move our eyeball with external forces (e.g., by pushing it with the finger), we perceive a movement of the visual surroundings although only the eyeball moves — but without eye movement command. These phenomena can be explained by the reafference principle. However, von Holst and Mittelstaedt (1950) assume that the efference copy and the reafference can be directly summed up by the CNS, thus they need no FM. This claim seems to be questionable, especially for as complex sensory data as from the visual sense.

More recent evidence for the cancelation of self-induced sensory effects has been presented in a study by Blakemore et al. (1999): In their experiments, subjects had to tickle themselves via a robotics interface. The experimenters manipulated the correspondence between the action of the subjects' left hand and the tactile stimulus on their right hand. One manipulation concerned the time delay between the action and the sensory effect, the other the movement direction of the tactile stimulus (applied by the robot) in relation to the movement direction of the left hand. By these manipulations, the difference between the normal self-elicited sensory effect and the real sensory effect was varied. As result, the subjects rated their tactile sensation as less tickly, pleasant, and intense, the smaller this difference was (smaller time delay, more similar hand and stimulus movement). This effect is ascribed to FMs whose output cancels the sensory effect of self-executed tickling out as long as the sensory prediction matches closely the experienced stimulus. In a follow-up study, Blakemore et al. (2000) employed a simplified version of their tickliness experiment to explore the brain regions which are dedicated to the prediction of sensory consequences. Based on their fMRI data, the authors concluded that "the cerebellum is involved in predicting the specific sensory consequences of movements and in providing the signal that is used to attenuate the somatosensory response to self-produced tactile stimulation" (Blakemore et al., 2000, p. R14). In addition, predictive sensory cancelation has also been observed in a very different species, in the electrosensory system of electric fish (Bell, 2001).

1.5.3 Visual perception

In psychology, the two main theoretical approaches to visual perception are the constructivist and the ecological one (for a thorough comparison, see Norman, 2002). In the constructivist approach, perception is viewed as an inferential process. The sensory signals are seen as inherently insufficient for unequivocal perception. Instead, it is assumed that the sensory information has to be processed on the basis of stored schemata and unconscious thought-like processes before perception can arise. In this aspect, the constructivist approach resembles strongly the classical AI approach to perception (see Sect. 1.3.1) which also starts from the assumption that perception arises from pure sensory processing without any reference to the motor system. On the contrary, the ecological approach (Gibson, 1979) is build around the conception of the so-called "direct information pickup". In this view, perception is an active process in which an active observer explores his environment by deliberately moving his eyes, his head, and his whole body. Perception extends over space and time, and objects are not perceived by the knowledge-driven interpretation of cues found in a single retinal image, but instead by directly detecting the *affordances* the objects offer to the observer. Gibson writes: "The affordances of the environment are what it offers the animal, what it provides or furnishes, either for the good or for the ill" (Gibson, 1979, p. 127). For example, surfaces can be "stand-on-able", "climb-on-able", or "sit-on-able". These affordances are closely related to the shape of the body of the observer and to his repertoire of motor actions. Basically, perception in the ecological approach is the direct perception of the behavioral meaning of the objects in the environment.

Möller (1996, 1999) suggested the "perception through anticipation" approach, which is related to the ecological view, but replaces the *direct* perception of affordances by a mental simulation process based on internal models. The main thesis of this approach is: "Perception of space and shape is based on the anticipation of the sensory consequences of actions that could be per-

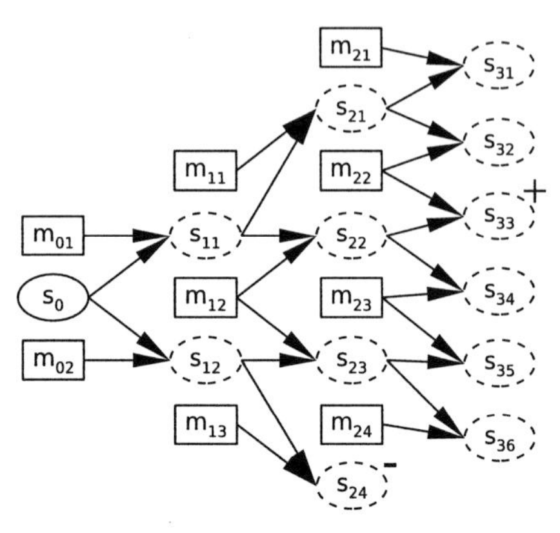

Figure 1.4 — Graphical illustration of the internal simulation process in the "perception through anticipation" approach (Möller, 1999). Starting from the current sensory state s_0, different movement sequences with motor commands m_{ij} are simulated by internal prediction (the predicted sensory states are depicted in dashed ellipoids). The predicted sensory state s_{24} is evaluated as a negative outcome, thus it is not used for further simulation. When the simulation process encounters the predicted sensory state s_{33} with a positive rating, the simulation is halted (at a simulation depth of three steps).

formed by the agent, starting from the current sensory situation. Perception and the generation of behaviour are two aspects of one and the same (neural) process" (Möller, 1999, p. 186). Starting from the current sensory situation, an IM suggests several motor actions. A corresponding FM predicts the sensory consequences of all suggested actions. On the basis of the predicted sensory situations, further motor actions are suggested, afterwards their consequences are predicted as well, and so on, until a maximum step size is reached or at least some simulated action sequences have led to sensory results with a clear positive or negative meaning to the agent (Fig. 1.4 illustrates the internal simulation process, omitting the IM and FM for clarity). In this way, a human agent can for example detect if an object is "sit-on-able", because at least one of the simulated movement sequences would result in a typical sitting posture with support for the body by the top surface of the object. Together with the affordances which have emerged from the other movement sequences simulated in parallel, this may result finally in the perception of a chair. Thus, in this approach, perception is an integrated sensorimotor process which relies on IMs, FMs, and the evaluation of sensory states. The main problem of the anticipation approach is how to restrict the number of simulated motor sequences to a feasible amount, despite the combinatorial explosion that occurs when a large number of motor commands is tested in parallel for each predicted sensory state. As we will see in the following, various solutions have been proposed in the literature.

Several robot studies have been published, which investigate the feasibility of the anticipation approach. Hoffmann and Möller (2004) used a mobile robot equipped with an omnidirectional camera, which had to move within a circle of obstacles. The sensory data was reduced to ten distance values, each indicating the distance of the robot to the closest obstacle in a different sector of the circle. These distance values were generated from the camera image, in this way the FM in this study may be interpreted as a simple visual FM. As motor input, the FM received the wheel velocities for a fixed small time interval. By internal simulation, the system was able to tell if the robot was positioned at the center of the circle or not (perceptual task), and to generate a sequence of motor commands to move from the starting position to a goal position which was defined by the desired activation of a single selected distance detector (motor task). The sequence of motor commands was generated by concatenating a chain of FMs. The resulting free parameters were the motor inputs for all FMs. These parameters were determined by the optimization method "fast simulated annealing" (Szu and Hartley, 1987). The optimization goal was to minimize the difference between the desired activation of the selected distance detector and the corresponding sensory output of the last FM in the chain. In this way, the main problem of the anticipation approach — how to restrict the number of simulated motor sequences to a feasible amount — was solved by transforming the internal simulation into an optimization problem. The study presented in Chapt. 7 relies on a similar idea.

In a further study with the same robot setup, Hoffmann (2007) refined the visual FM. The new FM was capable to predict downsampled images of the omnidirectional camera with a size of 40×40 pixels. To predict the intensity of each single pixel, an individual MLP was used. However, because of the lacking precision of the MLPs, noise crept very easily into the prediction process. To overcome this problem, an abstract recurrent neural network (similar to NGPCA in Sect. 3.4) was trained to represent the downsampled camera images which were encountered during training. This abstract recurrent neural network was used to project the predicted noisy image back onto the data manifold of "clean" camera images. The denoised image was the final output of the visual FM. This model was applied to a dead-end recognition task. The robot was positioned at the beginning of either a dead end or of a passage, depending on the arrangement of obstacles. By the mental simulation of a sequence of movement steps with prediction by the visual FM, the robot could distinguish between dead ends and passages. The movement steps were either generated by an obstacle avoidance algorithm or by a recursive search algorithm. Both variants worked well for the perception of dead ends by mental simulation as suggested by the anticipation approach.

Möller and Schenck (2008) went even a step further. In their study, a (simulated) mobile robot system first acquired a visuo-tactile FM (which also predicted collisions with the surrounding obstacles). This FM was used to generate learning examples for an IM. A simple movement strategy drove the mental simulation along different search paths, and the collision-free movement sequence with the smallest costs was selected (costs were caused by rotary movements and changes of the movement direction, thus straightforward movements were favored). The first motor command in the selected sequence was applied as training output to the IM. Both the FM and the trained IM were used for the subsequent dead end recognition task. From the current visual input, the system had to decide if it was at the entrance of a dead end or a passage. As in the previous study, dead end perception relied on a mental simulation process. Motor commands were generated by the IM, in this way the combinatorial explosion of a systematic search in motor space was avoided. However, to generate at least some motor sequences in parallel, the output of the IM was disturbed in a random fashion at certain movement steps. If none of the simulated movement sequences allowed a passage through the obstacles, the visual scene was interpreted by the system as showing a dead end. This classification task was solved successfully for many different obstacle arrangements. The authors concluded that "the agent [...] 'understands' the behavioral meaning of all situations by revealing their 'affordances' through a process of internal simulation" (Möller and Schenck, 2008).

However, the study by Möller and Schenck (2008) not only provides support for the anticipation approach to visual perception, but also illustrates a further possible function of FMs: They can be used for mental practice, here for acquiring an IM which favors collision-free straightforward movement trajectories. This corresponds to the well-known phenomenon from the psychomotor literature that pure mental practice can help to exercise movement tasks, e.g. in sports (e.g., Martin et al., 1999) or rehabilitation (e.g., Jackson et al., 2001).

The work by Gross et al. (1999) is also based on the anticipation approach. In their study, a mobile robot had to navigate through an arena with obstacles. For the internal simulation process, an FM was used which predicted the optical flow field at the next time step. To avoid an exhaustive search for the best movement sequence, they restricted the internal simulation to motor commands that are typical for the current situation. They suggested a neurophysiological model of the internal simulation process with special emphasis on the cortex (generation of motor commands), the basal ganglia (evaluation of sensory outcomes), and the cerebellum (sensory prediction). The robot studies by Mel (1988), Tani

(1996), and Ziemke et al. (2005) are also based on the idea of an internal prediction and simulation process, but their emphasis is not on visual perception. Instead, Mel (1988) aimed on motor planning, Tani (1996) on symbol grounding, and Ziemke et al. (2005) on the emergence of an "inner world".

O'Regan and Noë (2001) added a further aspect to the basic idea that visual perception relies on acting (although without an explicit reference to sensory prediction). According to their theory, "vision is a mode of exploration of the world that is mediated by knowledge, on the part of the perceiver, of what we call sensorimotor contingencies" (O'Regan and Noë, 2001, p. 940). Sensorimotor contingencies are the rules that govern the sensory changes produced by various motor commands. For example, the sensory activation on the retina which is caused by a straight line remains constant for certain eye movement directions. Eye movements in different directions will instead cause specific changes of the retinal activation. O'Regan and Noë understand visual perception as "the activity of exploring the environment in ways mediated by knowledge of the relevant sensorimotor contingencies" (O'Regan and Noë, 2001, p. 943). Following this line of thought, visual and auditory perception are different, for example, because the sensorimotor contingencies differ between these modalities.

However, the behavioral accounts to visual perception are not only supported by theoretical considerations and robot studies, but also by neurophysiological findings. Ungerleider and Mishkin (1982) presented evidence for two different cortical pathways in the brain of the monkey: the ventral stream leading from the occipital cortex to the inferior temporal cortex, and the dorsal stream leading to the posterior parietal cortex. These streams serve different purposes. In the interpretation of Goodale and Milner (1992), the ventral stream is the "what" system for object identification, whereas the dorsal stream is the "how" system which transforms visual information into an egocentric framework which allows the subject to *act* on the object. Norman (2002) puts forward the hypothesis that the ventral stream corresponds to the constructivist approach to visual perception, whereas the dorsal stream corresponds to the ecological approach. Such a unified theory could explain why both approaches are supported by experimental findings. Even more important, the experimental evidence for the existence and the function of the dorsal pathway provides a solid neurophysiological background for the action-related accounts to visual perception (see for example Grezes and Decety, 2002).

Furthermore, studies on visual imagery and motor imagery support the idea that perceiving is closely related to an internal simulation of motor sequences. Visual mental images "correspond to short-term memory representations that lead to the experience of 'seeing with the mind's eye' " (Kosslyn et al., 1993, p. 263). While the Behaviorist school of thought even questioned the very existence of visual mental images, neuroimaging studies meanwhile have shown that these images have clear neural correlates in rather low-level areas of the visual cortex (Kosslyn et al., 1993; Kosslyn, 1994). This shows that an internal simulation process, in which sensory states are predicted and not elicited by real stimulation, could be possible on a neural level. A similar argument follows from the studies on motor imagery. According to Jeannerod (1995), a motor image is a conscious motor representation which is related to intending and preparing movements. Motor imagery implies that the subject feels himself executing a given action, thus motor imagery "requires a representation of the body as the generator of the acting forces" (Jeannerod, 1995, p. 1420). Jeannerod (1995) reviewed several neuroimaging studies which show that the cortical activation during motor imagery is similar to the activation during intentionally executed actions (see also Jeannerod, 2001). Thus, the internal simulation of actions is closely related to the real execution of actions on a neural level. This supports the initial claim of the "perception through anticipation" approach, that "perception and the generation of behaviour are two aspects of one and the same (neural) process" (Möller, 1999, p. 186). However, one has to note that the findings on mental imagery concern conscious processes, while the hypothesized mental simulation during perception are supposed to take place on a subconscious level.

1.5.4 Cognition and consciousness

It is a long-standing idea that the motor apparatus might be involved in thinking and other cognitive abilities. According to Hesslow (2002), it traces back to the 19th century. In its modern form, this idea is closely linked to various forms of "simulation theory". The basic assumption is that cognition relies on the internal simulation of motor actions (or action-related concepts) without actually executing these actions. Although this assumption is not necessarily linked to internal models, it fits very well to the framework of FMs and IMs for sensorimotor processing. FMs might be identified here with general anticipatory mechanisms, and IMs with general mechanisms for the generation of motor commands.

1.5.4.1 Cognition as internal simulation

In the context of the "perception through anticipation" approach (Sect. 1.5.3), we already mentioned studies on visual imagery and motor imagery. These studies demonstrated that the neural correlates of visual and motor imagery correspond (partly) to the neural correlates of visual perception (Kosslyn et al.,

1993; Kosslyn, 1994) and of real actions (Jeannerod, 1995). These findings not only support simulation theories of perception, but also simulation theories of cognition. Hesslow (2002) hypothesized that the CNS might be able to generate long chains of simulated actions and perceptions, in this way not only enabling motor planning, but also genuinely cognitive tasks like problem solving in games like chess or the "Tower of London" (and imaging experiments revealed actually an activation of premotor areas for the latter task; Dagher et al., 1999). In a related experiment with a simulated mobile robot setup, Ziemke et al. (2005) demonstrated that the internal simulation of perception can replace an explicit representational world model for movement planning over hundreds of iteration steps. Schubotz (2007) went a step further by claiming that the sensorimotor system is even used to simulate events that cannot be produced or imitated like the rhythm of ocean waves or the flight of a mosquito. She suggested that the premotor cortex houses FMs for general transformations like rotations. These FMs could be applied both to motor actions and to external events. Related ideas were put forward by Bar (2007), who speculated that the brain is continuously busy generating predictions of the near future to facilitate perception and cognition.

With regard to the social domain, the discovery of the mirror neuron system in humans (Rizzolatti et al., 1996) (see also Sect. 1.3.4.2) established a close link between the motor system and social cognition (Jacob and Jeannerod, 2005). For example, it is suggested that we understand the behavior of other people by carrying out a simulation of their internal states; the activity of the mirror neurons serves to map the observed behavior to action plans, which in turn are associated with the supposed intentions and goals of the observed subject (e.g., Gallese and Goldman, 1998). Miall (2003) connected the concept of internal models with the mirror neuron system, in this way explaining how the brain might anticipate future sensory effects caused by the actions of others (see also Iacoboni, 2005). A related approach to social cognition by internal models is provided in the HMOSAIC model by Wolpert et al. (2003) (see Sect. 2.1.5 for more details).

1.5.4.2 Consciousness and attention

Cruse (2003) presented an approach to cognition which is also based on the idea that the same neural substrate serves for the generation of motor commands and for higher cognitive abilities. Instead of an iterative internal simulation process based on feedforward models, he proposed a recurrent sensorimotor body model as underlying structure (on basis of the MMC network architecture; Cruse and Steinkühler, 1993). The MMC network represents the underlying sensorimotor mechanisms and not only specific input-output relationships. Furthermore, the MMC network converges dynamically to its attractor states, a feature which is biologically plausible but completely missing in pure feedforward architectures. Cruse (2003) hypothesized that the content of the body model is subjectively experienced when the recurrent network has sufficiently relaxed to an attractor, in this way giving rise to consciousness.

Taylor (2006) proposed a control system for attention in the brain. This system is based on an IM to move the focus of attention and on an FM which predicts the attented state of the world. These models do not work with motor outputs/inputs, but instead with signals for attention control. Taylor (2006) suggested that the corollary discharge of the "attention movement" functions as control basis of consciousness. This activity is supposed to create the experience of an "owner" to whom the content of consciousness belongs.

Theoretical considerations like these are highly speculative at the current level of knowledge, but they show that the concepts of integrated sensorimotor processing and of internal models reach far beyond the level of motor control and basic perceptual skills.

Chapter 2
Adaptive Motor Control

2.1 Motor Control with Internal Models

Motor control is a complex task. For example, even for a simple arm movement the human central nervous system (CNS) has to coordinate the activation of a large number of muscles over an extended period of time. Within the framework of internal models, one assumes that motor control in the CNS is split into distinct modules. Each module processes a certain input-output relationship on the way from sensory data and movement goals towards the final motor commands which specify the activation of the muscles. In biological reality, several control loops exist on various levels of the CNS, e.g., the muscular, the spinal, the transcortical, and the cortical, thus forming a hierarchy of control loops (Mehta and Schaal, 2002). For this reason, one has to be aware that every "block model" of motor control is an abstraction and simplification. Moreover, the framework of internal models has its origin in control theory, and therefore the formalism and the mathematical methods from this area are widely used. This has proved to be a fruitful theoretical approach, capable of explaining a large amount of experimental data. Nevertheless, reverse engineering of the motor system is a difficult task, which requires to integrate experimental evidence from various sources (e.g., behavioral observations, controlled behavioral experiments, electrophysiological and neuroimaging studies) and data from mainly theory-driven work (like robot models). At the time of this writing, it seems fair to say that the CNS *might* use internal models and *might* employ control and learning strategies similar to those which have been developed in control theory in engineering (Shadmehr and Wise, 2005).

In this section, we concentrate on human arm movements (and to a smaller extent, on eye movements). This restriction is motivated by the fact that most research on internal models for human motor control concerns arm movements for pointing, reaching, and grasping. Therefore, these models are often tested on simulated or real robot arms. Moreover, they are developed on the basis of

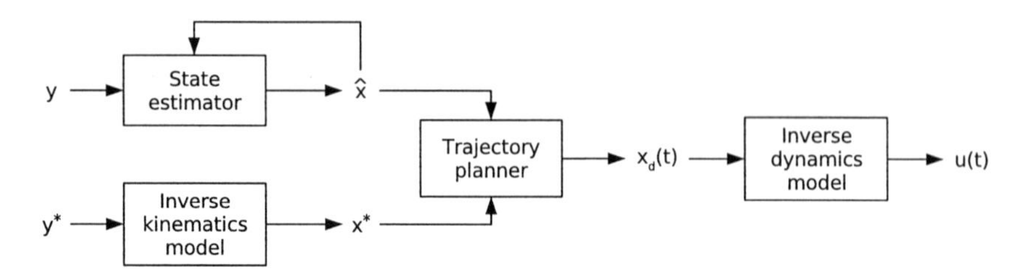

Figure 2.1 — Overview of the modules for motor control (for details see text).

robotics-related concepts, although these concepts can only rarely be applied directly to the human motor system. For example, the state of a robot arm can be described by a vector $\mathbf{x}(t) = \left(\boldsymbol{\theta}(t), \dot{\boldsymbol{\theta}}(t)\right)$. $\boldsymbol{\theta}(t)$ is the vector of joint angles, $\dot{\boldsymbol{\theta}}(t)$ the vector of joint angle velocities. Applied to the human motor system, this state description is an abstraction. To the best of the author's knowledge, there is no experimental study which directly shows that there is a neural correlate of the current joint angles and velocities anywhere in the CNS. The CNS *might* use representations of the physical state of the body and relevant parts of the environment, but one has to keep in mind that the exact nature of these representations is still unknown. It might be that these state descriptions are more related to the state of the muscles. Considering the motor output, a similar abstraction applies. For the dynamic control of robot arm movements, the motor command is usually a vector of torques $\boldsymbol{\tau}(t)$. In the human body, these torques are generated by the muscles which contract with a certain force depending on their state and on the activation of the attached motor neurons. In addition, muscles for joint movements usually appear in agonist/antagonist pairs; the exerted forces in the final posture determine the stiffness of the limb. Nevertheless, in accordance with the majority of the studies presented here, we ignore the stiffness aspect in the following and use $\mathbf{x}(t) = \left(\boldsymbol{\theta}(t), \dot{\boldsymbol{\theta}}(t)\right)$ as state description of the arm and $\boldsymbol{\tau}(t)$ as motor output in dynamic motor control. One just has to keep in mind that these are abstractions when they are applied to human motor control.

To structure motor control in a modular fashion, it is a convenient approach to distinguish between the following modules (see Fig. 2.1): state estimation, inverse kinematics, trajectory planning, and inverse dynamics. In the following, these modules are described in the context of a reaching task: The hand has to move from point A to point B.

State estimation In a real-world task with a biological agent in a dynamic environment, the state of the overall system cannot be described completely with a reasonable number of state variables. Therefore, it is necessary to restrict the state description to a set of variables which is most relevant for the control task. For arm movements, the joint angles and angular velocities form such a set: $\mathbf{x}(t) = \left(\boldsymbol{\theta}(t), \dot{\boldsymbol{\theta}}(t)\right)$. The CNS does not have direct access to this state, instead is has to be estimated from the sensory signals $\mathbf{y}(t)$ from proprioception (originating in the receptors imbedded in the joints, tendons, muscles, and skin) (Zimmermann, 2005). In control theory, for state estimation an "observer" is specified which processes the current sensory input and the last state estimate to generate a new state estimate (Dorf and Bishop, 2004). Here, we propose a similar "state estimator" which generates the estimated state $\widehat{\mathbf{x}}(t)$. At time t_0, when the arm's tip is at point A, this estimated state is $\widehat{\mathbf{x}}(t_0) = \left(\widehat{\boldsymbol{\theta}}(t_0), \widehat{\dot{\boldsymbol{\theta}}}(t_0)\right)$.

Inverse kinematics In a very general sense, the term "kinematics" refers to a transformation between coordinate systems (Jordan, 1996). For a reaching task, this could be the transformation from the position of the reaching target (provided in body-centered coordinates) to the corresponding arm posture. In Fig. 2.1, we use $\mathbf{y}^*$ to denote the position of the reaching target since it is closely related to visual sensory information (in a more generic way, one can think of $\mathbf{y}$ as the overall sensory inflow of the agent; depending on the specific module, a different subset of this information is selected and further processed before it is finally used as the module's input). In this example, the inverse kinematics model transforms the target coordinates of the desired movement end point $\mathbf{y}^*$ (point B) into the corresponding desired arm state $\mathbf{x}^* = (\boldsymbol{\theta}^*, \mathbf{0})$ (the desired final velocity should be zero in the reaching example). One has to note that the notation is slightly inconsistent at this point since the output of inverse models is usually a motor command $\mathbf{u}$. However, in the overall framework of Fig. 2.1, we consider the kinematic motor command $(\boldsymbol{\theta}^*, \mathbf{0})$ mainly in its role as desired state $\mathbf{x}^*$ of the system.

Trajectory planning Knowing the current state estimate $\widehat{\mathbf{x}}(t_0)$ and the desired final state $\mathbf{x}^*$, a trajectory between both points can be planned. This is the task of the trajectory planning module. The result is a function $\mathbf{x}_d(t)$ which is defined over a time interval $[t_0; t_{\text{final}}]$. The movement duration $t_{\text{final}} - t_0$ is also a result of the planning process. Knowing $\mathbf{x}_d(t)$ implies knowing $\boldsymbol{\theta}_d(t)$ and all of its derivatives; most interesting are the angular velocity $\dot{\boldsymbol{\theta}}_d(t)$ and the angular acceleration $\ddot{\boldsymbol{\theta}}_d(t)$.

Inverse dynamics The inverse dynamics module has to transform the trajectory $\mathbf{x}_d(t)$ into the correct motor commands $\mathbf{u}(t)$. The function $\mathbf{u}(t)$ describes the muscle activation during the movement (or on a more abstract level the torques $\boldsymbol{\tau}_d(t)$). Because of the physical characteristics of the system, not every trajectory $\mathbf{x}_d(t)$ can be realized by motor commands $\mathbf{u}(t)$ (e.g., because of impossible arm postures or accelerations which are too large to be achievable). Thus, the inverse dynamics has to be already considered during trajectory planning.

This modular approach is intended as guideline to structure the problems which are connected with motor control, to organize research approaches and results, and to discuss the role of internal models in these different contexts. It is *not* intended as model of the flow of information in the CNS during motor control. For example, trajectory planning and inverse dynamics are deeply intermingled. One could as well assume an inverse dynamics model (IDM) which takes $\widehat{\mathbf{x}}(t)$ and $\mathbf{x}^*$ as inputs and which directly generates motor commands $\mathbf{u}(t)$. Moreover, instead of the system states this IDM could use the sensory data $\mathbf{y}(t)$ and $\mathbf{y}^*$ as inputs (which would require most likely state estimation by an internal observer).

2.1.1 State estimation

The estimation of the current or of future states usually involves prediction. Within the framework of internal models, the role of the predictor is quite naturally assigned to forward models (FMs). In an influential study by Wolpert et al. (1995b), a Kalman filter model on the basis of an FM was proposed. The Kalman filter is an algorithm to estimate or predict the state of a process, in a way that approximately minimizes the mean of the squared error (Kalman, 1960).[1] It is often used in control theory as observer model. The new state estimate is computed recursively from the previous state estimate and the new sensory data.

Figure 2.2 shows the structure of the discrete Kalman filter for prediction (as outlined in Haykin, 2002, for linear systems). The filter has a predictor-corrector structure. In the upper pathway, the predictor (here an FM) generates a preliminary state estimate $\widehat{\mathbf{x}}_{t+1}^-$ for the next time step. As input, it receives the previous state estimate $\widehat{\mathbf{x}}_t$ and an external input $\mathbf{u}_t$ (in the context of motor

[1] Here, we do not differentiate between the standard Kalman filter for linear systems and the extended Kalman filter for non-linear systems; for linear systems, the Kalman filter results in a true minimization of the mean of the squared error of the state prediction.

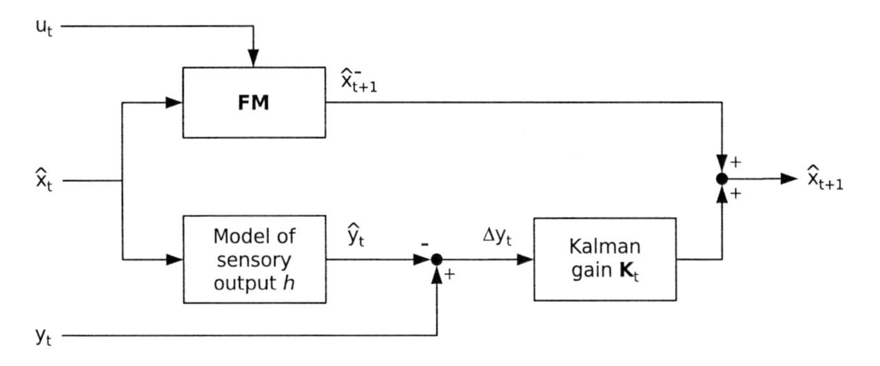

Figure 2.2 — Block diagram of the Kalman filter (for details see text).

control, this is generally a motor command). In the lower pathway, a correction of the last state estimate is computed: First, the sensory error is determined as $\Delta\mathbf{y}_t = \mathbf{y}_t - \widehat{\mathbf{y}}_t = \mathbf{y}_t - h(\widehat{\mathbf{x}}_t)$. The function h is a model of the measurement process and generates an estimate of the sensory input $\widehat{\mathbf{y}}_t$ from the estimated state $\widehat{\mathbf{x}}_t$. Afterwards, the sensory error is transformed back into the state domain by the Kalman gain matrix $\mathbf{K}_t$. Finally the results of the upper and lower pathway (predictor and corrector) are summed up to yield the final state prediction:

$$\widehat{\mathbf{x}}_{t+1} = \widehat{\mathbf{x}}_{t+1}^{-} + \mathbf{K}_t \Delta\mathbf{y}_t$$

This equation illustrates clearly that the magnitude of the elements in the Kalman gain matrix $\mathbf{K}_t$ determines the amount of correction of the original prediction $\widehat{\mathbf{x}}_{t+1}^{-}$.

The discrete Kalman filter includes an algorithm to compute the Kalman gain matrix $\mathbf{K}_t$ in a way that the mean of the squared error of the state prediction is approximately minimized. The basis of this computation is the noise in the state prediction process and the noise in the measurement process. Depending on the magnitude of these two sources of noise, the Kalman gain either gives more weight to the prediction $\widehat{\mathbf{x}}_{t+1}^{-}$ or to the correction $\Delta\mathbf{y}_t$. In this prediction scheme, the computation of $\mathbf{K}_t$ involves knowing the derivatives of the predictor in the current and in the previous time step (Haykin, 2002).[2]

In the study by Wolpert et al. (1995b), human subjects had to move one of their arms in the absence of visual feedback. At the end of the movement, subjects had to estimate the position of their thumb by moving a cursor projected in

[2] To be precise, this prediction scheme works only for systems with linear state transitions; we stick to this scheme since it is used in the study by Wolpert et al. (1995b).

the plane of the thumb along the movement line (via a trackball operated with the other hand). In doing so, the subjects showed a clear bias to overestimate the movement length (peaking around a movement duration of one second). Moreover, when an assistive force field was generated during the movement, the overestimation bias increased, while a resistive force field resulted in a reduced bias. Wolpert et al. (1995b) report the time course of the bias under the varying conditions for a movement time between 0.5 and 2.5 seconds. The main research question in their study is if the estimation of the thumb's position relies on motor information (efference copy), on sensory information (from proprioception), or on a mixture of both.

Wolpert et al. (1995b) proposed a Kalman filter model (like in Fig. 2.2) to determine which kind of processing yields the best fit to the experimental results. The predictor is a linear FM of the arm dynamics (with two state variables, the velocity and the position of the thumb). The motor input $\mathbf{u}_t$ is simplified to the sum of the externally and the internally generated force. The parameters of the model were fitted to the experimental data, especially the amounts of noise in the state prediction process (relying on motor data) and in the measurement process (relying on sensory data), which determine the Kalman gain. Only when both the state prediction and the sensory correction contributed to the state estimation, the time course of the estimation bias could be reproduced by the model under the varying conditions. Using efferent or proprioceptive information alone could not account for the experimental data. The authors conclude: "We feel that the results of this state estimation study provide evidence that a forward model is used by the CNS in maintaining its estimate of the hand location" (Wolpert et al., 1995b, p. 1882).

This study is important in two respects: First, it supports the FM hypothesis with experimental data, and second, it puts forward the idea that the state prediction by FMs in the CNS is corrected as soon as sensory data is available. For this predictor-corrector process, the Kalman filter provides a well-understood framework from signal processing.

The Kalman filter approach to state estimation fits well into the larger framework of Bayesian integration of sensorimotor information (Kording and Wolpert, 2006). According to Bayes's theorem, the probabilities $P(A)$ and $P(B)$ and the conditional probabilities $P(A|B)$ and $P(B|A)$ are linked in the following way:

$$\overbrace{P(A|B)}^{posterior} = \frac{\overbrace{P(B|A)}^{likelihood}\ \overbrace{P(A)}^{prior}}{P(B)}$$

$P(A|B)$ is often called the posterior belief, while $P(A)$ is the prior belief.

Bayes's theorem also applies to continuous probability densities. For state estimation, the continuous version yields:

$$p(\mathbf{x}|\mathbf{y}) = \frac{p(\mathbf{y}|\mathbf{x})p(\mathbf{x})}{p(\mathbf{y})} \tag{2.1}$$

$p(\mathbf{x})$ is the estimated probability density of the state *a priori* with the expected value $\widehat{\mathbf{x}}^-$. To update this state estimate, an observation $\mathbf{y}$ is made. Through Bayes's rule, the estimated probability distribution $p(\mathbf{x}|\mathbf{y})$ of the state *a posteriori* with the expected value $\widehat{\mathbf{x}}$ is determined, which takes the observation $\mathbf{y}$ into account. The likelihood $p(\mathbf{y}|\mathbf{x})$ has to be derived from the measurement process; this likelihood is the probability density of the observation $\mathbf{y}$ under varying values of $\mathbf{x}$. If all distributions are univariate Gaussians and the sensory output model h is the identity function, the computation of $p(x|y)$ is straightforward (Kording and Wolpert, 2006):

$$\begin{aligned} \widehat{x} &= \lambda y + (1-\lambda)\widehat{x}^- \\ \lambda &= \frac{\sigma_x^2}{\sigma_x^2 + \sigma_y^2} \end{aligned} \tag{2.2}$$

σ_x^2 is the variance of the prior distribution, σ_y^2 is the variance of the likelihood. The width of the posterior distribution is $\sigma = \lambda\sigma_y$.

Equation (2.1) can be generalized to multiple sources of sensory information. The resulting posterior distribution provides the optimum state estimate in which each source of information is weighted in proportion to its precision (the inverse of its variance) as in Eqn. (2.2) (Bays and Wolpert, 2007). The Kalman filter can be derived from the Bayesian framework as well (Barker et al., 1994).

Bays and Wolpert (2007) and Kording and Wolpert (2006) review experimental studies which support the view that the CNS uses Bayesian inference to minimize uncertainty and variability in sensorimotor control: These studies concern visual illusions, the integration of cues from different modalities, and strategies in motor control. The study by Vaziri et al. (2006) is of special interest because it deals with an oculomotor FM. When humans fixate a reach target but then look away, the CNS generates before the actual eye movement an estimate of the remapped peripheral location of the reach target (as it would appear after the eye movement) in fixation-centered coordinates (this is an interpretation of the studies on predictive remapping by Duhamel et al. (1992); Umeno and Goldberg (1997); Walker et al. (1995); see also Sect. 1.3.4.2). This remapping can be interpreted as FM which uses a copy of the oculomotor command as motor input. In this setting, subjects have two sources of target information for the reaching movement available after looking away. First, the remapped target position, and second, the target as it appears in peripheral vision. From a theoretical viewpoint, this could be advantageous because the actual peripheral position of the reach target on the retina might not be precise enough due to the low resolution to generate an adequate reaching movement. In a controlled experimental setting, Vaziri et al. (2006) examined the precision of reaching movements when only one of the two sources or both were available to the subjects. Their results indicate that the CNS combines both sources in a Bayesian way for optimal reaching performance. Vaziri et al. (2006) propose two different integration mechanisms: Both target representations coexist until the reaching movement is triggered (one representation in memory and the other in visual cortical areas), or there is only one target representation which is updated through a Kalman filter. In conclusion, this study provides further support for the existence of FMs for state estimation in the CNS.

2.1.2 Kinematic control

As already mentioned, the term "kinematics" refers to a transformation between coordinate systems (Jordan, 1996). A common example from engineering is the transformation between joint angles $\boldsymbol{\theta}$ of a robot arm and the position and orientation of its gripper tip in the world coordinate system. This relationship is often referred to as "forward kinematics" while the inverse relationship is designated as "inverse kinematics" (Spong and Vidyasagar, 1989). For motor control, the inverse kinematics is required to determine the correct posture for a given desired limb position. In the example of the reaching task, the target is first represented in retinal (eye-centered) coordinates. These coordinates have to be transformed first into a head-centered and finally into a body-centered representation (Battaglia-Mayer et al., 2003; Buneo et al., 2002; Carrozzo et al., 1999). From the body-centered coordinates, the final reaching posture of the arm (the joint angles) can be generated by an inverse kinematics model (IKM).

Actually, there is a large amount of experimental evidence from neurophysiological and psychophysical studies on humans and primates that these coordinate transformations take place in the CNS (for a review, see Battaglia-Mayer et al., 2003). Depending on the motor task and the available sensory information about target and hand, the most relevant reference frames seem to differ as the following two findings suggest. In support of the stepwise coordinate transformations as outlined above, Snyder (2000) presents the finding that in some cortical regions the locally represented retinal position is modulated by the population code of the gaze direction (this modulation has been termed "gain fields"). On the other hand, the results by Buneo et al. (2002) from a neurophysiological study on monkeys indicate that the posterior parietal cortex (PPC) can also directly transform eye-centered target coordinates to hand-centered target coordinates if both the target and the hand are visible. So far, these results refer to the input of hypothetical IKMs in the CNS. Moreover, in a study by Graziano et al. (2002) on monkeys, the stimulation of certain motor cortex neurons lead to hand locations independent of the initial arm posture. This corresponds to the output of an IKM. Another candidate for an IKM in the CNS is the oculomotor map found in the superior colliculus. Here, retinotopic target positions are closely linked to eye viewing directions (Leigh and Zee, 1999). On a higher level, the lateral intraparietal area in the PPC seems to serve as a sensorimotor "interface" for the production of saccades. Buneo and Anderson write: "By interface we mean a shared boundary between the sensory and motor systems where the meanings of sensory and motor-related signals are exchanged" (Buneo and Andersen, 2006, p. 2595).

Very often, the input-output relationship of the inverse kinematics is a one-to-many mapping. For example, to grasp for a cup on the desk before them, humans can use a large variety of different final arm postures. Nevertheless, many human movements and final postures are rather stereotype, thus the CNS seems to prefer certain solutions of the inverse kinematics (Cruse et al., 1990; Grea et al., 2000). In robotics, one distinguishes between redundant robot manipulators with a continuous set of solutions for the inverse kinematics and non-redundant ones for which only several distinct solutions exist (Spong and Vidyasagar, 1989). The existence of one-to-many mappings is a challenge for many adaptive learning strategies in motor control (see Sect. 2.2.2).

The "knowledge" model of Rosenbaum et al. (1995) is a kinematic model of motion planning for planar arm movements. One of its principal aims is to explain how the human motor system solves the redundancy problem. The solution which is offered by the model relies on a memory for stored postures. To generate a movement, all stored postures are evaluated with respect to their spatial error costs and travel costs. Spatial error costs rely on the difference between the position of the movement target in the world coordinate system and the position of the specific contact point (usually, the hand). Travel costs are related to the difference in the joint angles between the stored posture and the current posture. Depending on their costs, a weight term is computed for all stored postures. The final posture is determined as weighted sum of all stored postures in joint angle space. Whenever the spatial error costs of the final posture are too large, a special mechanism called "feedforward correction" is applied to generate new postures with smaller spatial errors. This model predicts successfully several experimental results, for example the effects of starting positions on final postures (for an overview, see Rosenbaum et al., 2001). Subsequently, Rosenbaum et al. (2001) extended the model to grasping and obstacle avoidance and offered a more flexible way of accounting for costs through a constraint hierarchy. The authors put forward several reasons why human motion planning is based on the final posture of the movement. Most important, the variability of end positions is generally smaller than the variability of movements to those end positions (Desmurget et al., 1995). Moreover, the memory for final positions is better than the memory for movements (Baud-Bovy and Viviani, 1998). In summary, the IKM in the knowledge model is based on a lookup table. The posture which is finally retrieved depends on the movement costs.

The research by Cruse and colleagues (Cruse et al., 1990, 1993) provides further support for the idea that final postures of the redundant human arm are determined on the basis of cost functions. In their approach, neither the spatial error nor the travel costs are considered, but instead the "comfort" of keeping the arm in a certain posture. In different psychophysical experiments on reaching movements in a horizontal plane, Cruse et al. (1990) discovered u-shaped cost functions for the shoulder, elbow, and wrist. The minimum cost (maximum comfort) was found approximately at the middle of the range of joint movement for different subjects. In the model of Cruse et al. (1990), the overall cost of an arm posture is the sum of the cost values for shoulder, elbow, and wrist. This model offers no mechanism for an IKM, but explains how an IKM could arrive at a certain arm posture despite the redundancy.

In contrast to dynamical control tasks where even the optimal output of the controller can only decrease the difference between the desired and the current state over time, there is always a "one-shot" solution available in kinematic problems. Kinematics involves only the relation between two different coordinate systems without any dynamic aspect.

2.1.3 Trajectory planning

The trajectory planning module has to specify a trajectory $\mathbf{x}_d(t)$ with $t \in [t_0; t_{\text{final}}]$ from a known starting point $\widehat{\mathbf{x}}(t_0)$ to a desired final point $\mathbf{x}^*$. Here, our presentation suggests that the trajectory is planned in state space, consisting of joint angles and velocities. Although every trajectory can be specified in this space, their is actually a lot of controversy about the level which is most relevant for trajectory planning. Two main classes of models exist: kinematic and dynamic models. Kinematic models start from the assumption that the trajectory is planned in (body-centered) workspace coordinates, while dynamic models refer to dynamic variables like torques, forces, or muscle activations.

Point to point arm movements by human subjects show roughly straight hand paths and bell-shaped velocity profiles (Abend et al., 1982) in workspace coordinates. Every trajectory formation model has to account for this finding.

Many models start from a cost function which has to be minimized (often with the mathematical methods of optimal control theory). The final trajectory is the one with minimum cost. One of the most influential kinematic models is the minimum jerk model (Flash and Hogan, 1985). Jerk is the derivative of acceleration. The cost function for planar arm movements in this model is:

$$C = \frac{1}{2} \int_{t_0}^{t_{\text{final}}} \left(\left(\frac{d^3 x}{dt^3} \right)^2 + \left(\frac{d^3 y}{dt^3} \right)^2 \right) dt$$

Here, x and y denote the coordinates of the hand in the 2D plane. The resulting trajectory is computed in 2D workspace coordinates. Accordingly, in this model the IDM which finally computes the torques has also to solve the inverse kinematics problem (at least implicitly). But in the first place, the optimal trajectory is determined completely independently of the dynamical quantities such as additional payloads, torques, or external forces. This appeared to be rather implausible and led to the development of dynamic trajectory formation models like the minimum torque-change model (Uno et al., 1989). Here, the cost function is:

$$C = \frac{1}{2} \int_{t_0}^{t_{\text{final}}} \sum_{i=1}^{n} \left(\frac{d\tau_i}{dt} \right)^2 dt$$

n is the number of joints, and τ_i is the torque applied to joint i.

At the first glance, the minimum torque-change model seems to be a very elegant solution because the final trajectory is directly specified in torques $\tau_i(t)$, thus no further motor controller is required. But internally, the trajectory formation process requires an IKM, a state-space model of the arm dynamics, and advanced mathematics to find the optimal trajectory. To overcome these difficulties, Uno et al. (1989) presented an iterative learning scheme which generates an approximate solution for the trajectory. In contrast to the minimum jerk model, the minimum torque-change model succeeds in predicting the hand path of human subjects when an external force is applied. However, more recent studies have shown that subjects adapt their movements in dynamic environments over many practice trials, finally resulting in a straightening of the hand paths (Shadmehr and Mussa-Ivaldi, 1994). Thus, a straight movement path in workspace coordinates seems to be the preferred solution. This finding conforms rather to the hypothesis of kinematic trajectory planning.

Further support for the kinematic approach stems from a study by Wolpert et al. (1995a). In this study, subjects had to perform two-joint planar arm movements. The visual feedback was disturbed to suggest increased curvature of the movements. In accordance with the prediction of the minimum jerk model, subjects adjusted their movements so that the perceived movement became nearly straight again. This contradicts with the minimum torque-change model which predicts no adjustment in such a setting. Wolpert et al. (1995a) conclude that their results are incompatible with purely dynamics-based models. They suggest that "spatial perception — as mediated by vision — plays a fundamental role in trajectory planning" (Wolpert et al., 1995a, p. 460).

The minimum-variance theory by Harris and Wolpert (1998) offers an integration of the kinematic and the dynamic approach. The foundation of their theory is the assumption that the neural signals for motor control are corrupted by noise, whose variance increases with the size of the control signal. The main principle of trajectory formation is the minimization of the variance of the final arm (or eye) position for a specified movement duration, or equivalently the minimization of the movement duration for a specified final positional variance. This theory predicts smooth trajectories with bell-shaped velocity profiles; moreover, it accounts for Fitt's law (Fitts, 1954), which basically states that the speed of human movements is reciprocally related to their accuracy. A further advantage of the minimum-variance theory is that it provides a reasonable explanation in itself: Relying on the minimum-variance criterion, the CNS would achieve the most precise trajectories it can generate with a noisy motor system. On the contrary, it is not clear why the CNS should apply optimization criteria like minimum jerk or minimum torque-change. On the downside, the minimum-variance model requires a state-space model of the plant and advanced mathematical computations (see Harris and Wolpert, 2006, on saccadic eye movements) which is not biologically plausible.

Cruse and colleagues (Cruse and Brüwer, 1987; Cruse et al., 1993) proposed a different mechanisms to explain that hand paths sometimes deviate from a straight line. They assume that "the control system might use a compromise between a straight line in joint space and a straight line in the work space" (Cruse et al., 1993, p. 138). A global strategy aims on achieving a straight path in joint space, a local strategy aims on a straight path in the work space. The local strategy works by computing the pseudoinverse of the Jacobian of the plant which maps joint angles to workspace coordinates. The pseudoinverse is used to determine the changes of the joint angles for a small movement step in the workspace. Moreover, the resulting joint angle changes are weighted by the corresponding cost functions which assign comfort values to joint angles (Cruse et al., 1990, 1993). Thus, the local strategy combines pseudoinverse control with a minimum cost principle. However, although Dean and Porrill (1998) suggested an adaptive biological model for pseudoinverse control of eye movements, it is still unclear how the CNS could carry out the necessary computations for pseudoinverse control of the arm.

So far, we have considered simple point-to-point movements. To define

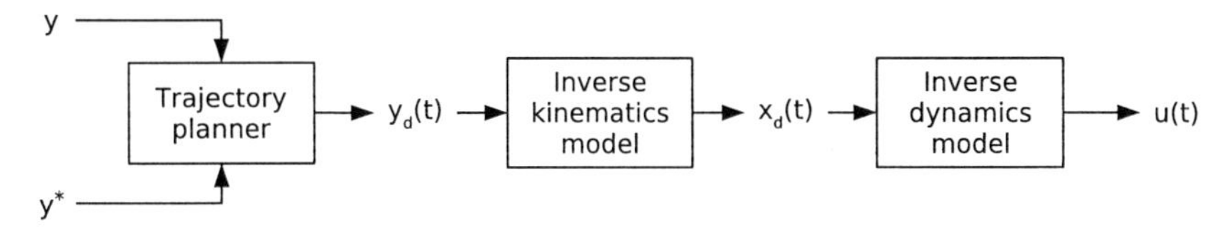

Figure 2.3 — Modification of the framework in Fig. 2.1 for the minimum jerk model (for details see text).

the trajectories of complex and curved arm movements, it is often assumed that the CNS uses via-points. These are points in the work space with a time stamp. Along the trajectory, the hand has to pass through each via-point at the specified time. Usually, the overall trajectory is smooth without stopping at the via-points (except for movement reversals). Wada and Kawato (1993) proposed the "forward-inverse relaxation model" (FIRM) for trajectory formation with via-points with a fixed time stamp. The FIRM model uses the minimum torque-change criterion to generate the overall trajectory. Wada and Kawato (1993) report that it can reproduce complicated human motion trajectories precisely. In an extension of their work, Wada and Kawato (2004) modified the FIRM model so that it only needs the spatial position of the via-points to start with and generates the time stamps within the optimization process.

Although the model by Rosenbaum et al. (2001) (see Sect. 2.1.2) is purely kinematic, it offers a mechanism for obstacle avoidance and trajectory planning via a single via-point. In this model, each movement to a target point is a superposition of a first movement to a via-point and to a second movement from the via-point to the final goal. Candidate postures for the goal and via-point are evaluated according to a certain constraint hierarchy. One constraint implies that collisions should be avoided along the trajectory. Collision detection is carried out by simulating movements on basis of the simplified (and basically wrong) assumption that trajectories follow straight-line movements through joint space. In contrast to the other aforementioned models for trajectory generation, the model by Rosenbaum et al. (2001) does not *minimize* costs, instead it searches for appropriate postures until the movement constraints are *satisfied*. The Rosenbaum model ignores the dynamics in this process, but instead relies on a simplified mechanism for trajectory generation. In this respect, the dynamic trajectory planning models and the Rosenbaum model supplement each other.

When we reconsider the framework in Fig. 2.1, the minimum jerk model fits in quite well, although some modifications are needed. A simplified and modified version of the framework is depicted in Fig. 2.3. The new framework leaves

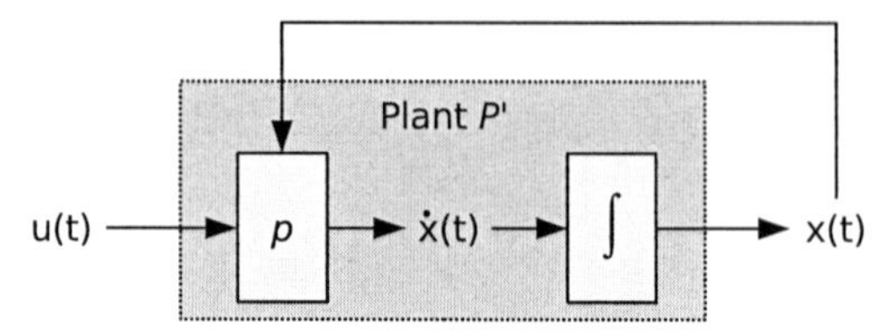

Figure 2.4 — The plant P' as transfer element, composed from the serial connection of the state change function p and an integrator.

out state estimation, and the sensory variables $\mathbf{y}$ and $\mathbf{y}^*$are now identified with the position of the hand and the reaching target in the workspace. Although these variables are directly connected to the visual information from the retina, their representation in body-centered workspace coordinates requires state estimation. Thus, this notation is a simplification. According to the minimum jerk model, the trajectory planner generates the trajectory $\mathbf{y}_d(t)$ in workspace coordinates. An IKM has to transform $\mathbf{y}_d(t)$ into the corresponding joint angle trajectory $\mathbf{x}_d(t) = \boldsymbol{\theta}_d(t)$. As in Fig. 2.1, $\mathbf{x}_d(t)$ is the input for the IDM which generates the final motor commands.

2.1.4 Dynamic control

The task of dynamic control is to generate the necessary torques or muscle activations for a desired movement trajectory. In the dynamic domain, it is more convenient to switch to a time-continuous notation. In time-continuous state-space systems, the plant dynamics is usually defined by the state equation, a set of differential equations:

$$\dot{\mathbf{x}}(t) = p(\mathbf{x}(t), \mathbf{u}(t)) \tag{2.3}$$

In contrast to the time-discrete notation, in which the plant P' is just a simple next state function, P' is now a transfer element which can be depicted by the serial connection of the state change function p and an integrator (see Fig. 2.4). To determine $\mathbf{x}(t)$, it is necessary to solve the state equation. For a multi-joint arm with the state vector $(\boldsymbol{\theta}, \dot{\boldsymbol{\theta}})$, the state equation is usually defined as follows (Kawato, 1990):

$$\begin{aligned} d\boldsymbol{\theta}/dt &= \dot{\boldsymbol{\theta}} \\ d\dot{\boldsymbol{\theta}}/dt &= f(\boldsymbol{\theta}, \dot{\boldsymbol{\theta}}, \boldsymbol{\tau}) \end{aligned} \tag{2.4}$$

Equation (2.4) expresses that the changes of the angular velocities (the angular accelerations $\ddot{\boldsymbol{\theta}}$) depend on the current joint angles $\boldsymbol{\theta}$, the current joint velocities $\dot{\boldsymbol{\theta}}$, and on externally applied torques $\boldsymbol{\tau}$. For multi-joint manipulators, the function f is usually non-linear since it contains sine and cosine terms of θ_i (for an example, see Eqns. (2.7-2.8)). The task of the IDM is to transform a trajectory $\boldsymbol{\theta}_d(t)$ (which implies $\dot{\boldsymbol{\theta}}_d(t)$ and $\ddot{\boldsymbol{\theta}}_d(t)$) into the torques $\boldsymbol{\tau}_d(t)$. In the following, two opposing control strategies, feedforward control and feedback control, and a mixture of both are presented.[3]

2.1.4.1 Feedforward control

In feedforward control, the IDM is applied as shown in Fig. 2.1. The IDM implements a solution of Eqn. (2.4) with respect to $\boldsymbol{\tau}$:

$$\boldsymbol{\tau}_d(t) = \mathrm{IDM}(\boldsymbol{\theta}_d(t), \dot{\boldsymbol{\theta}}_d(t), \ddot{\boldsymbol{\theta}}_d(t)) \tag{2.5}$$

Ignoring static friction, Eqn. (2.4) is linear with respect to the torques $\boldsymbol{\tau}_d(t)$ because of the general relationship $\tau = I\ddot{\theta}$ (with I being the moment of inertia). Other than this, τ has no direct impact on joint angles or velocities. Thus, there exists an explicit analytical form of the IDM in Eqn. (2.5).

Nevertheless, pure feedforward control through the IDM is little bit risky. While the IDM represents a functional relationship (a many-to-one mapping), the inverse of the IDM does not. Accordingly, a certain control signal $\boldsymbol{\tau}(t)$ can have a very different impact depending on the actual state $(\boldsymbol{\theta}(t), \dot{\boldsymbol{\theta}}(t))$ of the system. Only if the IDM is really a perfect counterpart of the plant, and if the actual starting state $(\boldsymbol{\theta}(t_0), \dot{\boldsymbol{\theta}}(t_0))$ is equal to $(\boldsymbol{\theta}_d(t_0), \dot{\boldsymbol{\theta}}_d(t_0))$, the actual trajectory $\boldsymbol{\theta}(t)$ follows exactly the desired trajectory $\boldsymbol{\theta}_d(t)$.

In practice, feedforward control is not well suited for arm movements where a lot of state variables come into play and even a slightly inprecise IDM can lead to an undesired outcome. Moreover, feedforward control cannot cope with noise in the sensory and motor channels and cannot react at all to unexpected external forces or changes in the payload.

Nevertheless, there are models which propose feedforward control for certain tasks of the human motor system. One of these models is the model by Robinson (1981) for the vestibulo-ocular reflex (VOR). The VOR compensates for head motion by moving the eyes in the opposite direction with opposite speed. In the VOR model by Robinson (1981), this is achieved by an inverse model of the oculomotor plant which is used for open-loop feedforward control.

[3] Depending on the control strategy, the desired state for multi-joint arm movements is sometimes defined as $\mathbf{x}_d(t) = (\boldsymbol{\theta}_d(t), \dot{\boldsymbol{\theta}}_d(t))$ (matching the system state; mainly in feedback control) or as $\mathbf{x}_d(t) = (\boldsymbol{\theta}_d(t), \dot{\boldsymbol{\theta}}_d(t), \ddot{\boldsymbol{\theta}}_d(t))$ (extending the system state by $\ddot{\boldsymbol{\theta}}_d(t)$; mainly in feedforward control).

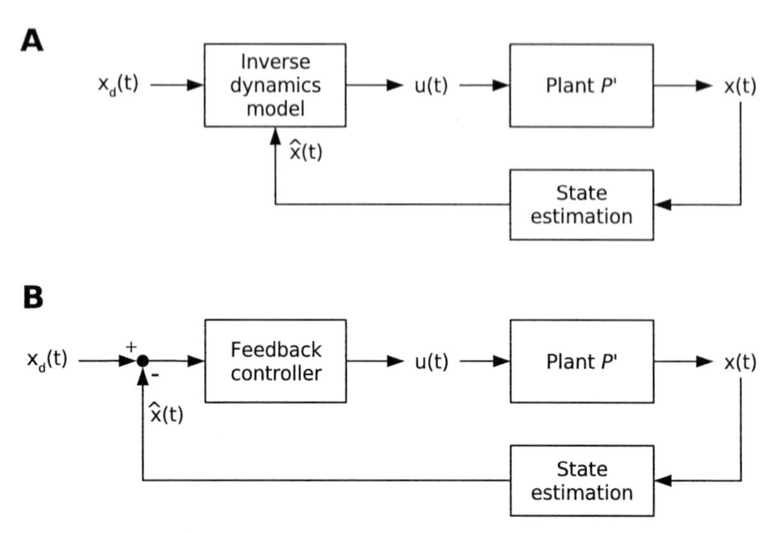

Figure 2.5 — Simplified feedback control schemes, leaving out any reference to the sensory variables $\mathbf{y}(t)$. A: Predictive control with an IDM. B: Error-correcting feedback control.

2.1.4.2 Feedback control

In feedback control, not only the desired state or the desired output of the plant are fed as input to the controller, but also the actual output of the plant or an estimate of its current state (see Fig. 2.5). If the controller is well-designed, it can overcome all the difficulties of pure feedforward control. On the downside, feedback control faces the problem of time delays in the feedback loop. But this is a topic on its own which is covered in Sect. 2.1.4.6.

Jordan (1996) distinguishes between two types of feedback control. The first belongs to the category of so-called "predictive control". Here, the controller is an IDM (see Fig. 2.5a). In contrast to the IDM for feedforward control, this IDM gets both the desired state $\mathbf{x}_d(t)$ and the current state estimate $\widehat{\mathbf{x}}(t)$ as input. The feedback IDM generates (ideally) the optimal motor ouput $\mathbf{u}(t)$ to decrease the difference between $\mathbf{x}_d(t)$ and $\widehat{\mathbf{x}}(t)$ as fast as possible (or according to another optimality criterion). Since $\mathbf{x}_d(t)$ and $\widehat{\mathbf{x}}(t)$ are functions over time, their difference will change dynamically. Overall, the feedback IDM is basically an optimal controller. In addition, a feedback IDM can be viewed as a kind of trajectory planner because it generates a trajectory "online" from the starting point of the movement towards the desired goal position (assuming that this position is fixed). This shall illustrate that the overall sketch in Fig. 2.1 is only a framework to guide the presentation, and that there are smooth transitions between the different modules.

The feedforward IDM can be interpreted as a feedback IDM where the desired and the current state are infinitesimally close together. Returning to the example of arm control, where $\mathbf{x}(t) = (\boldsymbol{\theta}(t), \dot{\boldsymbol{\theta}}(t))$, this infinitesimal difference is expressed by the angular acceleration $\ddot{\boldsymbol{\theta}}(t)$. A direct comparison of the output equations of the feedforward $\mathrm{IDM_{ff}}$ and the feedback $\mathrm{IDM_{fb}}$ shows this relationship:

$$\begin{aligned}\boldsymbol{\tau}_d(t) &= \mathrm{IDM_{ff}}(\boldsymbol{\theta}_d(t), \dot{\boldsymbol{\theta}}_d(t), \ddot{\boldsymbol{\theta}}_d(t)) \\ \boldsymbol{\tau}_d(t) &= \mathrm{IDM_{fb}}(\boldsymbol{\theta}_d(t), \dot{\boldsymbol{\theta}}_d(t), \widehat{\boldsymbol{\theta}}(t), \widehat{\dot{\boldsymbol{\theta}}}(t))\end{aligned}$$

Using a feedforward IDM belongs also to the category of predictive control (Jordan, 1996).

The second type of feedback control is "error-correcting feedback control" (see Fig. 2.5b). Here, the controller receives only the difference between $\mathbf{x}_d(t)$ and $\widehat{\mathbf{x}}(t)$ as input. Jordan (1996) emphasizes as main difference between predictive and error-correcting control that the latter can only react *after* the error has already occured. Furthermore, feedback controllers (FCs) are mostly constructed heuristically according to some qualitative knowledge of the plant, but not as its optimal counterpart like an IDM. Often, this approach works quite well and can provide a starting point to identify a real IDM as outlined in the next sections.

2.1.4.3 Composite control systems

Feedforward and feedback control have complementary strengths and weaknesses. For this reason, it is a reasonable approach to integrate both in a composite control scheme. A common and generally successful approach is to add both control signals as shown in Fig. 2.6. Here, a feedforward IDM and an error-correcting FC work together. Because of the feedforward IDM, the system is less sensitive to time delays in the feedback loop, and on the other hand, the FC helps to overcome unexpected disturbances, noise, and possible inaccuracy of the IDM (Jordan, 1996).

2.1.4.4 Feedback-error learning

Feedback-error learning (FEL) is a learning strategy for feedforward IDMs which has been proposed by Kawato et al. (1987) (see also Gomi and Kawato, 1993; Kawato, 1990). It is based on the composite control system in Fig. 2.6.

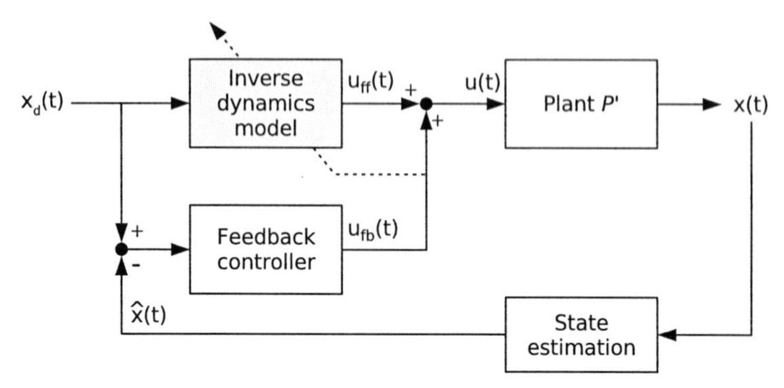

Figure 2.6 — Combination of feedforward and feedback control. The dashed arrow indicates the error signal in feedback-error learning. For simplicity, any reference to the sensory variables $\mathbf{y}(t)$ is left out.

It has been developed in the context of dynamical systems. For this reason, it is presented here. The following section on motor learning (Sect. 2.2) is mostly dedicated to kinematic control and discusses FEL only in this context.

In adaptive control systems, the motor controller is not "pre-wired" but instead acquired during the interaction of the controller with the plant. Often, neural networks are used as adaptive controllers. In the FEL strategy, the corrective output of the FC is used as training signal for the adaptive feedforward IDM. At each time $\tilde{t}$, the input $\mathbf{x}_d(\tilde{t})$ and the output $\mathbf{u}(\tilde{t}) = \mathbf{u}_{\text{ff}}(\tilde{t}) + \mathbf{u}_{\text{fb}}(\tilde{t})$ form a learning example for supervised neural network training. In the context of arm movements, Kawato (1990) suggests the following FC (omitting the time variable for simplicity):

$$\tau_{\text{fb}} = K_P(\boldsymbol{\theta}_d - \boldsymbol{\theta}) + K_V(\dot{\boldsymbol{\theta}}_d - \dot{\boldsymbol{\theta}}) + K_A(\ddot{\boldsymbol{\theta}}_d - \ddot{\boldsymbol{\theta}}) \tag{2.6}$$

This is a proportional controller with three different gain factors for position, velocity, and acceleration. Kawato (1990) shows that the resulting learning scheme is a valid approximation of a Newton-like method in functional space. A more thorough stability analysis is developed in Nakanishi and Schaal (2004) on the basis of nonlinear adaptive control theory.

Kawato and colleagues (Kawato, 1990; Kawato and Gomi, 1992a,b) have proposed a cerebellar feedback-error learning model (CBFELM). According to this model, distinct microzones within the cerebellar cortex act as feedforward IDMs for different motor tasks. Each microzone has two different sources of afferent input, mossy fibers and climbing fibers. The CBFELM proposes that the mossy fibers carry sensory information and information about the desired

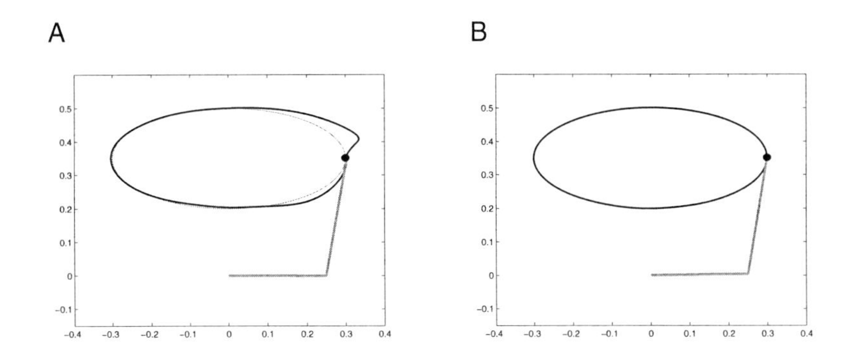

Figure 2.7 — The desired elliptical trajectory is shown as fine line, the performed trajectory as bold line. The starting point is marked with a black dot. In addition, the simulated arm is depicted in light gray. A: First round during FEL. B: 120th round.

trajectory, while the climbing fibers transmit a motor error signal from crude FCs in the brain stem or spine. The mossy fibers form excitatory synapses with the granule cells (first layer of the cerebellar cortex). The Purkinje cells (second cortical layer) integrate the activation carried via the parallel fibers (axons of the granule cells) and the climbing fibers. At this integration point, adaptation according to the feedback-error signal in the climbing fibers is supposed to take place.

Wolpert and Kawato (1998) summarize the results of several neurophysiological studies on ocular-following responses (slow tracking movements of the eyes evoked by movements of large-field visual stimuli). These studies provide support for the CBFELM by data from single-cell recordings comparing the complex spikes and simple spikes of Purkinje cells (the former elicited by climbing fiber input, the latter by parallel fiber input).

A small experiment on FEL In the following paragraphs, a small experiment on FEL for a simulated two-link manipulator for planar arm movements

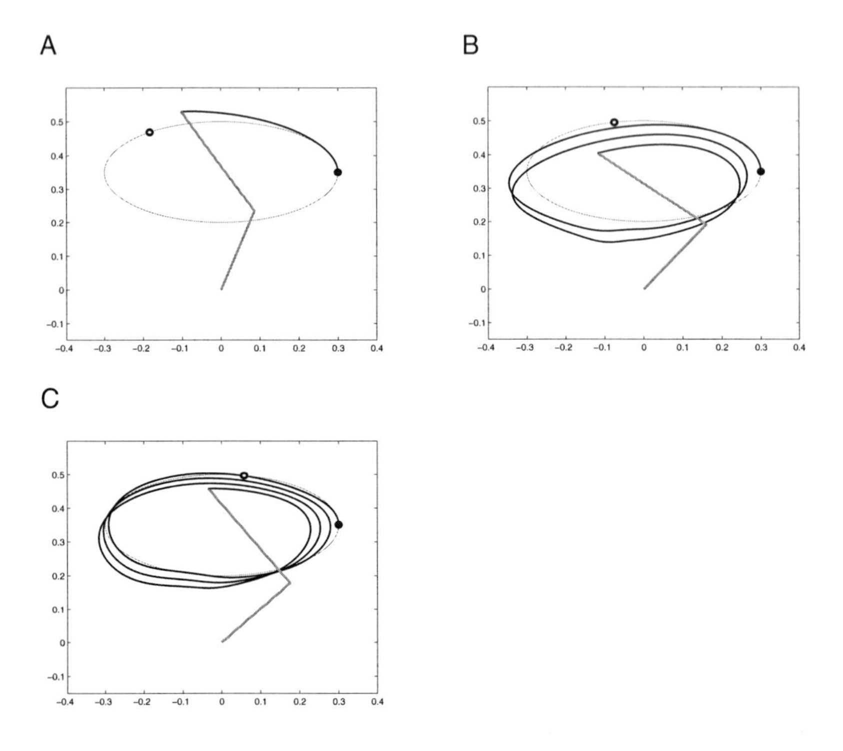

Figure 2.8 — The small circle on the desired trajectory marks the position at which the simulation of the feedforward IDM is stopped; for further explanation of the diagrams see Fig. 2.7. A-C: Performance of feedforward IDMs without FC after different types of training. A. Training by standard FEL. B. Training by FEL with random ordering of learning examples. C. Training by a strategy similar to direct inverse modeling.

is presented. The dynamics for the arm are defined by the following equations:

$$\begin{aligned} \tau_1 &= (I_1 + I_2 + 2M_2L_1S_2\cos\theta_2 + M_2(I_1)^2)\ddot{\theta}_1 \\ &\quad + (I_2 + M_2L_1S_2\cos\theta_2)\ddot{\theta}_2 \\ &\quad - M_2L_1S_2(2\dot{\theta}_1 + \dot{\theta}_2)\dot{\theta}_2\sin\theta_2 + b_1\dot{\theta}_1 \qquad (2.7) \\ \tau_2 &= (I_2 + M_2L_1S_2\cos\theta_2)\ddot{\theta}_1 + I_2\ddot{\theta}_2 \\ &\quad + M_2L_1S_2(\dot{\theta}_1)^2\sin\theta_2 + b_2\dot{\theta}_2 \qquad (2.8) \end{aligned}$$

These equations and the parameter values M_i (mass), L_i (length), S_i (distance from the center of mass to the joint), I_i (rotary intertia around the joint), and

b_i (coefficient of viscosity) are taken from the paper by Uno et al. (1989).[4] They approximately model the dynamics of the human forearm and upper arm for planar movements. As feedforward IDM, a standard multi-layer perceptron (MLP) (see Sect. 3.1) with ten hidden units is used. The FC is defined like in Eqn. (2.6) with $K_P = 1.2$, $K_V = 0.8$, and $K_A = 0.01$. The MLP is adapted by plain gradient descent with a learning rate of $\eta = 0.004$ (see Sect. 3.1.2). The motor task is to move the arm's tip along an elliptical trajectory (the fine line in Fig. 2.7a). During each round, 1000 learning cycles are carried out (the time-continuous system is approximated with Euler's method; each discrete step is used for one learning cycle).

Figure 2.7a shows the simulated trajectory (bold line) during the first round. The poor performance of the forward IDM and the necessary correction by the FC are clearly visible in the beginning. After 120 rounds, the performance of the composite system is nearly flawless (Fig. 2.7b). Nevertheless, the performance without FC is less convincing as shown in Fig. 2.8. In this figure, the performance of different feedforward IDMs is presented. The simulated trajectory is halted as soon as the Euclidean difference between the arm's tip and the current position on the desired trajectory (marked with a small circle) exceeds $0.1\ \mathrm{m}$. Figure 2.8a shows the performance of the feedforward IDM trained by FEL. The performed trajectory deviates very early from the desired one, thus the lack of the FC results in a considerable performance drop.

For comparison, the last $100,000$ learning examples with the structure $[\boldsymbol{\theta}_d, \dot{\boldsymbol{\theta}}_d, \ddot{\boldsymbol{\theta}}_d \longrightarrow \boldsymbol{\tau}]$ (with $\boldsymbol{\tau} = \boldsymbol{\tau}_{\mathrm{ff}} + \boldsymbol{\tau}_{\mathrm{fb}}$) which have been generated during FEL were used to train another MLP, but this time the learning examples were presented in random order during network training (each example exactly once). The performance of this feedforward IDM is depicted in Fig. 2.8b. The simulated trajectory continues much longer for more than two rounds. This result shows that FEL with MLPs is prone to suffer from "catastrophic interference"[5] because the learning patterns occur in a temporal order during training. Learning one part of the trajectory in one part of the input space causes worse performance in other parts of the input space. For this reason, it is advisable to use supervised online learning algorithms for FEL which adapt only locally (e.g., "supervised growing neural gas" by Fritzke, 1998). For the third feedforward IDM in the comparison, the learning examples $[\boldsymbol{\theta}_d, \dot{\boldsymbol{\theta}}_d, \ddot{\boldsymbol{\theta}}_d \longrightarrow \boldsymbol{\tau}]$ are replaced by learning examples $[\boldsymbol{\theta}, \dot{\boldsymbol{\theta}}, \ddot{\boldsymbol{\theta}} \longrightarrow \boldsymbol{\tau}]$, also collected during the last 100 rounds

[4] Parameter values in SI base units: $M_1 = 0.9; M_2 = 1.1; L_1 = 0.25; L_2 = 0.35; S_1 = 0.11; S_2 = 0.15; I_1 = 0.065; I_2 = 0.100; b_1 = 0.08; b_2 = 0.08$.

[5] The term "catastrophic interference" is often used to describe that artificial neural network tend to "forget" what they have already learned as soon as they have to adapt to new training data.

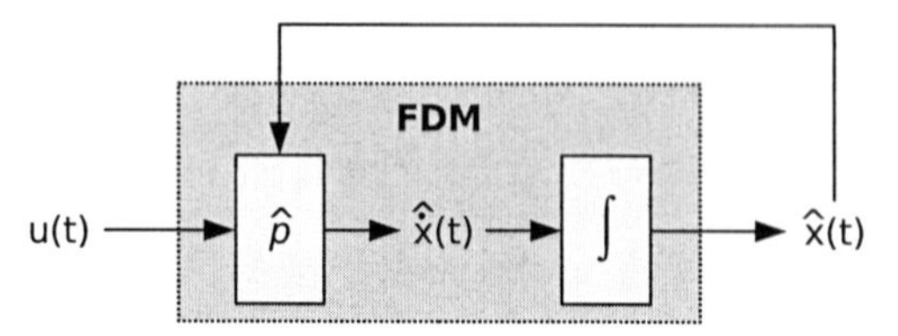

Figure 2.9 — The forward dynamics model (FDM) as transfer element, composed from the serial connection of the approximated state change function $\widehat{p}$ and an integrator.

of FEL and presented in random order during training. These learning examples perfectly represent the inverse of the plant (basically, using such learning examples is similar to "direct inverse modeling", which is described in detail in Sect. 2.2.5 for kinematic problems; for a related dynamic approach see Miller et al., 1990). The MLP which is trained with these learning examples yields the best performance (see Fig. 2.8c) with more than three rounds. This ranking order of the different types of feedforward IDMs could be replicated during several training attempts with different parameter settings, thus there is more to it than just anecdotical evidence.

In conclusion, this small experiment shows that FEL works as expected. However, depending on the implementation of the feedforward IDM, FEL may suffer from catastrophic interference. Moreover, a modification of FEL towards direct inverse modeling can improve the final performance.

2.1.4.5 Feedforward control with an internal feedback loop

For the composite control systems that are presented in this and in the next section, it is necessary to clarify the concept of the forward model in the time-continuous dynamical domain. These models are often called "forward dynamics models" (FDM). Since one cannot differentiate between the current and the next time step in the time-continuous domain, an FDM cannot directly predict a future system state. Instead, FDMs mimic the dynamics of the plant. Thus, the FDM has its own state equation (like Eqn. (2.3) for the plant) based on an approximated state change function $\widehat{p}$:

$$\widehat{\dot{\mathbf{x}}}(t) = \widehat{p}(\widehat{\mathbf{x}}(t), \mathbf{u}(t))$$

If FDMs are acquired through learning, it is basically this function which has to be learned. The overall FDM is a transfer element, composed from the approximated state change function $\widehat{p}$ and an integrator, similar to the plant P' (see Fig. 2.9). The internal state of the FDM is the state estimate $\widehat{\mathbf{x}}(t)$ of the real plant state $\mathbf{x}(t)$. At distinct points in time, here denoted as t_0, the state estimate of the FDM can be set to the real state of the system (or to a better estimate obtained from other sources): $\widehat{\mathbf{x}}(t_0) = \mathbf{x}(t_0)$. An FDM can be used to generate an internal simulation of the plant's behavior over a certain time interval $[t_0; t_0 + \Delta t]$: One just has to feed it with the planned motor output $u(t)$ for this time interval. In this way, an FDM can be used to predict states at arbitrary future points in time. A discretization of the FDM (for example via the Euler method with a step size $h \in \mathrm{IR}$) leads directly to the FM in time-discrete notation, demonstrating their equivalence:

$$\widehat{\mathbf{x}}_{t+1} = \widehat{\mathbf{x}}_t + h\widehat{p}(\widehat{\mathbf{x}}_t, \mathbf{u}_t) = \mathrm{FM}(\widehat{\mathbf{x}}_t, \mathbf{u}_t)$$

In the time-discrete form, $\widehat{\mathbf{x}}_t$ as input can be replaced by the real state $\mathbf{x}_t$ (if available). In the time-continuous form, this makes only sense at distinct points in time as outlined above (otherwise one would end up with $\widehat{\mathbf{x}}(t) = \mathbf{x}(t)$ and therefore not need an FDM at all). In the following, we will use the term FDM only for the formal description of time-continous systems, otherwise we will stick to the more general term FM.

The FDM can be used for a composite control system, in which the feedforward IDM is replaced by an internal feedback loop with an FDM instead of the plant (see Fig. 2.10). The estimated state of the system $\widehat{\mathbf{x}}(t)$ (generated by the FDM) is used instead of the real state of the system $\mathbf{x}(t)$ to compute the difference to the desired state $\mathbf{x}_d(t)$. The difference $\mathbf{x}_d(t) - \widehat{\mathbf{x}}(t)$ serves as input for an FC which generates the motor command $\mathbf{u}(t)$. At distinct points in time, especially at the beginning of the movement, the state estimate $\widehat{\mathbf{x}}(t)$ can be set to values from an external source, e.g. the real system state: $\widehat{\mathbf{x}}(t_0) = \mathbf{x}(t_0)$. If the real system state is only available with a certain time delay, this approach is reasonable: At the beginning of the movement, after a moment of rest, the real system state is known, during the movement it is not and has to be replaced by the estimated state. Since precise feedforward control relies on the precondition that the initial desired state is equal to the initial real state of the system, one can as well apply $\widehat{\mathbf{x}}(t_0) = \mathbf{x}_d(t_0)$. How well this type of feedforward control approximates the desired trajectory depends on the precision of the FDM and on the choice of the gain parameters in the FC. Moreover, this approach suffers from external disturbances and noise like all feedforward control systems.

Composite control systems like this are presented by Jordan (1996) and Miall et al. (1993). These systems are closely related to neurophysiological models which propose that the cerebellar cortex implements FMs which are used in motor control. For example, Keeler puts forward that "the cerebellar cortex combines its input sensory information to build an internal model of the world that allows prediction of the dynamics of the sensory-motor system" (Keeler, 1990,

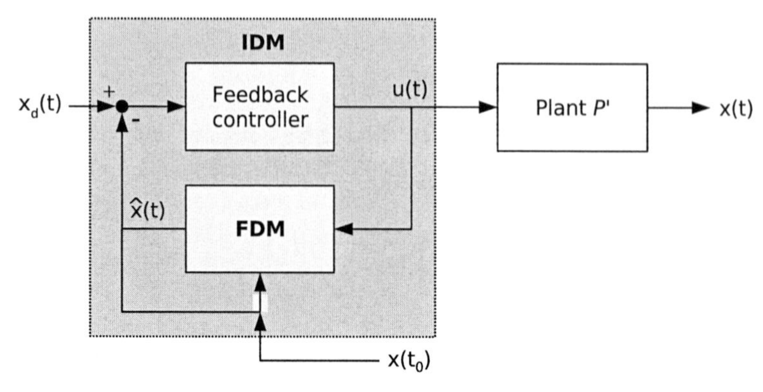

Figure 2.10 — Composite control system with an FDM in a negative feedback loop which replaces a feedforward IDM (gray box) (for details see text).

p. 409). Ito (1984) suggested a model in which the cerebellar cortex serves as a predictive, adaptive filter in an FC.

2.1.4.6 Control systems with time delay

Feedback control systems like in Fig. 2.5 are sensitive to delays in the feedback loop. If considerable delays are present, the gain parameters in the FC need to be small to avoid instabilities. But small gains result in slow adjustments and overall larger differences beween the desired and the actual trajectory. Thus, time delays are an important problem for motor control, especially in biological organisms. Here, delays occur on the way from the sensors to the brain (e.g., 50 ms from retina to visual cortex; Miall et al., 1993), during the generation of the motor commands, during the transmission of the efferent commands to the muscles (axonal delays), and finally in the response of the limb (muscle latencies). For the total feedback-loop in the human motor system, Miall et al. (1993) present the following figures: 130 ms for oculomotor control, 110-150 ms for proprioceptive control, and 200-250 ms for visuomotor control. Since fast arm movements can last less than 200 ms, feedback control alone does not seem to be feasible for this purpose.

To avoid the time delay problem, one could rely on pure feedforward control. However, feedforward control has its own disadvantages: requirement of very precise feedforward IDMs, high noise sensitivity, and moreover no means to deal with external disturbances (see Sect. 2.1.4.1). In Sect. 2.1.4.3, composite control systems consisting of a feedforward IDM and an FC (Fig. 2.6) have been suggested to compensate for the complementary weaknesses of feed-

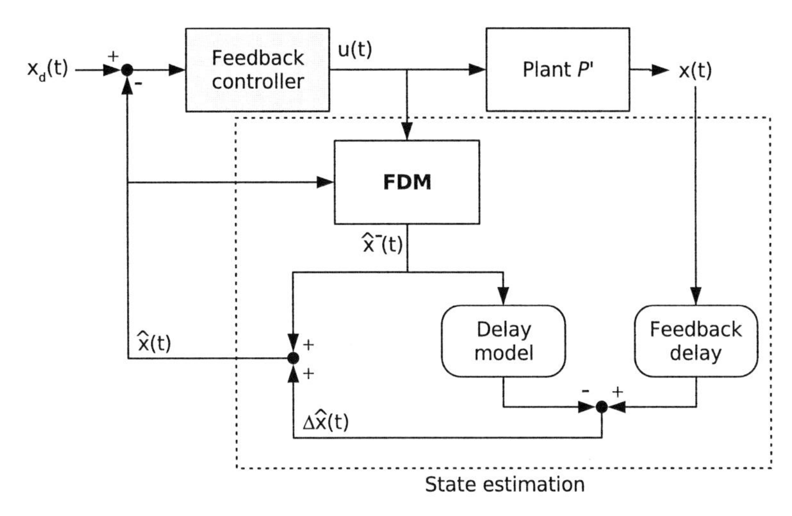

Figure 2.11 — The Smith predictor. An FDM and a delay model are used for state estimation in a system with time delay in the feedback loop (for details see text).

forward and feedback control. In these systems, the time delay problem of feedback control is not solved, but it is less prevalent because the feedforward IDM does the main work in generating the motor output: The FC has only to compensate for the remaining deviations between the desired and the performed trajectory, therefore it can work with small gain parameters (necessary because of the time delay) without sacrificing too much of the overall performance of the control system. Nevertheless, time delays also deteriorate the performance of such composite control systems. The Smith predictor, which is presented in the following, offers a better way to deal with time delays in the feedback loop.

Miall et al. (1993) suggested the Smith predictor as a model how the human motor system could overcome the time delays. The Smith predictor was first proposed by Smith (1959) for factory processes with long transport delays, but the idea applies as well to other control processes with long loop delays. The flow chart of the Smith predictor for motor control is shown in Fig. 2.11. The state estimator (large dashed box) is based on an FDM which generates a preliminary state estimate $\widehat{\mathrm{x}}^-(t)$ (the superscript minus sign indicates that this is an *uncorrected* state estimate). This state estimate is corrected by an adjustment term $\Delta\widehat{\mathrm{x}}(t)$ which is computed as difference of the delayed real state of the

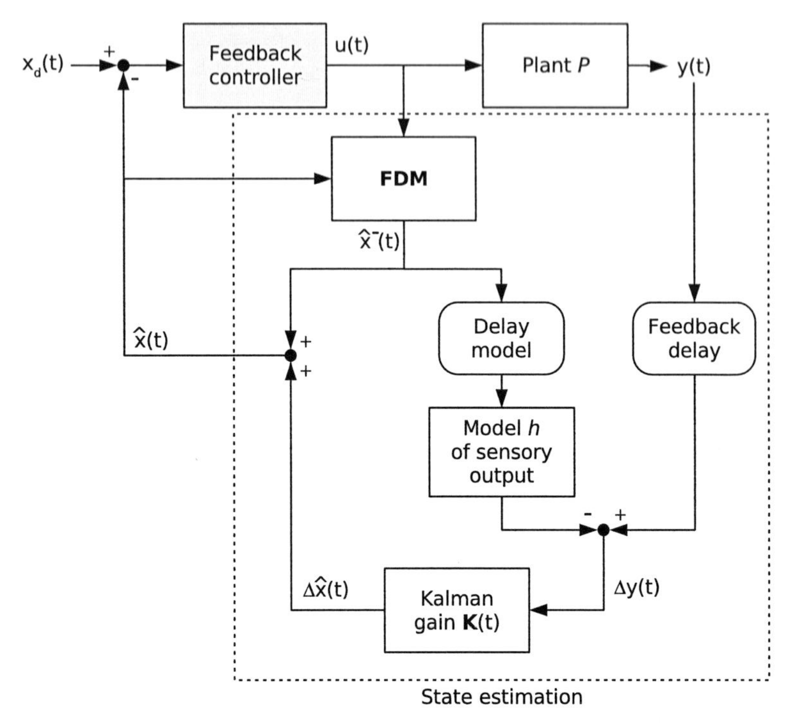

Figure 2.12 — Modification of the Smith predictor into a time-continuous Kalman filter with time delay (for details see text).

system $\mathbf{x}(t - t_{\mathrm{fb}})$ and the delayed preliminary state estimate $\widehat{\mathbf{x}}^{-}(t - \widehat{t}_{\mathrm{fb}})$:

$$\begin{aligned}\Delta\widehat{\mathbf{x}}(t) &= \mathbf{x}(t - t_{\mathrm{fb}}) - \widehat{\mathbf{x}}^{-}(t - \widehat{t}_{\mathrm{fb}}) \\ \widehat{\mathbf{x}}(t) &= \widehat{\mathbf{x}}^{-}(t) + \Delta\widehat{\mathbf{x}}(t)\end{aligned}$$

$\widehat{\mathbf{x}}(t)$ is the final state estimate which is used for comparison with the desired state $\mathbf{x}_d(t)$. Besides the FDM, the Smith predictor features a delay model which is used to estimate the time delay t_{fb} in the feedback loop. Its output is the estimate $\widehat{t}_{\mathrm{fb}}$. Since the FDM is placed in the feedback loop instead of the plant (resulting in neglible time delays), larger gain parameters are usable without sacrificing stability compared to a feedback loop with time delays. Only unexpected differences between the estimated state and the real events are subject to long time delays before they contribute to the input of the FC.

The control scheme in Fig. 2.11 (as proposed by Miall et al., 1993) can be modified to incorporate a Kalman filter (with time delay). The extended control system is illustrated in Fig. 2.12. There is no direct access to the plant state $\mathbf{x}(t)$ anymore, but instead the system incorporates the delayed sensory output $\mathbf{y}(t)$ of the plant. $\mathbf{y}(t)$ is used to correct the preliminary state estimate $\widehat{\mathbf{x}}^{-}(t)$ which is generated by the FDM. The amount of correction is determined by the Kalman gain $\mathbf{K}(t)$. Like in the original Smith predictor, an FDM and a delay model are required. This demonstrates how close the concepts of the Smith predictor and the Kalman filter are related to each other. Gerdes and Happee (1994) suggested a similar Kalman filter model for fast goal-directed arm movements (without an explicit delay model). Their simulated results show a good fit with experimental data from human subjects.

The proponents of the Smith predictor hypothesis (Miall et al., 1993; Miall and Wolpert, 1996; Wolpert et al., 1998) suggest that the cerebellum is the location of the FMs and the delay models. They cite functional imaging, clinical, and neurophysiological studies which provide at least indirect evidence for this claim. For the delay model, Miall et al. (1993) list several putative physiological mechanisms: The parallel fibers in the cerebellum or chains of pontine nuclear cells could act as a "tapped delay line" for short time intervals. For long time delays, Miall et al. (1993) propose a model which predicts backwards instead of a model which implements a time delay.

The weakness of the Smith predictor model is that it requires two predictive models (the FM and the backprediction or time delay model) which are trained simultaneously during learning. They share the same training signal (the real state $\mathbf{x}(t)$; see Fig. 2.11), thus a credit assignment problem occurs. Miall et al. (1993) expresses the view that the delay model could be learned first by carrying out single motor commands and awaiting the change in the afferent sensory response. Over many trials, this could result in an estimate of the time delay. Alternatively, in a simulation of a two-joint planar arm Miall and Wolpert (1995) trained both predictive models simultaneously, but with different learning rates. This proved to be successful as well.

2.1.4.7 Internal models vs. single-point equilibrium hypothesis

So far, dynamic control has been discussed on the basis of internal models. A very different approach is the single-point equilibrium hypothesis. It is based on the observation that muscles and peripheral reflex loops have spring-like properties. This viscoelasticity pulls joints back to their equilibrium position by generating a restoring force against external perturbations. The CNS could exploit the viscoelasticity by commanding a series of stable equilibrium positions along the desired trajectory (e.g., Flash, 1987). In this approach, only IKMs are needed, but neither IDMs nor any of the control schemes with internal models.

However, this theory predicts that the viscoelastic forces increase as the movements gets faster, while control by internal models can realize fast movements with low viscoelastic forces. At least for well-trained movements, experimental observations of relatively low stiffness of the limbs support the existence of internal models (Gomi and Kawato, 1996; Morasso and Schieppati, 1999).

2.1.5 Forward-inverse coupling: MOSAIC

Motor control in the real world faces the problem that the dynamics of the plant changes depending on the context. For example, while lifting an object, the dynamics of the arm changes depending on the payload. In theory, one could overcome this problem with monolithic and complex IDMs with a lot of additional input lines which specify the perceived context. However, a more elegant solution is offered by a modular approach in which a multitude of different IDMs serves for different contexts. Wolpert and Kawato state the following three advantages of a modular approach: "First, the world is essentially modular, in that we interact with multiple qualitatively different objects and environments. [...] Second, the use of a modular system allows individual modules to participate in motor learning without affecting the motor behaviors already learned by other modules. [...] Third, many situations which we encounter are derived from combinations of previously experienced contexts, e.g. novel conjoints of manipulated objects and environments" (Wolpert and Kawato, 1998, p. 1318).

Based on the theoretical framework of sensorimotor coordination with internal models, Wolpert and colleagues (Haruno et al., 1999, 2001; Wolpert and Kawato, 1998) proposed the MOSAIC model ("modular selection and identification for control") as modular approach for motor control. MOSAIC overcomes the two basic challenges of modular systems, the module selection problem (which module to select in which context) and the module learning problem (how to adapt each module). Within each module, an FM and an IDM are combined. They are matched to the same context: The more precise the FM predicts in the current context, the better the IDM is currently suited for motor control. Accordingly, depending on the prediction accuracy of the FM in the current context, the module gets more or less responsibility (denoted as λ_t^i; t is the time step, i the module index; $\sum_i \lambda_t^i = 1$). The overall motor output of the system is the sum of the motor outputs of the IDMs in all modules, weighted by λ_t^i. Thus, module selection is carried out on the basis of the prediction accuracy of the FMs.

Module learning has to address the adaptation of the FMs and of the IDMs. The FMs in all modules are adapted to the same learning examples, which are sampled during the movements of the agent, but the learning rate of each FM is multiplied with λ_t^i. Thus, the best FM in the current context is adapted the most. This results in a specialization of the different modules to different contexts. The motor error for the adaptation of the IDMs is determined by FEL like in Sect. 2.1.4.4. It is also multiplied with λ_t^i, so that the IDM in the module with the most precise FM is adapted the most. In this way, it is ensured that the FM and the IDM in every module become specialists for the same context.

Up to this point, the outlined architecture can switch between modules and distribute the error signals for the adaptation of the FMs and IDMs after the first motor command has been generated and the prediction by the FMs has started. To generate responsibility estimations in advance in a feedforward fashion, each module gets an additional responsibility predictor. These predictors learn how much responsibility their respective module receives in the current context. The final responsibility estimate is a multiplicative combination of the output of the responsibility predictor and of the responsibility which is derived from the accuracy of the FM. This final estimate serves also as the desired output for the adaptation of the responsibility predictor.

The MOSAIC model allows several interesting predictions. The most important is that FMs have a primary role in motor learning. The FMs have to learn to predict before the IDMs learn to control. Experimental evidence for this adaptation order has been found in studies on grip-force modulation (e.g., Flanagan et al., 2003) and for the learning of a visuomotor task completely novel to the test subjects (Sailer et al., 2005). Furthermore, the MOSAIC model predicts that context-dependent IDMs are learned. This claim has been experimentally confirmed in a study by Wada et al. (2003) in which subjects had to adapt their arm movements to two different external force fields, cued by blue or red color on a screen. Finally, Davidson and Wolpert (2004) demonstrated that inverse models for grasping objects with different weights are additively combined. In principle, a superposition of module outputs is also predicted by the MOSAIC model. However, MOSAIC can only interpolate between the output of different modules while the subjects in the study by Davidson and Wolpert (2004) extrapolated new motor output by adding up the motor output of two modules. Thus, the MOSAIC model needs further modification to incorporate results like this.

Wolpert and Kawato (1998) are proponents of the view that the cerebellum is used as neural substrate for internal models. They suggest that the cerebellum "implements" both FMs and IDMs, integrating earlier theories that the cerebellum it either the location of FMs or of IDMs (see Sect. 2.1.4.4). As extension of their work, Wolpert et al. (2003) proposed a hierarchical version of the MOSAIC model: Higher-level modules in this extended model might be used to understand the actions of others, for imitation learning, and for social interac-

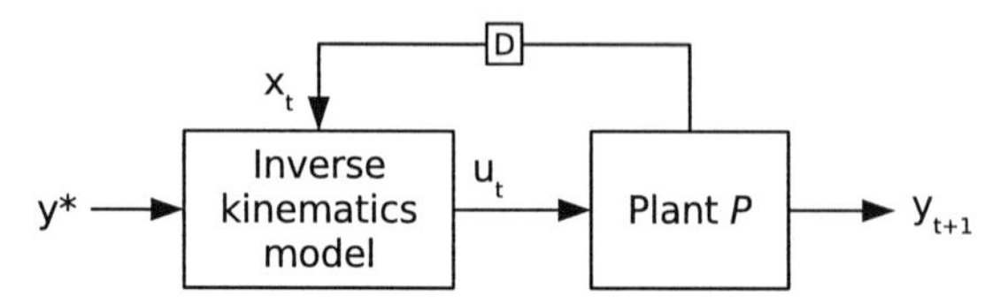

Figure 2.13 — Simplified motor control scheme (the internal state loop of the plant as in Fig. 1.2 is omitted for simplicity). The box labeled D indicates a delay by one time step. In kinematic problems without strict separation of sensory and state variables, $\mathbf{x}$ can also serve as sensory context information.

tion. This approach fits to other theories which suppose that the motor system is the basis for cognitive abilities (see Sect. 1.5.4).

2.2 Kinematic Motor Learning

2.2.1 Overview of inverse models

In the previous section on motor control, three different types of inverse models have been presented, the feedforward inverse dynamics model (IDM), the feedback IDM, and the inverse kinematics model (IKM).

The feedforward IDM is usually a many-to-one mapping. Although it is possible to define plants with redundant dynamics, this redundancy can be mostly resolved by introducing a more precise state description. For example, muscles for limb movements appear in agonist/antagonist pairs. To generate a certain torque, a continuum of different muscle activations is available. But if one introduces the stiffness of the limb as additional state variable, this ambiguity vanishes and for a given torque and stiffness only a precisely defined pair of muscle activations applies (from a biological perspective, this is still a severe simplification). In summary, feedforward IDMs are "simple" functions which drive the plant along a predefined way.

Feedback IDMs correspond to optimal controllers as known from control theory. Since they are used to control plants with internal dynamics, they have to generate the optimum motor command over a certain time interval to achieve the desired state $\mathbf{x}_d(t)$. Moreover, the "correct" motor output of the IDM depends on the applied optimality criterion.

The IKM is quite different from its dynamic cousins. It is the counterpart of a "kinematic plant" (see the simplified control scheme in Fig. 2.13; the internal state loop of the plant as in Fig. 1.2 is omitted in the following for simplicity). Ideally, such a plant has the form $\mathbf{y} = P(\mathbf{x}, \mathbf{u})$ without any reference to distinct time steps, since both the forward and the inverse relationship are just a mapping between the respective input and output space in the kinematic domain. However, with certain restrictions we also include plants of the form $\mathbf{x}_{t+1} = P'(\mathbf{x}_t, \mathbf{u}_t)$ with $\mathbf{y}_{t+1} = h(\mathbf{x}_{t+1})$ (see Sect. 1.5.1) in the kinematic domain, although these plants define a time-discrete dynamical system as soon as the loop between $\mathbf{x}_t$ and $\mathbf{x}_{t+1}$ is closed. These restrictions are: First, with zero motor input, the state of the system does not change; second, an IKM can always achieve the desired system output $\mathbf{y}^*$ in a single time step (as long as this is possible at all). Kinematic control does not require a strict separation of sensory output and state variables, thus it is also possible to interpret $\mathbf{y}$ as a set of variables which are used to define the desired plant output and $\mathbf{x}$ as (sensory) context information. This suggests that IKMs are easier to handle than IDMs, but on the downside the inverse of a kinematic plant is often a one-to-many mapping. This has to be considered in the implementation and adaptation of IKMs.

This section on adaptive motor learning focuses on IKMs since the studies in this thesis deal solely with kinematic problems. Learning of feedforward IDMs is only briefly addressed in Sect. 2.3 (in addition to Sect. 2.1.4.4 on feedback-error learning), learning of feedback IDMs in Sect. 2.4.1.

2.2.2 Problems of motor learning

Motor learning is required for both biological organisms and adaptive robotic systems. While the former need to control their musco-sceletal system, the latter have to command their artificial actuators. Motor learning is goal-oriented and aims on a certain effect in the external world or on the agent itself. This effect has to be specified and measured in the sensory domain. For example, visual and tactile information indicate the failure or success of a human grasping movement. Nevertheless, although this sensory feedback is available, the correct motor command remains unknown. This example shows that motor learning has to rely on the sensory error, the deviation between the desired and actual sensory outcome after a movement. The main problem of motor learning is the transfer of this sensory error to an error signal in the motor domain. A correct mapping from sensory to motor error space might be very complex. In the example, high-dimensional visual and tactile data has to be mapped to the activation of a large set of motor neurons.

In general, the mapping from sensory to motor error is unknown to the agent. A very simple solution to this problem would be the random exploration of sensorimotor space in the search for motor commands which accomplish the task at hand. Unfortunately, through the high dimensionality of sensorimotor

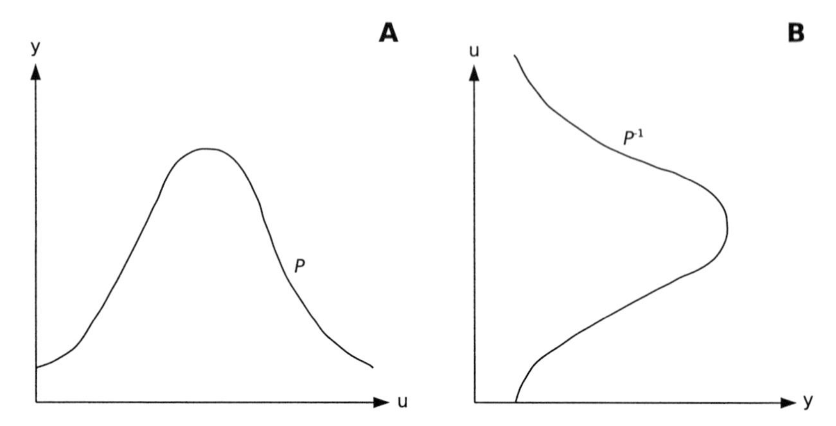

Figure 2.14 — A. Plant P as mapping between motor and sensory space. B. Inverse plant. In contrast to P, P^{-1} is not a function.

space in natural tasks, such an approach is way too expensive in terms of time and energy.

For the learning of kinematic motor control, we start from the simplified motor control scheme in Fig. 2.13. Instead of the term "inverse kinematics model", the shorter term "controller" is used in the following in the text (abbrev.: C). The following general equations apply:

$$\mathbf{u}_t = C(\mathbf{x}_t, \mathbf{y}^*)$$
$$\mathbf{y}_{t+1} = P(\mathbf{x}_t, \mathbf{u}_t)$$

Ideally, the controller is the inverse of the plant (with regard to $\mathbf{u}$ and $\mathbf{y}$). For an untrained adaptive controller, there remains a residual $\mathbf{\Delta y} = \mathbf{y}^* - \mathbf{y}_{t+1}$ after a movement. This sensory error is accessible while the corresponding $\mathbf{\Delta u}$ for which $\mathbf{\Delta y} = \mathbf{0}$ is unknown. This is the first problem of motor learning, the "problem of the missing teacher signal", as mentioned above.

The second problem arises if the inverse plant is a one-to-many mapping. Figure 2.14a shows a very simple plant mapping from a one-dimensional motor space u to a one dimensional sensory space y (omitting the state x). $y = P(u)$ is a smooth function. A properly working controller should implement P^{-1}, thus $u = C(y) = P^{-1}(y)$. As shown in Fig. 2.14b, P^{-1} is not a function. This is a serious challenge to any approach to motor learning using a function approximator (like the MLP) as adaptive controller. The problem can be illustrated solely refering to P as well (see Fig. 2.15a). To reach y^*, two different motor commands u_1^* and u_2^* are suitable. This causes an ambiguity in the motor error

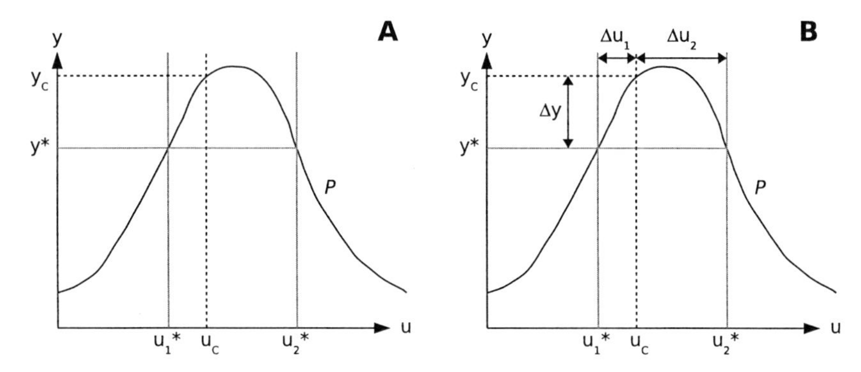

Figure 2.15 — A. Two motor commands u_1^* and u_2^* are applicable to obtain the desired sensory output y^*. B. Ambiguity of motor error signals. Both Δu_1 and Δu_2 can be used to correct the controller output u_C.

as well. If we consider an untrained controller which produced $u_C = C(y^*)$ leading to $y_C \neq y^*$ (Fig. 2.15b), two different corrections of its motor output, Δu_1 and Δu_2 are possible. Thus, a motor learning procedure either has to guarantee that the controller output converges to one of the possible solutions, or the controller architecture itself has to be capable of storing multiple outputs for one and the same input. The MLP and other function approximators converge instead to the average of u_1^* and u_2^*. The resulting motor command has obviously not the desired sensory effect y^*. Only if the solution sets $\{u|P(u) = y^*\}$ are convex, function approximators can deal with one-to-many mappings. When we speak in the following of the "problem of one-to-many mappings", we usually address one-to-many mappings with non-convex solution sets.

In the literature regarding internal models and adaptive controller learning, several approaches have been proposed to circumvent these problems. The most popular are "feedback-error learning" (FEL) (Kawato, 1990), "distal supervised learning" (DSL) (Jordan and Rumelhart, 1992), and "direct inverse modeling" (DIM) (e.g., Kuperstein, 1988). Moreover, Kröse et al. (1990) and van der Smagt (1995) suggested "learning by input adjustment" (LbI). In the following, we provide a review of these approaches to motor learning in the context of internal models, and show the relationship between them. Moreover, we present a novel learning procedure called "learning by averaging" (LbA) (Schenck and Möller, 2004, 2006). In Chapter 4, the performance of FEL, DSL, DIM, and LbA is compared on two different learning tasks, concerning active vision and the control of a planar robot arm.

FEL and DSL are closely related; they use a local linear approximation for

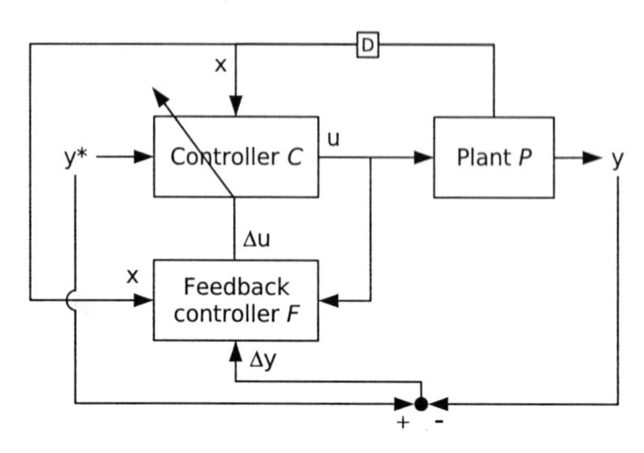

Figure 2.16 — Feedback-error learning scheme for kinematic problems. The box labeled D indicates a delay by one time step (for details see text).

the unknown mapping from sensory error to motor error space and solve the one-to-many problem by converging to one of the possible solutions. DIM and LbI rely on a reformulation of the learning problem. They offer no genuine solution for the one-to-many problem, but the usage of abstract recurrent neural networks as adaptive controller (Hoffmann and Möller, 2003; Möller and Hoffmann, 2004) can overcome this shortcoming. LbA employs an MLP as adaptive controller and actually exploits its averaging capabilities in the learning process. Learning examples are generated by a heuristic search process. LbA can only cope with the one-to-many problem as long as additional learning constraints are defined. It will be presented in a staged and a continuous version.

To reduce the complexity of figures and equations, we omit the time step indices t and $t + 1$ in the following for the most part as long as it is possible without any loss in generality or clarity.

2.2.3 Feedback-error learning

FEL has been developed in the context of dynamic motor control (see Sect. 2.1.4.4), but it can also be applied to kinematic problems. The control scheme is shown in Fig. 2.16. In addition to the controller C and the plant P, a feedback controller F is displayed. As input, it gets the sensory error $\mathbf{\Delta y}$. The feedback controller has the task to convert $\mathbf{\Delta y}$ into the motor error $\mathbf{\Delta u}$ which is used as error signal for the adaptation of C. In contrast to the dynamic version, $\mathbf{u}$ and $\mathbf{x}$ are additional inputs to F that provide the necessary context information. Usually, F is a linear function (see also Eqn. (2.6)):

$$\Delta \mathbf{u} = F(\boldsymbol{\Delta}\mathbf{y}, \mathbf{u}, \mathbf{x}) = \mathbf{G}_{\mathbf{u},\mathbf{x}} \Delta \mathbf{y}$$

$\mathbf{G}_{\mathbf{u},\mathbf{x}}$ is a gain matrix depending on $\mathbf{u}$ and $\mathbf{x}$. $\mathbf{G}_{\mathbf{u},\mathbf{x}}$ can be computed by the following approach as long as P is an analytically known continuously differentiable function. As basis for the local linear approximation, the Jacobian $\mathbf{J}_{\mathbf{u},\mathbf{x}}$ of the plant at the position $(\mathbf{u}, \mathbf{x})$ in motor and state space is used. Note that $\mathbf{J}_{\mathbf{u},\mathbf{x}}$ is not the full Jacobian; it only contains the columns concerning motor space:

$$\begin{aligned} \Delta \mathbf{y} &= \mathbf{J}_{\mathbf{u},\mathbf{x}} \Delta \mathbf{u} \\ \Rightarrow \Delta \mathbf{u} &:= \mathbf{J}^{+}_{\mathbf{u},\mathbf{x}} \Delta \mathbf{y} \end{aligned} \tag{2.9}$$

Employing the pseudoinverse $\mathbf{J}^{+}_{\mathbf{u},\mathbf{x}}$, the solution $\Delta \mathbf{u}$ is either the least squares solution to Eqn. (2.9) (if the problem is overdetermined) or the minimum length solution (if the problem is underdetermined). Thus, using the pseudoinverse $\mathbf{J}^{+}_{\mathbf{u},\mathbf{x}}$ of the motor part of the Jacobian of the plant as gain matrix $\mathbf{G}_{\mathbf{u},\mathbf{x}}$ is a general and straightforward approach. In addition, it is useful to introduce a gain factor η to control learning speed:

$$\mathbf{G}_{\mathbf{u},\mathbf{x}} = \eta \mathbf{J}^{+}_{\mathbf{u},\mathbf{x}} \tag{2.10}$$

The appropriate value range for η for convergence of the learning process depends on the motor task. Usually, a function approximator like an MLP is used as adaptive controller, and η implicitly adjusts the learning rate. FEL does not provide a perfect mapping from sensory error to motor error space. Instead, the feedback controller provides a coarse local linear approximation. By adapting along these small local corrections during the course of learning, the MLP output converges to one of the possible solutions for $\mathbf{u}$ (in case of a one-to-many problem) as we explore in Sect. 4.2. Because of this stepwise linear adaptation, FEL is basically an online training procedure. It is not possible to collect a set of training examples for batch learning. For rather simple sensorimotor relationships, one may determine the gain matrix of the feedback controller heuristically. This approach is used for the saccade controller example in Sect. 4.1.

FEL for kinematic problems is rarely used in the literature. Exemplary applications are from the field of biologically inspired active vision systems (Bruske et al., 1997; Dean et al., 1991) where FEL is used for adaptive oculomotor control. If the plant is too complex to define a gain matrix heuristically, FEL is no longer a really adaptive learning scheme since one needs to know the plant analytically to compute $\mathbf{G}_{\mathbf{u},\mathbf{x}}$. Thus, for the biological modeling of kinematic control, FEL is only attractive for rather simple plants like the ones found in oculomotor control. Prewired feedback controllers for such simple plants could have been acquired during evolutionary development.

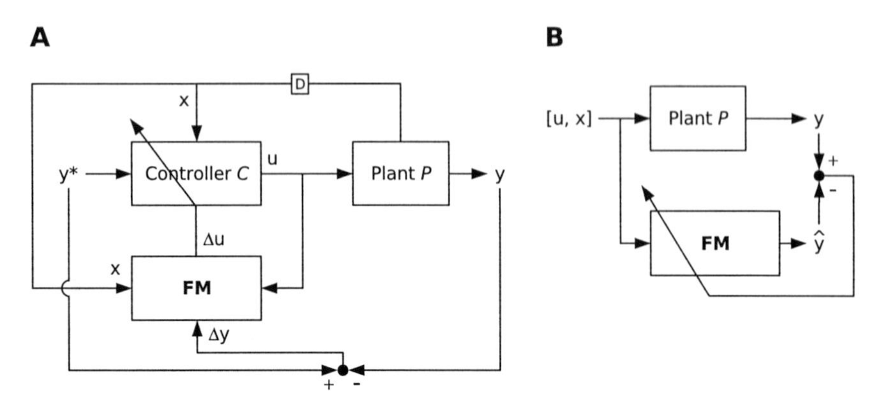

Figure 2.17 — A. Distal supervised learning scheme. The box labeled D indicates a delay by one time step. B. Learning scheme for the FM (for details see text).

2.2.4 Distal supervised learning

DSL was introduced by Jordan and Rumelhart (1992) for the kinematic control of a planar arm. The controller learning scheme of DSL is very similar to FEL as shown in Fig. 2.17a. The feedback controller F is replaced by a forward model FM. The FM itself approximates the plant (see Fig. 2.17b): FM $= \widehat{P}$. It receives a motor command $\mathbf{u}$ and state $\mathbf{x}$ and predicts the sensory outcome $\mathbf{y}$. For DSL, the FM has to be implemented by an MLP. Training data for the FM is generated by collecting the plant's response to random motor commands.

For controller training, DSL uses the trained FM in reverse direction (Fig. 2.17a): The sensory error $\mathbf{\Delta y}$ becomes the input, the motor error $\mathbf{\Delta u}$ becomes the output. This conversion of error signals is possible by error backpropagation (see Sect. 3.1.2) without weight change, as we show in the following.[6]

Backpropagation implements gradient descent on MLPs to minimize the error E of the activation of the units of the output layer. For the sensory output $\mathbf{y}$, the error is defined as $E = \frac{1}{2}\left\|\mathbf{\Delta \widehat{y}}\right\|$ with $\mathbf{\Delta \widehat{y}} = \mathbf{y}^* - \mathbf{\widehat{y}}$. $\mathbf{\widehat{y}}$ is the output of the FM (used in its normal direction). In the following, N_{out} denotes the dimension of the output space (with index j), while N_{in} is the dimension of the motor input space (with index i). The backpropagated error signal δ_i for each motor unit of

[6] Although Jordan and Rumelhart (1992) derived DSL analytically, they did not elaborate on this step.

the input layer is computed as follows:

$$\delta_i = -\frac{\partial E}{\partial u_i}$$

According to the chain rule:

$$\frac{\partial E}{\partial u_i} = \sum_{j=1}^{N_{\text{out}}} \frac{\partial E}{\partial y_j} \frac{\partial y_j}{\partial u_i}$$

Substituting for the partial derivatives we obtain:

$$\begin{aligned}
\frac{\partial E}{\partial y_j} &= -\left(y_j^* - \widehat{y}_j\right) \\
\frac{\partial y_j}{\partial u_i} &= \widehat{\mathbf{J}}_{\mathbf{u},\mathbf{x}(ji)} = \widehat{\mathbf{J}}^t_{\mathbf{u},\mathbf{x}(ij)} \\
\Rightarrow \frac{\partial E}{\partial u_i} &= -\sum_{j=1}^{N_{\text{out}}} \left(\widehat{\mathbf{J}}^t_{\mathbf{u},\mathbf{x}(ij)} \cdot \left(y_j^* - \widehat{y}_j\right)\right)
\end{aligned}$$

$\widehat{\mathbf{J}}_{\mathbf{u},\mathbf{x}}$ is the motor part of the Jacobian of the FM (which approximates the plant). Gradient descent in motor space for error minimization with a step size $\Delta E = -\eta$ yields:

$$\begin{aligned}
\Delta u_i &= \frac{\partial E}{\partial u_i} \Delta E = -\eta \frac{\partial E}{\partial u_i} \\
&= \eta \sum_{j=1}^{N_{\text{out}}} \left(\widehat{\mathbf{J}}^t_{\mathbf{u},\mathbf{x}(ij)} \cdot \left(y_j^* - \widehat{y}_j\right)\right) \\
\Leftrightarrow \mathbf{\Delta u} &= \eta \widehat{\mathbf{J}}^t_{\mathbf{u},\mathbf{x}} \mathbf{\Delta \widehat{y}}
\end{aligned} \tag{2.11}$$

Equation (2.11) shows that backpropagation through the FM results in a local linear approximation of the motor error with $\widehat{\mathbf{J}}^t_{\mathbf{u},\mathbf{x}}$ as gain matrix. However, there is still one downside: The learning algorithm would try to change the motor commands $\mathbf{u}$ in a way that the output of the FM finally equals the desired sensory output. But we are actually interested in a close match with the sensory output of the plant, which might differ from the output of the FM. To overcome this problem, Jordan and Rumelhart (1992) replaced $\mathbf{\Delta \widehat{y}}$ with $\mathbf{\Delta y}$, the difference between the desired output and the real plant output. Thus, Eqn. (2.11) changes to

$$\mathbf{\Delta u} = \eta \widehat{\mathbf{J}}^t_{\mathbf{u},\mathbf{x}} \mathbf{\Delta y} \,. \tag{2.12}$$

Considering Eqns. (2.10) and (2.12), DSL is equivalent to FEL with a gain matrix $\mathbf{G}_{u,x} = \eta \widehat{\mathbf{J}}^t_{u,x}$.[7] The adaptive motor controller (also an MLP) is trained online by small approximated motor error signals. For one-to-many problems, the controller output converges to one of the possible solutions (Jordan and Rumelhart, 1992).

DSL is truely adaptive because no analytical knowledge about the plant is needed beforehand. Moreover, even with a rather imprecise FM, successful controller learning is possible as Jordan and Rumelhart (1992) emphasize. It is also possible to learn both the FM and the controller online simultaneously from scratch. The main drawback of DSL for biological modeling is the requirement of backpropagation which is in itself not a biologically plausible neural learning mechanism. Although DSL could be realized in theory with other decomposable learning systems, we are not aware of any implementation of DSL which is really biologically plausible.

2.2.5 Direct inverse modeling

DIM is based on a reformulation of the motor learning problem (Fig. 2.18). Instead of searching for the right motor command $\mathbf{u}$ for a certain desired sensory output $\mathbf{y}^*$ and state $\mathbf{x}$, random motor commands $\mathbf{u}^*$ are generated. For each $\mathbf{u}^*$, the resulting sensory output $\mathbf{y}^*$ of the plant is recorded. Afterwards, $\mathbf{y}^*$ is interpreted as desired output for which $\mathbf{u}^*$ is actually a perfect motor response. The combination $[\mathbf{x}, \mathbf{y}^* \longrightarrow \mathbf{u}^*]$ forms a perfect learning example for the training of the motor controller. Due to this characteristic, DIM works both for online and batch learning.

DIM has been used in various applications and simulations of robot arms for visuomotor coordination (Mel, 1988; Kuperstein, 1987, 1988, 1990). Kuperstein motivated DIM by the "circular reaction" Piaget (1952) observed in children during their development: Children carry out explorative actions in a rather random fashion and observe the sensory effects. In this way, they find out which actions are best suited to obtain these effects. This corresponds very closely to the DIM learning strategy.

Although this psychological motivation seems to be plausible, DIM has several shortcomings. First, DIM cannot cope with the one-to-many problem as long as the controller is implemented by a function approximator. This was illustrated by Jordan and Rumelhart (1992) for a planar arm (see also

[7] FEL can also be used with a gain matrix $\mathbf{G}_{u,x} = \eta \mathbf{J}^t_{u,x}$. In this way, the computation of the pseudoinverse can be omitted. Although DSL and FEL look nearly identical from this perspective, DSL has still the advantage over FEL that no analytical knowledge of the plant is required. Comparative results for a planar arm are shown in Sect 4.2.

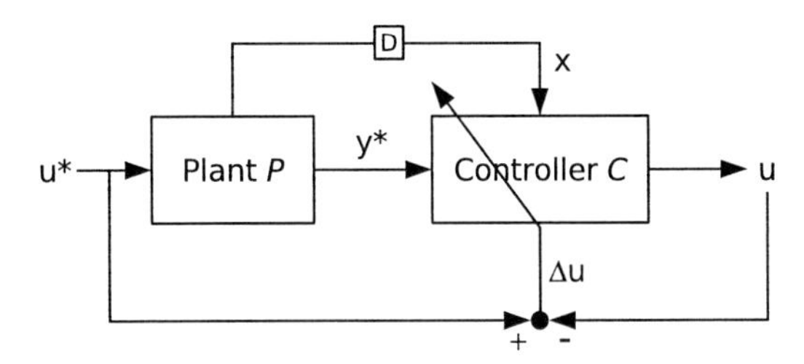

Figure 2.18 — Direct inverse modeling: Random motor commands u* are generated for controller training. The box labeled *D* indicates a delay by one time step (for details see text).

Sect. 4.2.1.1). To overcome this shortcoming, Hoffmann and Möller (2003) used abstract recurrent networks as controller. This type of network approximates high-dimensional sensorimotor data manifolds by a mixture of local PCAs (see Sect. 3.4 on NGPCA). Using such networks, the controller is actually able to reproduce all motor outputs of the one-to-many mapping as long as an appropriate recall procedure is used (Hoffmann, 2004; Hoffmann et al., 2005; Schenck et al., 2003). The learning strategy which combines DIM with NGPCA is called DIM_NGPCA in the following.

Second, DIM is not goal-oriented. Random sampling in motor space may elicit various sensory effects, but may rarely hit the ones which are later used as desired sensory outcomes. Therefore, the resulting controller has to extrapolate the motor output in the region of sensory space containing the desired outcomes and will most likely exhibit bad performance. However, this criticism also applies to a certain extent to DSL with regard to FM training, because the learning examples for the FM are generated in the same non-goal-directed way. But an FM which is inaccurate in the region of sensory space containing the desired outcomes may still allow slow but nevertheless successful training of the controller, because only the coarse direction of the local linear approximation needs to be correct. Thus, usually one would expect DIM to suffer considerably more from the lacking goal-directedness than DSL.

Third, on the neural implementation level, DIM seems to be biologically rather implausible at the first glance. During learning, the input units of the controller receive the real sensory signals, later during the usage of the controller, these units receive the desired sensory outcome. This switch requires a "rewiring" of the connections to the input units which does not seem to take place in the CNS (Kawato, 1990). However, it is questionable if such a rewiring is really necessary. One could also hypothesize that the real and the desired

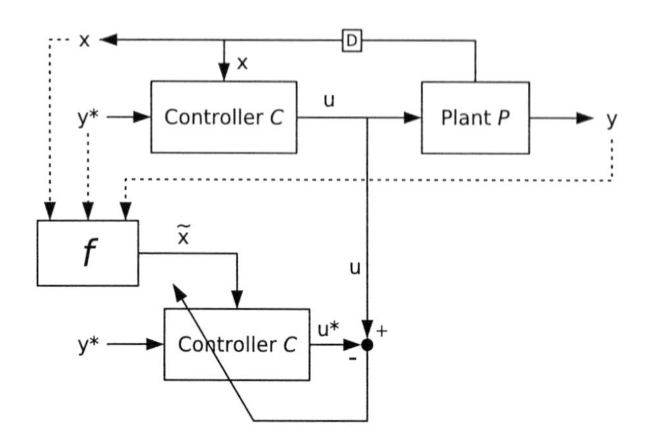

Figure 2.19 — Learning by input adjustment: Generated learning examples map $(\mathbf{x}^*, \mathbf{y}^*)$ to $\mathbf{u}$. The box labeled D indicates a delay by one time step (for details see text).

sensory outcome are represented by the activation of the same neural substrate like in the model by Kuperstein (1987, 1988).

2.2.6 Learning by input adjustment

LbI was applied by Kröse et al. (1990) for visually guided grasping with a robot arm. Like DIM, LbI works by a reformulation of the learning problem. The learning scheme is shown in Fig. 2.19. It consists basically of three steps for the generation of one learning example for the motor controller.

First, for a given state $\mathbf{x}$ und desired sensory output $\mathbf{y}^*$, the controller generates an output $\mathbf{u}$. This motor command evokes a plant response $\mathbf{y}$ which is different to $\mathbf{y}^*$ for an untrained controller. In the second step, LbI does not attempt to solve the problem of the missing teacher signal directly, but instead to determine a state $\widetilde{\mathbf{x}}$ for which the plant equation is fulfilled in the desired way with $\mathbf{y}^* = P(\widetilde{\mathbf{x}}, \mathbf{u})$. $\mathbf{u}$ is the motor command that has been actually generated by the controller before. In combination, $\widetilde{\mathbf{x}}$, $\mathbf{y}^*$, and $\mathbf{u}$ form a valid learning example $[\widetilde{\mathbf{x}}, \mathbf{y}^* \longrightarrow \mathbf{u}]$ for controller training. Thus, LbI does not correct the controller output $\mathbf{u}$, but instead the state input $\mathbf{x}$. For the state correction, an input adjustment function f is required: $\widetilde{\mathbf{x}} = f(\mathbf{x}, \mathbf{y}^*, \mathbf{y})$. This function has to be determined analytically through knowledge of the plant characteristics. In the third step, the resulting learning example $[\widetilde{\mathbf{x}}, \mathbf{y}^* \longrightarrow \mathbf{u}]$ is used for controller training (depicted in the lower part of Fig. 2.19). Both online and batch training are possible, although the latter requires a really precise input adjustment function f to obtain a set of good learning examples.

In the original LbI scheme (Kröse et al., 1990; van der Smagt, 1995), a controller with a fixed desired sensory output $\mathbf{y}^*$ is used. The control task is to move a robot arm which operates above a table surface. A camera is mounted at the tip of the arm which records the image of the table surface. The controller has to position the robot arm in a way that a target object on the table appears exactly in the center of the camera image. The sensory output $\mathbf{y}_t$ of the system in time step t is defined as the image coordinates of the target object. Thus, the desired sensory output $\mathbf{y}^*$ is constant. The controller output is specified as change of the joint angles of the robot arm: $\mathbf{u}_t = \Delta\boldsymbol{\theta}_t$. The state $\mathbf{x}_t$ is defined as $\mathbf{x}_t = (\mathbf{y}_t, \boldsymbol{\theta}_t)$. In this application, the input adjustment function computes $\widetilde{\mathbf{x}}_t = (\widetilde{\mathbf{y}}_t, \boldsymbol{\theta}_t)$, thus basically only $\widetilde{\mathbf{y}}_t = f(\mathbf{y}_t, \mathbf{y}_{t+1})$. Compared to the general presentation in the previous paragraph without time indices, $\mathbf{y}^*$ is completely omitted, $\mathbf{y}_t$ replaces $\mathbf{x}$, and $\mathbf{y}_{t+1}$ replaces $\mathbf{y}$. In this special case, the input adjustment function f works only with image coordinates as parameters. f is defined on the basis of the known geometry of the imaging process, but without explicit analytical knowledge of the complete plant. On this background, Kröse and colleagues claim that LbI "does not need a model of the system" (Kröse et al., 1990, p. 201). However, this claim seems to be too strong since at least a part of the system has to be known precisely enough for the definition of a reasonable input adjustment function f.

With regard to the one-to-many problem, LbI does not offer any inherent solution if a function approximator is used as adaptive controller. Only in the online version of LbI, one can hope that the learning process converges to one of the possible solutions of the inverse kinematics. But this has not been discussed by the original authors (Kröse et al., 1990; van der Smagt, 1995) and has yet to be tested. For the batch version, one might use abstract recurrent neural networks for the learning of one-to-many mappings as suggested for DIM.

LbI is the least general and least adaptive approach among the presented ones. It always requires the analytical construction of a function f for input adjustment. For this reason, we do not pursue LbI any further and do not include it in the simulation study in Sect. 4.

2.2.7 Learning by averaging

The development of LbA started with the simple observation that unfavorable results in motor performance are often scattered around the desired outcome in the sensory domain. E.g., in throwing a ball to a certain target, the throw might be too close or too far. In this task, the sensory outcome depends heavily on the force generated by the muscles, thus it seems to be a good guess to apply the average force of a throw which is too close and of a throw which is too far. If a throw is way off (e.g., accidentally backwards), it should be discarded completely and not considered at all for learning.

This illustrates the basic idea of LbA: to collect learning examples which are neither too bad nor necessarily perfect, and to adapt the motor controller to their average motor output during training. In this way, the problem of the missing teacher signal is solved: Instead of directly converting the sensory into the motor error, averaging over non-perfect learning examples takes place. To accomplish this, one needs a controller architecture capable of averaging over learning examples. For example, the MLP fulfills this requirement.

Pure averaging alone is not sufficient to train precise motor controllers. For this reason, controller performance has to be further enhanced. This is achieved by incrementally improving the quality of the learning examples used for controller training. Such improved learning examples are obtained by searching in motor space in the region around the motor output of the already trained but still inprecise controller for even better motor output. This search can be accelerated by an evolutionary optimization method. In the following, we present LbA as staged version for batch learning and as continuous version for online learning.

2.2.7.1 Staged version

"Staged learning by averaging" (SLbA) works by repeatedly generating a set of learning examples and subsequently training a controller with this set (see Fig. 2.20). Learning examples are included in the training set only if they exceed a certain quality threshold $\widetilde{Q}$. The quality Q of a learning example $[\mathbf{x}, \mathbf{y}^* \longrightarrow \mathbf{u}]$ or of a controller output $\mathbf{u}$ in response to the input $(\mathbf{x}, \mathbf{y}^*)$ depends on the desired output $\mathbf{y}^*$ and the resulting plant output $\mathbf{y} = P(\mathbf{x}, \mathbf{u})$. It is computed by a function $Q : (\mathbf{y}, \mathbf{y}^*) \rightarrow Q(\mathbf{y}, \mathbf{y}^*)$. This function has the following property: the smaller the deviation between $\mathbf{y}$ and $\mathbf{y}^*$, the larger Q. Here, we assume that the maximum of Q is 1.0. The deviation between $\mathbf{y}$ and $\mathbf{y}^*$ can be expressed by their Euclidean distance, but alternative distance measures which are meaningful for the motor learning task at hand are usable as well. Moreover, the quality function Q can be used to incorporate additional constraints in the learning task as we will show for the planar arm (see Sect. 4.2).

In stage k, a single learning example $[\mathbf{x}, \mathbf{y}^* \longrightarrow \mathbf{u}]$ is generated by the following steps:

1. $\mathbf{x}$ and $\mathbf{y}^*$ are created at random within their operating range.

2. In the first stage ($k = 1$), without an existing controller, a random vector $\mathbf{u}_0$ is drawn. In later stages ($k > 1$), $\mathbf{u}_0 = C_{k-1}(\mathbf{x}, \mathbf{y}^*)$; C_{k-1} is the

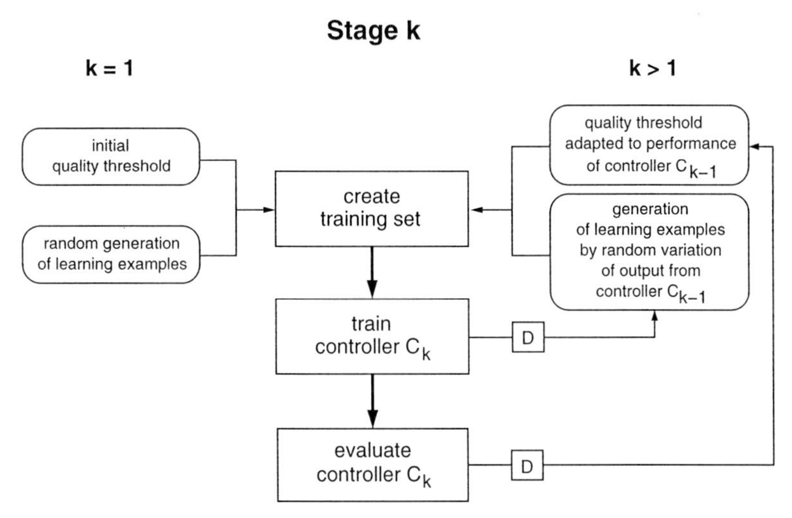

Figure 2.20 — Basics steps of staged learning by averaging. The boxes labeled D indicate a delay by one learning stage (adapted from Hoffmann et al., 2005, © Springer).

controller trained in the preceding stage.

3. The quality threshold $\widetilde{Q}_k$ is determined: In the first stage ($k = 1$), $\widetilde{Q}_k$ may be a constant or may depend on $\mathbf{x}$, $\mathbf{y}^*$, and $\mathbf{u}_0$. In later stages ($k > 1$), $\widetilde{Q}_k$ may depend on the overall quality of the preceding controller C_{k-1} as well.

4. $\mathbf{u}$ is repeatedly computed as random variation of $\mathbf{u}_0$ until $Q(P(\mathbf{x}, \mathbf{u}), \mathbf{y}^*) > \widetilde{Q}_k$. In its simplest form, the random variation is realized by adding noise to $\mathbf{u}_0$. Alternatively, one can apply an evolutionary optimization process instead (see Sect. 4.2.3.1 for the planar arm).

The generated learning examples are accumulated in a training set of a predefined size. This set is used for the training of the controller C_k afterwards. Ideally, with a controller implementation which is capable of averaging, two learning examples $[\mathbf{x}, \mathbf{y}^* \longrightarrow \mathbf{u}_1]$ (with quality q_1) and $[\mathbf{x}, \mathbf{y}^* \longrightarrow \mathbf{u}_2]$ (with quality q_2) result in a controller response $\mathbf{u}_C = \frac{\mathbf{u}_1+\mathbf{u}_2}{2}$ to the input $(\mathbf{x}, \mathbf{y}^*)$. Successful learning only takes place when the controller response is at least better than the inferior learning example of the two. Thus, SLbA requires as necessary precondition that $Q(P(\mathbf{x}, \frac{\mathbf{u}_1+\mathbf{u}_2}{2}), \mathbf{y}^*) > \min(q_1, q_2)$ for $\mathbf{u}_1 \neq \mathbf{u}_2$. This is illustrated in Fig. 2.21: For a certain $\mathbf{x}$ and $\mathbf{y}^*$, two different combinations of P and

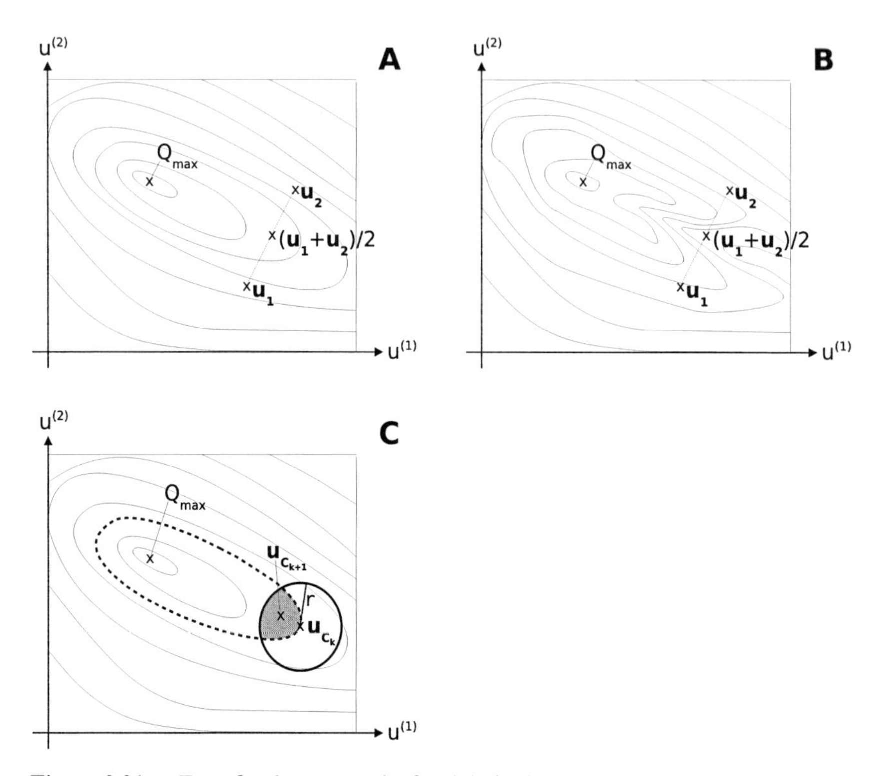

Figure 2.21 — For a fixed state x and a fixed desired sensory outcome y*, the quality of motor commands u in a two-dimensional motor space is shown as contour plot. Maximum quality is achieved at the point marked Q_{max}. A. Convex equi-quality curves. B. Non-convex equi-quality curves. C. Illustration of SLbA. For further explanation see text.

Q result in different equi-quality curves in a two-dimensional motor space. In Fig. 2.21a, these curves enclose convex subsets in motor space; for this reason, the SLbA precondition is fulfilled. On the contrary, in Fig. 2.21b, the quality curves enclose non-convex subsets; in this setting, it is easily possible that the average of two motor commands $\mathbf{u}_1$ and $\mathbf{u}_2$ has lower quality than each of the commands (as shown in the figure).

Fig. 2.21c illustrates the functioning of the staged procedure: in this example, $\widetilde{Q}_{k+1} = Q(P(\mathbf{x}, \mathbf{u}_{C_k}), \mathbf{y}^*)$ (marked by the bold dashed curve). Random learning examples are created by adding uniform noise with radius r to $\mathbf{u}_{C_k}$. Only the learning examples whose quality exceeds $\widetilde{Q}_{k+1}$ are included in the training set of stage $k+1$ (gray area). Training controller C_{k+1} with this set results in a controller output $\mathbf{u}_{C_{k+1}}$ with better quality than $\mathbf{u}_{C_k}$. Instead of just adding noise, for the simulations with the planar arm in this study a simple evolutionary strategy (see Sect. 4.2.3.1) was applied to explore the region around $\mathbf{u}_{C_k}$ in the search for learning examples exceeding $\widetilde{Q}_{k+1}$.

While SLbA replaces the unknown mapping from sensory to motor error with averaging and a goal-directed search in sensorimotor space, it does not offer a straightforward solution for the one-to-many problem. Usually, a one-to-many mapping implies a multi-modal quality function in motor space for fixed values of $\mathbf{x}$ and $\mathbf{y}^*$ (like in Fig. 2.21a, but with multiple peaks equal to $Q_{\max}$ for the multiple solutions of the inverse kinematics and "valleys" between these peaks). Accordingly, the equi-quality curves no longer enclose convex subsets of the motor space, and it is not longer guaranteed that averaging over learning examples results in improved motor output. Whenever a one-to-many mapping spoils learning by averaging in this way, one has to include additional constraints in the quality function Q to disambiguate the learning task. SLbA for the planar arm in Sect. 4.2 relies on such constraints. In previous studies, SLbA has been applied to saccade control with a robot camera head (Hoffmann et al., 2005; Schenck et al., 2003; Schenck and Möller, 2004).

2.2.7.2 Continuous version

Continuous learning by averaging (CLbA) works similar to the staged version with one important difference: Instead of collecting a set of learning examples before training, each example is immediately taken for the training of one single controller. The learning process is organized in training cycles j instead of stages k. In each cycle, one learning example is created in four steps as described for SLbA (substituting stages k for cycles j; moreover, instead of multiple controllers C_k, there is only one controller C). As fifth step, controller C is trained with the resulting learning example. Like SLbA, CLbA requires additional constraints in the quality function to circumvent the one-to-many problem.

Fig. 2.22 shows the structural similarity of CLbA to FEL and DSL. Instead of a feedback controller or reversed FM, a "quality enhancer" serves to generate an improved motor command $\mathbf{u}_{QE}$ by which the motor error $\mathbf{\Delta u}$ is computed. But in contrast to FEL and DSL, this motor error is not result of a local linear approximation, but of a heuristic search in motor space. The generated motor errors average out in controller training to guide learning in the right direction. Results for the performance of CLbA on saccade control have been published by Schenck and Möller (2006). Moreover, CLbA has been used to train the saccade controller in the study by Schenck et al. (to appear) (see also Sect. 6.3).

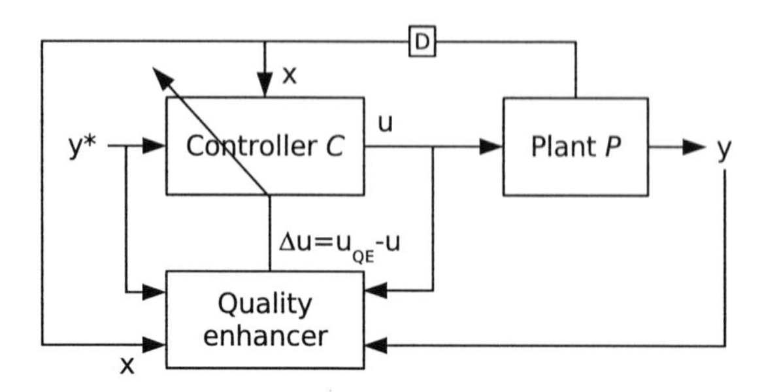

Figure 2.22 — Continuous learning by averaging: A quality enhancing mechanism serves to generate motor error signals Δu. The box labeled *D* indicates a delay by one time step (for details see text).

2.2.8 Summary on learning strategies

Table 2.1 summarizes and compares the most important properties of FEL, DSL, DIM, LbI, and LbA. In conclusion, no learning strategy is clearly preferable. Nearly all of them are fully adaptive insofar as no analytical knowledge about the plant is required (except for LbI; FEL only for simple plants), thus, they are suited for adaptive motor control in the kinematic domain. Being fully adaptive is also a precondition for biological plausibility. Moreover, as long as there is no convincing counterargument, we assume that a learning mechanism is biologically plausible. Following this reasoning, FEL and LbA are the most promising candidates for biological plausibility (but also with restrictions). FEL is only adaptive for simple plants, and the staged version of LbA (SLbA) requires the storage of a set of learning examples before the actual training takes place. Batch learning like this is always biologically less plausible than online learning because it seems to be rather unlikely that the CNS is capable of storing hundreds or thousend of learning examples with the necessary precision over a long period of time. With regard to DIM and DSL, one should not reject their biological plausibility too early (for this reason, Table 2.1 shows the entry “questionable” at the respective positions). Maybe neural rewiring is not necessary at all (with regard to DIM), maybe a biologically plausible variant of backpropagation will be discovered in the future (with regard to DSL).

FEL and LbA are the only learning strategies which are clearly goal-directed towards the desired sensory outcomes. DIM is not goal-directed at all and DSL and LbI suffer from limitations: While the controller learning in DSL is goal-directed, the preceding or accompanying adaptation of the FM is not. LbI is goal-directed with regard to the desired sensory outcome, but one cannot ex-

	FEL	DSL	DIM	LbI	LbA
Batch learning	No	No	Yes	Yes	SLbA only
Online learing	Yes	Yes	Yes	Yes	CLbA only
Usable for one-to-many problems	Yes, converges to one solution	Yes, converges to one solution	Only with abstract recurrent networks	Only with abstract recurrent networks (batch version)	Only with additional constraints
Goal-directed	Yes	(Yes)	No	(Yes)	Yes
Fully adaptive	Only for simple plants*	Yes	Yes	No	Yes
Biologically plausible	Only for simple plants*	Questionable (because of backpropa-gation)	Questionable (because of neural rewiring)	No (because of analytical input adjustment function)	(Yes)
Applicable to dynamic domain (see Sect. 2.3)	Yes	Yes	Yes	No	No

* complex plants require analytical knowledge

Table 2.1 — Comparison of learning strategies. A "yes" in round brackets indicates a restricted affirmation.

clude that the input adjustment produces only state inputs which later on, during the application of the controller, are only rarely encountered, resulting in bad performance.

With regard to one-to-many mappings, DIM seems to be one of the worst candidates (similar to LbI). But this is only true if DIM is used with a function approximator as adaptive controller. If DIM is implemented on the basis of abstract recurrent neural networks, the resulting controller is capable of storing all solutions simultaneously and not only one (as it is the case for FEL, DSL, and LbA). In Chapter 4, these theoretical claims will undergo experimental examination.

2.3 Learning of Feedforward Inverse Dynamics Models

Some of the described learning strategies for adaptive IKMs are also suitable for feedforward IDMs. First of all, FEL has been first developed for the dynamic domain as outlined in Sect. 2.1.4.4 (Kawato et al., 1987; Kawato, 1990). Generally, a dynamic feedback controller (FC) like in Eqn. (2.6) can be specified for the control of manipulators. In a similar way, DSL can be used for the adaptation of feedforward IDMs (Jordan and Rumelhart, 1992). Instead of the FC used in FEL, a reversed dynamic FM is used to generate the motor error along the desired trajectory. Otherwise, the control loop stays the same. With regard to the training of the dynamic FM, Jordan and Rumelhart (1992) point out that it is generally not feasible to produce arbitrary random control signals in dynamic environments because the resulting trajectories are most likely just jitter movements. Instead, they recommend to produce random equilibrium positions for the arm instead of random torques, or to generate learning examples along the target trajectories by a preliminary PD feedback controller. However, FEL is by far the most popular learning strategy for IDMs. Although DSL is even more adaptive without the need to specify the FC in advance, DSL is barely used for applications or modeling. The main reason for rejecting DSL seems to be that backpropagation of the error signal through the FM is biologically not plausible (e.g., Kawato and Gomi, 1992a,b).

DIM could be used in principle for dynamic problems. However, the missing goal orientation of DIM is a severe problem in the dynamic domain. By "motor babbling", many training examples can be generated, but there is no guarantee that these examples are relevant for the final trajectories the feedforward IDM has to control afterwards (this is basically the same problem as raised by Jordan and Rumelhart (1992) for FM learning in the dynamic domain). Nevertheless, at least for the two-joint planar arm in Sect. 2.1.4.4, this problem could be solved by using a PID controller in a negative feedback loop to generate approximately the desired elliptical trajectory and to collect training examples along the way. With this training set, a feedforward IDM with good performance could be learned. Miller et al. (1990) used a different approach to the real-time dynamic control of a five-axis industrial robot with DIM. Their learning scheme is very similar to FEL, but the feedforward IDM was trained online with the learning examples $[\boldsymbol{\theta}, \dot{\boldsymbol{\theta}}, \ddot{\boldsymbol{\theta}} \longrightarrow \boldsymbol{\tau}]$ instead of $[\boldsymbol{\theta}_d, \dot{\boldsymbol{\theta}}_d, \ddot{\boldsymbol{\theta}}_d \longrightarrow \boldsymbol{\tau}]$, thus applying a mixture of DIM and FEL.

In real-world applications, inverse dynamics problems often involve high-dimensional input spaces: For every degree of freedom of a robot setup, its position, velocity, and acceleration have to be provided as input. In the study by Vijayakumar et al. (2005), a humanoid robot with 30 degrees of freedom was trained to draw a planar figure 8 with one of its end effectors. The adaptive feedforward IDM had to learn the mapping from a 90-dimensional input space to a 30-dimensional output space (the torques). Standard function approximator techniques like the MLP do not perform very well on such high-dimensional learning problems. For this reason, Vijayakumar and Schaal (2000) developed "locally weighted projection regression" (LWPR), a function approximator technique especially suited for high-dimensional input spaces. It works on the basis of partial least squares (PLS) regression (e.g., Wold, 1985), which identifies the (orthogonal) directions in the input space with the largest correlations to the output data and restricts the regression to these directions. This reduces the dimensionality of the learning problem. However, for non-linear functions these relevant directions vary from position to position in the input space, therefore LWPR combines a multitude of PLS regressions which are centered at different points in the input space and which are locally learned and applied. In this way, LWPR can deal with high-dimensional function approximation tasks. LWPR was successfully applied to the inverse dynamics problem with the 90-dimensional input space for drawing the figure 8 (Vijayakumar et al., 2005). Learning examples were collected in a DIM-like fashion. In the earlier study by Vijayakumar and Schaal (2000), LWPR was tested as well on the basis of a DIM-like learning scheme for a 50-dimensional inverse dynamics task. Furthermore, in the study by Vijayakumar et al. (2002) LWPR was used for a dynamic biomimetic gaze stabilization task with FEL as learning strategy. However, in a diploma thesis under the author's supervision (Gerstung, 2006) we explored how easy LWPR is applicable in practice to various synthetical learning tasks. Our findings suggest that LWPR is very sensitive with regard to its adjustable parameter settings. The best settings seem to vary strongly from task to task, and successful learning requires a lot of test runs in advance.

LbI is not suited for the adaptation of feedforward IDMs because they have no additional state information as input which could be adjusted after the movement. The best and only adjustment for feedforward IDMs is the usage of the real or estimated state of the plant after the movement as training input, and this means basically to apply DIM.

The application of LbA to dynamic motor control has not been tested yet. Since LbA relies on making several test steps in motor space from a given controller output, and because the state of the system changes during these test steps, LbA seems to be rather unsuited for dynamic problems. Generally, it is not possible to revert the state of the plant to the point at which the original controller output has been generated first. Thus, any comparative quality judgement with regard to the test steps becomes quickly impossible, and therefore the generation of new and better training examples.

In conclusion, FEL, DSL, and DIM are applicable to the training of feed-forward IDMs. With regard to feedback IDMs, things get more complicated. To learn a feedback IDM basically means to learn an optimal control strategy (which implicitly includes to learn how to generate trajectories). Tasks like this belong to the field of reinforcement learning (see next section).

2.4 Alternative Routes to Motor Learning

2.4.1 Reinforcement learning

The motor learning strategies in Sect. 2.2 belong to the paradigm of supervised learning which is based on providing an explicit correction of the output of the adaptive learning system (although the "supervisor" is another compound of the learning system). In contrast, reinforcement learning (RL) (Sutton and Barto, 1998) works by just providing a specific reward to each output. The learning process is driven by the maximization of the reward. For motor learning with continuous state and motor variables, the temporal difference (TD) family of RL algorithms is often used (Doya et al., 2001). In this framework, a learning system consists of an "actor" and a "critic". The actor is equivalent to the controller or inverse model. It implements a control law

$$\mathbf{u}_t = G(\mathbf{x}_t) + \boldsymbol{\nu}_t \ .$$

G it the control policy, $\boldsymbol{\nu} \in \mathrm{IR}^m$ denotes noise for exploration, m is the dimensionality of the motor space. The state change via the plant is defined as usual:

$$\mathbf{x}_{t+1} = P(\mathbf{x}_t, \mathbf{u}_t)$$

The reward in time step t is computed as:

$$r_t = R(\mathbf{x}_t, \mathbf{u}_t)$$

The critic has to predict the cumulated future reward V (a scalar value), given the current state of the system:

$$V(\mathbf{x}_t) = E[r_t + \gamma r_{t+1} + \gamma^2 r_{t+2} + ...]$$

The parameter γ is the "discount factor" with $0 \leq \gamma \leq 1$. The error signal δ_t of RL is the prediction error of the critic. It is called "temporal difference error" (TD error):

$$\delta_t = r(t) + \gamma\widehat{V}(\mathbf{x}_{t+1}) - \widehat{V}(\mathbf{x}_t)$$

If the policy and the critic are implemented as adaptive neural networks (e.g., as function approximators), their parameters are updated by gradient ascent proportional to δ_t (for details, see Doya et al., 2001). Therefore, the predicted cumulated future reward V is reduced for negative values of δ_t and vice versa. Over many learning trials, the critic is adapted in this way towards better prediction performance. The adaptation of the policy G takes place in a way that its motor output is moved either in the direction of the last exploration noise vector $\boldsymbol{\nu}_t$ in motor space or exactly in the opposite direction. This changes the policy towards motor output with higher rewards. The desired state of the system $\mathbf{x}^*$ has to be hard-coded in the reward function R. A reasonable approach is to choose larger values for R the smaller the distance between $\mathbf{x}_t$ and $\mathbf{x}^*$ is.

Wolpert and Flanagan (2003) point out that optimal control theory and RL are mathematically equivalent. The former focuses on systems with known dynamics and known cost functions, while the latter have to learn both the dynamics and the costs through experience. Referring to the different types of inverse models, the IKM, the feedforward IDM, and the feedback IDM (see Sect. 2.2.1), RL can be used to acquire the latter, while the former two are best learned with supervised learning strategies. However, RL for feedback IDMs faces the following problem: If the desired state $\mathbf{x}^*$ is time-dependent ($\mathbf{x}_t^*$ in time-discrete or $\mathbf{x}_d(t)$ in time-continuous notation), TD learning as described above does not converge because R is not stationary. To overcome this problem, one could provide $\mathbf{x}_t^*$ as additional input to the policy G and the critic V. During learning, one should keep $\mathbf{x}_t^*$ constant over longer time intervals or change it only slowly because otherwise the computation of the TD error would be meaningless. Moreover, the function approximators for the policy and the critic must not suffer from catastrophic interference, because learning in one part of the input space would result in forgetting in other parts otherwise. However, we are not aware of any study which attempts to learn a feedback IDM with varying desired state input on the basis of RL. It might be an interesting and challenging research project to develop a TD based learning algorithm for the training of a feedback IDM which generates arm trajectories between any two points in the workspace according to the minimum-variance principle (Harris and Wolpert, 1998). In the learning architecture by Shimansky et al. (2004), a feedback IDM for a simulated two-joint planar arm was trained which is at least capable to generate trajectories towards eight different target zones in the workspace. Their work is similar to RL to some extent, but instead of an actor and a critic they use a forward dynamics model, an internal model of the movement cost rate, and an internal model of the minimal movement costs in addition to the feedback IDM. Unfortunately, the generated velocity profiles are not bell-shaped, questioning the biological plausibility of their learning architecture.

While RL is usually restricted to time-discrete systems, Doya (2000b) developed a variant of TD learning for time-continuous systems which is even better suited for smooth motor control. He showed the successful application of his algorithm for a pendulum swing-up task with limited torque and a cart-pole swing-up task. These are rather low-dimensional systems in state and motor space which illustrates one of the weaknesses of RL: RL algorithms cannot be easily applied to large-scale problems. For this reason, the recent trend in RL research is to use modularization and hierarchical architectures for realistic and large-scale problems (Miyamoto et al., 2004).

With regard to the neurophysiological foundation of RL, Doya (2000a) reviews several studies which support the view that the basal ganglia are the neural substrate of reward-based learning, while the cerebellum serves for error-based (supervised) learning. Especially the activity of dopamine neurons in the basal ganglia seems to resemble the output of the critic, predicting future reward. To complete the picture, Doya (2000a) ascribes unsupervised learning to the cerebral cortex.

2.4.2 Learning by imitation

A very different approach to motor learning is learning by imitation. For humans, imitation provides an important means to acquire new skills, both on a basic motor level and on the level of complex social interactions (see for example Tomasello et al., 1993). Imitation-like behavior is also observed in nonhuman primates and other species (e.g., Bard, 2007), although there is disagreement about the equivalence of human and nonhuman imitative behavior (Byrne and Russon, 1998). Regardless of this debate, one of the most important neurophysiological findings with regard to imitation was the discovery of "mirror neurons" in the premotor cortex of monkeys, which discharge both when the monkey performs an action and when he observes a similar action made by another monkey (Rizzolatti et al., 1996) (see also Sect. 1.3.4.2). These neurons are thought to be part of the neural foundation of learning by imitation (Billard and Schaal, 2006).

Compared to the supervised learning of internal models, imitation learning has many additional requirements: pose estimation and tracking, body correspondence, coordinate transformation from external to egocentric space, matching of observed against previously learned movements, suitable movement representations for imitation, etc. (Schaal et al., 2003). In return, imitation offers a speedup of learning because the troublesome conversion from sensory to motor error is less difficult: The search space for the right motor command is significantly reduced through the observation of the correct behavior. In addition, imitation offers a fast way to transfer the optimal behavior under certain task conditions from one individuum to the next. In robotics, learning by imitation gathers a lot of interest, motivated both from the cognitive modeling and from the technical application perspective (Billard and Schaal, 2006). Especially for humanoid robots with a large number of degrees of freedom, learning by imitation might offer a flexible means for human-robot interaction.

Chapter 3

Computational Methods

This section summarizes the theoretical background of the neural network and optimization techniques used throughout the thesis. Neural network techniques can be classified along different characteristics. One important distinction is between feedforward and recurrent networks, another between supervised and unsupervised learning rules. Feedforward networks are usually used for function approximation while recurrent networks are capable of representing high-dimensional data manifolds or spatio-temporal data. Usually, feedforward networks are coupled with supervised learning, requiring a "teacher" which corrects their output. Recurrent networks, on the other hand, are mainly trained by presenting learning examples without explicit correction of any output. By this unsupervised learning regime, recurrent networks learn to model the distribution of the data.

A classical feedforward network is the multi-layer perceptron (MLP). In this thesis, it is used for adaptive motor control and for the learning of FMs. Moreover, for the latter purpose radial basis functions networks (RBFN) are applied.

Recurrent networks are often used for pattern completion. They receive an incomplete pattern as input, and after several iterative cycles in which the activation of the network's units is updated, they settle down to a final state which is supposed to be the complete pattern (e.g., the Hopfield network works in this way; Hopfield, 1982, 1984). In this thesis, the focus lies on so-called "abstract" recurrent neural networks. They are based on vector quantization methods (Gersho and Gray, 1991) and do not require an iterative activation update for pattern recall (therefore the term "abstract"). From this class of networks, NGPCA (a combination of neural gas and local principal component analysis; Möller and Hoffmann, 2004) is used in the context of motor learning. However, vector quantization is not only used for recurrent neural networks, but also in combination with supervised learning techniques, e.g. RBFNs.

Furthermore, the optimization method Differential Evolution (DE) (Storn

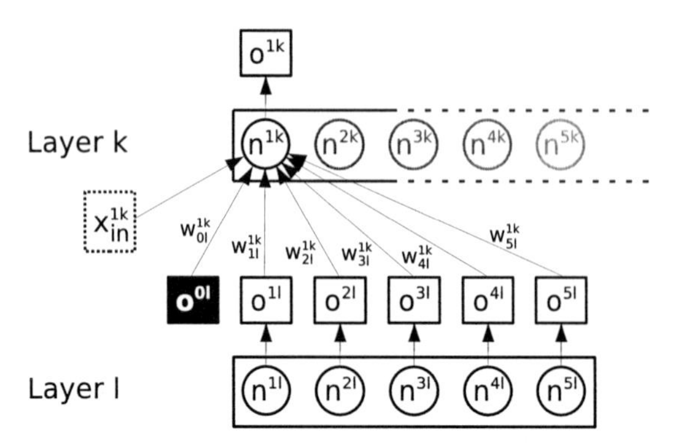

Figure 3.1 — General structure of the MLP, shown for a single unit n^{1k} (first unit of layer k). In this figure, the preceding layer l has five units. Their outputs are the weighted inputs for unit n^{1k}. Moreover, unit n^{1k} receives a weighted input from the bias unit of the layer l (with output $o^{0l} = 1.0$), and an external input x_{in}^{1k}. For clarity, only two layers and only the connections to a single unit are shown.

and Price, 1997) is used for the studies in Chapts. 4 and 7. In the following, all these methods are briefly described.

3.1 Multi-Layer Perceptron

The MLP belongs to the class of feedforward neural networks. It was proposed first by Werbos (1974), and became very popular among researchers with the publication by Rumelhart et al. (1986). It is mainly used for function approximation. We first describe the general structure of MLPs, afterwards the backpropagation algorithm for network adaptation is presented.

3.1.1 Topology

The basic component of MLPs are units $\{n^{ik}\}$. Each unit is an artificial neuron (in analogy to biological neurons, but on a very abstract level) with input and output links by which the unit is connected to other units in the network. The units are arranged in layers $\{k\}$ (the index i of n^{ik} counts units within a layer; $k = 1..K$, $i = 1..N_k$; K is the number of network layers, N_k is the number of units in layer k). Each unit can receive inputs from units in preceding layers, and can provide an output to units in succeeding layers. Within a layer, no connections between units exist.

Each unit n^{ik} has an activation a^{ik}. This activation is transformed into the unit's output o^{ik} by $o^{ik} = f_{act}(a^{ik})$. f_{act} is the so-called "activation function", which can take different forms. For the units of the first and the last network layer, often linear functions are used, for the layers in-between (the "hidden layers") squashing functions like hyperbolic tangent. The standard formula for the activation is

$$a^{ik} = \sum_{l \in L_k} \left[\sum_{j=0}^{N_l} (w_{jl}^{ik} o^{jl}) \right] + x_{\text{in}}^{ik} . \tag{3.1}$$

L_k is an index set referencing all layers l from which layer k receives input with $\bigvee_{l \in L_k} l < k$. The term x_{in}^{ik} is an external input to unit i of layer k; in the standard MLP, these inputs only exist in the first layer, but in a more general form, any unit may receive an external input. The parameter w_{jl}^{ik} is the weight assigned to the connection between unit j in layer l and unit i in layer k. The weights between two layers l and k are often collected in a matrix $\mathbf{W}^{lk}$. For the first layer ($k = 1$), Eqn. (3.1) can be simplified to $a^{ik} = x_{\text{in}}^{ik}$. The inner sum in Eqn. (3.1) starts with an index $j = 0$ although the units are counted beginning with 1. The unit index 0 in each layer is reserved for the so-called "bias unit" with a constant output of 1.0.

The main purpose of MLPs is the transformation of the real-valued inputs $\{x_{\text{in}}^{ik}\}$ into the real-valued outputs $\{o^{ik}\}$. In the standard MLP with three layers, in which the activation functions of the units in the single hidden layer are squashing functions, the external inputs $\{x_{\text{in}}^{i,1}\}$ $(i = 1, ..., N_1)$ of the first layer are transformed into the output values $\{o^{j,3}\}$ $(j = 1, ..., N_3)$ of the last layer. Moreover, for the standard MLP we have $L_k = \{k - 1\}$ for $k \in \{2, 3\}$. Hornik et al. (1989) showed for this standard type that it is an universal approximator for real-valued Borel measurable functions.[1] Actually, this topology is the minimum specification for MLPs. An MLP can be larger (more layers, additional "shortcut" connections between layers), but it cannot be smaller. A two-layered network is only capable to represent linearly separable input-output relationships and belongs to the class of perceptrons (Minsky and Papert, 1969).

3.1.2 Backpropagation

An important feature of MLPs is their ability to learn input-output relationships. They can adapt their output by adjusting the parameters w_{jl}^{ik}. This adaptation usually takes place by presenting examples of the input-output relationship the MLP is supposed to learn. For the standard MLP with three layers, there is an

[1] For practical applications, it is most important that continuous real-valued functions which map compact sets to compact sets are Borel measurable (Yen and Vaart, 1966).

input vector $\mathbf{x}_{\text{in}}$ with elements x^i_{in} (assigned to the external inputs $x^{i,1}_{\text{in}}$ of the first layer) and an output vector $\mathbf{x}_{\text{out}}$ with elements x^j_{out} which are the desired values for the network outputs $o^{j,3}$. The task of the MLP is to learn the empirical data in the set X:

$$(\mathbf{x}^{(1)}_{\text{in}}, \mathbf{x}^{(1)}_{\text{out}}), ..., (\mathbf{x}^{(P)}_{\text{in}}, \mathbf{x}^{(P)}_{\text{out}}) \in X \, .$$

In the process of learning, training patterns $(\mathbf{x}^{(p)}_{\text{in}}, \mathbf{x}^{(p)}_{\text{out}})$ $(p = 1..P)$ are presented to the MLP. The input $\mathbf{x}^{(p)}_{\text{in}}$ is transformed into the corresponding network output $\mathbf{o}_p$ which usually deviates considerably from the desired output $\mathbf{x}^{(p)}_{\text{out}}$ in the beginning of the learning process.

The following error measure quantifies this deviation for a single pattern p:

$$E_p = \frac{1}{2} \| \mathbf{x}^{(p)}_{\text{out}} - \mathbf{o}_p \|^2$$

The overall error for the whole training set is:

$$E = \sum_{p=1}^{P} E_p \tag{3.2}$$

The task of any training algorithm for the adaption of MLPs is to minimize this error by computing appropriate weight changes Δw^{ik}_{jl}. The popular backpropagation algorithm (Werbos, 1974; Rumelhart et al., 1986) for MLPs is based on gradient descent. The weight changes Δw^{ik}_{jl} are computed by

$$\Delta w^{ik}_{jl} = -\eta \frac{\partial E}{\partial w^{ik}_{jl}} \, . \tag{3.3}$$

η is the learning rate.

From this starting point, the backpropagation algorithm is derived as shown in Rumelhart et al. (1986). Here, we present only the final algorithm. Online backpropagation (with weight changes after each pattern presentation p) comprises the following steps:

1. Compute an error signal $\delta^{ik}_p = \frac{\partial E}{\partial o^{ik}_p}$ for each unit n^{ik} with $k > 1$:

$$\delta^{ik}_p = \begin{cases} f'_{act}(a^{ik}_p)(x^{i\,(p)}_{\text{out}} - o^{ik}_p) & \text{if layer } k \text{ is the output layer} \\ f'_{act}(a^{ik}_p) \sum_{j=0}^{N_{k+1}} \delta^{j,k+1}_p w^{j,k+1}_{ik} & \text{otherwise} \end{cases} \tag{3.4}$$

The error signals are determined first for the output layer with $k = K$, afterwards for the layer with $k = K - 1$, and so on. In this way, the error signals are "back propagated" through the network.

2. Compute the weight changes:

$$\Delta_p w_{jl}^{ik} = -\eta o_p^{jl} \delta_p^{ik}$$

Equation (3.4) can be extended to a more general form in which it applies to MLPs with multiple hidden layers and shortcut connections, where every unit can be an output unit.

In the batch version of backpropagation, the weight changes for the whole training set are summed up before applying the change. Batch learning is derived directly from Eqn. (3.3), while the online version has an additional heuristic component. It belongs to the class of optimization methods labeled "stochastic gradient descent". A presentation of a single pattern for learning is often called "learning cycle", while the presentation of the whole training set is called "learning epoch". Before the first learning cycle or epoch, the weights w_{jl}^{ik} are initialized to random values from the interval $[w_{min}; w_{max}]$.

Parameter	Description
η	Learning rate
w_{min}	Lower bound for the weight initialization range
w_{max}	Upper bound for the weight initialization range

Table 3.1 — Learning parameters for MLPs trained with the backpropagation method.

3.2 Radial Basis Function Networks

Radial basis function networks (RBFN) are used for function approximation and interpolation (Moody and Darken, 1989). Similar to the MLP, the learning task is to adapt to the empirical data in the set X:

$$(\mathbf{x}_{\text{in}}^{(1)}, \mathbf{x}_{\text{out}}^{(1)}), ..., (\mathbf{x}_{\text{in}}^{(P)}, \mathbf{x}_{\text{out}}^{(P)}) \in X \, .$$

RBFNs are two-layered networks. The first layer consists of N_1 Gaussian units with index $i = 1, ..., N_1$. The activation of a Gaussian unit is computed as:

$$\phi_i(\mathbf{x}_{\text{in}}) = \exp\left(-\frac{\|\mathbf{c}_i - \mathbf{x}_{\text{in}}\|^2}{2\sigma_i^2}\right)$$

$\mathbf{c}_i$ is the center of unit i, σ_i^2 its variance.

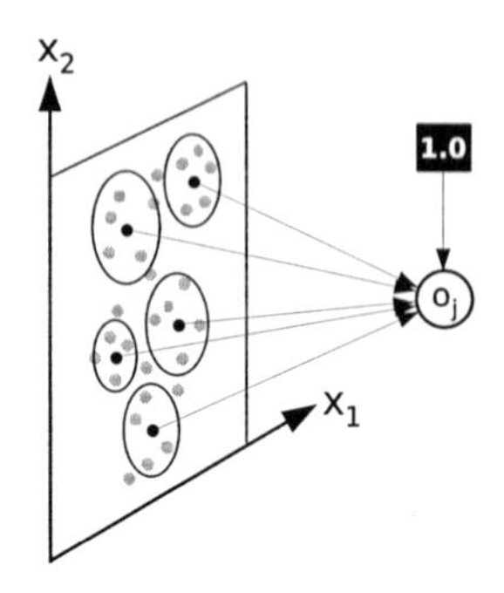

Figure 3.2 — Basic structure of an RBFN for a two-dimensional input space (spanned by x_1 and x_2). Data points $\{\mathbf{x}_{\text{in}}\}$ are shown as gray dots. Gaussian units ϕ_i are depicted as circles in the data plane. A single output unit o_j is shown which receives weighted input from all Gaussian units and from the bias unit with a constant activation of 1.0.

The second layer is very similar to the output layer of standard MLPs. It consists of N_2 units. The output of unit j with $j = 1, ..., N_2$ is determined by:

$$o_j = \sum_{i=0}^{N_1} w_i^j \phi_i(\mathbf{x}_{\text{in}}) \tag{3.5}$$

The unit with index $i = 0$ in this equation is a bias unit with a constant activation of $\phi_0 = 1.0$.

For the adaptation of RBFNs, usually the following steps are carried out (Moody and Darken, 1989; Fritzke, 1998):

- Position the centers $\{\mathbf{c}_i\}$ in the input data space by using a vector quantization method for the vectors $\{\mathbf{x}_{\text{in}}^{(p)}\}$ $(p = 1, ..., P)$; for this purpose, we use the K-means algorithm (see Sect. 3.3.1); alternatively, the NG method (see Sect. 3.3.2) might be applied as well.

- Determine the variances $\{\sigma_i^2\}$; it is suggested to choose σ_i proportionally to the mean distance d_i of unit i to the input data points that have been assigned to this unit in the preceding step (here we use: $\sigma_i = \alpha_\sigma d_i + \beta_\sigma$).

- Compute the weights w_i^j by regression techniques.

The last step requires further explanation. In vector form, and for a specific pattern p, Eqn. (3.5) can be written as

$$\mathbf{o} = \mathbf{W}\boldsymbol{\phi}(\mathbf{x}_{\text{in}}^{(p)})$$

with $\mathbf{o} = (o_1 ... o_{N_2})^T$ and the matrix $\mathbf{W}$ containing the weights w_i^j (i indexing columns, j indexing rows). Substituting the actual network output $\mathbf{o}$ with the desired output $\mathbf{x}_{\text{out}}^{(p)}$ yields

$$\mathbf{x}_{\text{out}}^{(p)} = \mathbf{W}\phi(\mathbf{x}_{\text{in}}^{(p)}) .$$

Since there are P training patterns in the empirical data set, P different equations of this form are available. This system of equations can be rewritten as

$$\mathbf{X}_{\text{out}} = \mathbf{W}\mathbf{\Phi} \tag{3.6}$$

with $\mathbf{X}_{\text{out}} = \left(\mathbf{x}_{\text{out}}^{(1)} \cdots \mathbf{x}_{\text{out}}^{(P)}\right)$ and $\mathbf{\Phi} = \left(\phi(\mathbf{x}_{\text{in}}^{(1)}) \cdots \phi(\mathbf{x}_{\text{in}}^{(P)})\right)$.

The system in Eqn. (3.6) is either underdetermined (if $N_1 > P$; very unlikely), exactly solvable, or overdetermined (if $N_1 < P$; the normal case). A general approach to its solution is the computation of the pseudoinverse $\mathbf{\Phi}^+$ of $\mathbf{\Phi}$. Then, the weights of the output layer can be determined by

$$\mathbf{W} = \mathbf{X}_{\text{out}}\mathbf{\Phi}^+ . \tag{3.7}$$

If Eqn. (3.6) is overdetermined, the solution from Eqn. (3.7) is not exact but conforms to a least squared error criterion (Golub and van Loan, 1996).

Instead of determining $\mathbf{W}$ with help of the pseudoinverse of $\mathbf{\Phi}$, one can use the same gradient descent technique as used in the backpropagation algorithm (based on the error function in Eqn. (3.2)). Since there are no hidden layers involved, it even simplifies to the standard delta rule (Widrow and Hoff, 1960). Moreover, one can generalize the backpropagation method to finetune the parameters $\{\mathbf{c}_i\}$ and $\{\sigma_i\}$ by gradient descent (Zell, 1997). This yields three different learning rates for RBFN networks: η, η_{c}, and η_σ (for explanation, see Table 3.2)

Parameter	Description
η	Learning rate for the weights w_i^j
η_{c}	Learning rate for the unit centers $\mathbf{c}_i$
η_σ	Learning rate for the unit variances σ_i^2
w_{min}	Lower bound for the weight initialization range
w_{max}	Upper bound for the weight initialization range
$\alpha_\sigma, \beta_\sigma$	Factor and offset for determining σ_i

Table 3.2 — Learning parameters for RBFNs.

3.3 Vector Quantization

The goal of vector quantization is to represent a set of data points $\{\mathbf{x}^{(p)}\}$ ($p = 1, ..., P$) in n-dimensional space with a smaller set of so-called "codebook vectors" $\{\mathbf{c}_i\}$; $i = 1, ..., N$ is the unit index. $P(i|\mathbf{x}^{(p)})$ is the probability that $\mathbf{x}^{(p)}$ belongs to unit i.

There are two basic categories of vector quantization methods: Hard-clustering methods assign only binary values (0 or 1) to $P(i|\mathbf{x}^{(p)})$ so that each data point is assigned to exactly one unit. On the contrary, soft-clustering methods work with continuous values for $P(i|\mathbf{x}^{(p)})$, thus each data point can be assigned to multiple units with different probabilities.

The optimal positions $\{\mathbf{c}_i\}$ are usually determined by minimizing the following error measure:

$$E = \sum_{i,p} P(i|\mathbf{x}^{(p)}) \|\mathbf{x}^{(p)} - \mathbf{c}_i\|^2$$

Generally, E has many local minima and it is hard to find the optimum solution. Therefore, many different vector quantization methods have been developed, both in the hard- and the soft-clustering category. In the following, the pattern index p is mostly omitted for clarity.

3.3.1 K-means algorithm

The K-means algorithm (Lloyd, 1982; Moody and Darken, 1989) is a widely used hard-clustering method. In the online version, a single data point $\mathbf{x}$ is presented in each learning step. The codebook vector $\mathbf{c}_i$ with the smallest Euclidean distance to $\mathbf{x}$ is determined. Afterwards, this codebook vector is updated according to the following rule:

$$\mathbf{c}_i \leftarrow \mathbf{c}_i + \epsilon\,(\mathbf{x} - \mathbf{c}_i)$$

ϵ is the learning rate. The update of the codebook vectors continues until convergence is reached. In the batch version, all training patterns $\{\mathbf{x}^{(p)}\}$ are presented first without changing the positions $\{\mathbf{c}_i\}$. Only the assignments $P(i|\mathbf{x}^{(p)})$ are determined. Afterwards, each position $\mathbf{c}_i$ is set to the mean value of all vectors assigned to unit i: $\{\mathbf{x}^{(p)}|P(i|\mathbf{x}^{(p)}) = 1\}$. This procedure is repeated until convergence is reached. The K-means algorithm is prone to end in local minima. The final result depends heavily on the initial choice of the codebook vectors $\{\mathbf{c}_i\}$.

3.3.2 Neural gas

Martinetz et al. (1993) presented a vector quantization method called "Neural Gas" (NG). NG belongs to the class of soft-clustering methods with "soft competition" between units. First, the units $\{\mathbf{c}_i\}$ are initialized by assigning randomly selected vectors from the training set to them. Afterwards, in each training step one vector $\mathbf{x}$ of the training set is drawn at random. The squared Euclidean distance

$$d_i(\mathbf{x}) = \|\mathbf{x} - \mathbf{c}_i\|^2$$

is computed for each unit; the vector of these distances is $\mathbf{d}$. A rank $r_i(\mathbf{d}) = 0, ..., N-1$ is assigned to each unit: A rank of 0 indicates the closest and a rank of $N-1$ the largest distance to the vector $\mathbf{x}$. After the ranking, the units $\{\mathbf{c}_i\}$ are updated according to the following rule:

$$\mathbf{c}_i \leftarrow \mathbf{c}_i + \epsilon \cdot h_\rho \left[r_i(\mathbf{d})\right] \cdot (\mathbf{x} - \mathbf{c}_i) \tag{3.8}$$

The function $h_\rho(\mathbf{d}) = \mathrm{e}^{-r/\rho}$ ensures that not only the best-matching unit is updated, but every unit with a factor exponentially decreasing with their rank (this is an important difference to the "hard-clustering" method K-means). The parameter ρ determines the neighborhood range. The second parameter ϵ is a global learning rate. Both ρ and ϵ decrease exponentially from initial positive values ($\rho(0)$, $\epsilon(0)$) to smaller final positive values ($\rho(T)$, $\epsilon(T)$), where T denotes the index of the final training step. By this means learning starts with the adaptation to the global structure of the training data and becomes more local during the training process.

3.4 NGPCA

NGPCA combines the vector quantization method NG with local principal component analysis (PCA) (Hoffmann and Möller, 2003; Möller and Hoffmann, 2004). As with vector quantization, the goal is to represent a high-dimensional data manifold by a model with only a small number of parameters (compared to the amount of training data).

PCA is a method of dimension reduction in multivariate data. The high-dimensional pattern space (with n dimensions) is approximated by a subspace of lower dimension. This subspace is spanned by the first m principal eigenvectors of the data covariance matrix (with $m \leq n$). In the following, $\mathbf{W}$ denotes the matrix of estimated eigenvectors $\mathbf{w}_j$, $j = 1, ..., m$ (one vector per column). $\mathbf{W}$ is of size $n \times m$. The eigenvalue λ_j of $\mathbf{w}_j$ is equal to the variance of the data distribution in this eigendirection. The $m \times m$ matrix $\mathbf{\Lambda}$ is a diagonal matrix

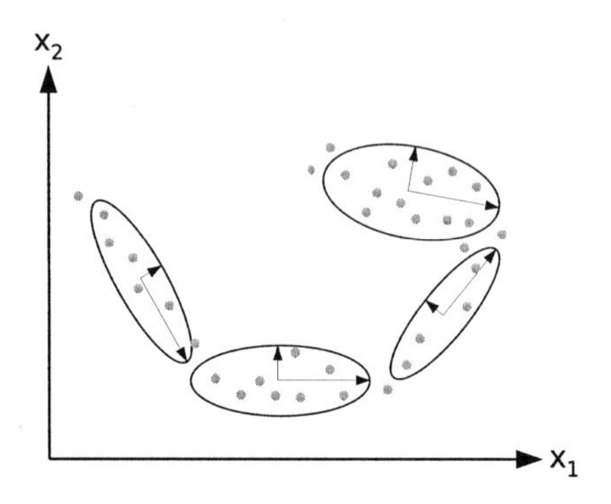

Figure 3.3 — NGPCA model with four local PCA units, depicted as ellipsoids in a two-dimensional data space. The arrows within each ellipsoid indicate the principal components. The data points $\{\mathrm{x}\}$ belonging to the data manifold are shown as gray dots.

containing the values λ_j. If the data has not been centered before the PCA, an additional vector $\mathbf{c}$ of dimension n is necessary. $\mathbf{c}$ is the mean vector of the data distribution and the center of the PCA. Altogether, the multivariate data is just represented by the matrices $\mathbf{W}$ and $\mathbf{\Lambda}$ and the vector $\mathbf{c}$.

Both NG and PCA have certain strengths and weaknesses. In NGPCA, the weaknesses of one method are compensated by the other, while the strengths are preserved. NG suffers from dimensions which contain mainly noise. Many units are needed to fill these dimensions. This is quite inefficient. PCA, on the other hand, needs just a single eigenvector to represent a noise dimension. But PCA is only a linear method and not capable to describe data manifolds if components have non-linear dependencies. NGPCA overcomes this drawback by describing the data manifold with a sufficiently large number of local PCAs which are placed at different positions $\{\mathbf{c}_i\}$. Each local PCA only adapts to the structure of the data distribution in the neighborhood of $\mathbf{c}_i$. In this way, curved data distributions are described by a local linear approximation. The positions of the units $\{\mathbf{c}_i\}$ are determined by an algorithm closely related to NG. Thus, NGPCA overcomes the inefficiency of NG with regard to noise dimensions and of PCA with regard to non-linear dependencies in the data by combining both methods.

A complete NGPCA model consists of N local PCAs, each described by the tupel $\{\mathbf{c}_i, \mathbf{W}_i, \mathbf{\Lambda}_i, \lambda_i^*\}$, with $i = 1, ..., N$. λ_i^* represents an estimate of the eigenvalue in each of the remaining $n - m$ minor eigendirections. In geometrical terms, each unit i is a hyperellipsoid centered at $\mathbf{c}_i$. In the $n - m$ minor eigendirections, it has the form of a hypersphere.

3.4.1 Learning

Adaptation of the centers $\{\mathbf{c}_i\}$ is carried out in the same way as in the NG algorithm, only the distance measure for the ranking of the units is modified (Möller and Hoffmann, 2004):

$$d_i(\mathbf{x}) = \left(\boldsymbol{\xi}_i^T \mathbf{W}_i \boldsymbol{\Lambda}_i^{-1} \mathbf{W}_i^T \boldsymbol{\xi}_i + \frac{1}{\lambda_i^*} \left(\boldsymbol{\xi}_i^T \boldsymbol{\xi}_i - \boldsymbol{\xi}_i^T \mathbf{W}_i \mathbf{W}_i^T \boldsymbol{\xi}_i \right) \right) V^{2/n} \tag{3.9}$$

$\boldsymbol{\xi}_i$ is the difference between the unit center $\mathbf{c}_i$ and the presented data point $\mathbf{x}$: $\boldsymbol{\xi}_i = \mathbf{x} - \mathbf{c}_i$. V is proportional to the volume of the hyperellipsoid and computed as $V = \sqrt{|\boldsymbol{\Lambda}| \lambda_i^{*n-m}}$.

Equation (3.9) is the sum of a normalized Mahalanobis distance plus reconstruction error. It is a modified form of the equation found in (Möller and Hoffmann, 2004, p. 308); in the original equation, the ranking measure $d_i(\mathbf{x})$ depends on the volume of the ellipsoid. In Hoffmann (2004), Eqn. (3.9) is presented which handles all ellipsoids for the ranking as if they had the same constant volume. According to Hoffmann (2004), this can be advantageous for the learning process.

λ_i^* in Eqn. (3.9) is determined as

$$\lambda_i^* = \frac{\sigma_i^2}{n - m} .$$

σ_i^2 is the residual variance of unit i. It is computed by an iterative update:

$$\sigma_i^2 \leftarrow \sigma_i^2 + \alpha_i \cdot \left(\boldsymbol{\xi}_i^T \boldsymbol{\xi}_i - \boldsymbol{\xi}_i^T \mathbf{W}_i \mathbf{W}_i^T \boldsymbol{\xi}_i - \sigma_i^2 \right)$$

According to Möller and Hoffmann (2004), the choice of the starting value of the iteration $\sigma^2(0)$ is not critical. α_i is a unit-specific learning rate: $\alpha_i = \epsilon \cdot h_\rho \left[r_i(\mathbf{d}) \right]$ (as in Eqn. (3.8) for NG). As in the NG method, the neighborhood range ρ and the learning rate ϵ decrease exponentially.

In addition to the centers $\{\mathbf{c}_i\}$, the matrices $\{\mathbf{W}_i\}$ and $\{\boldsymbol{\Lambda}_i\}$ have to be updated with each pattern presentation $\mathbf{x}$. For this purpose, an online PCA method is applied:

$$\mathbf{W}_i, \boldsymbol{\Lambda}_i \leftarrow \text{PCA}\{\mathbf{W}_i, \boldsymbol{\Lambda}_i, \boldsymbol{\xi}_i, \alpha_i\}$$

Throughout the studies in this thesis, the online PCA method presented in Möller (2002) is used. It works by the interlocking of RRLSA (Ouyang et al., 2000), a neural method for PCA based on the recursive least squares method, and the Gram-Schmidt method for orthonormalization (Golub and van Loan, 1996). Since this interlocked online PCA method does not guarantee perfect orthogonality, after every T_{ortho} learning steps an explicit Gram-Schmidt orthonormalization is carried out for all units.

In the beginning of the learning process, the centers $\{\mathbf{c}_i\}$ are set to randomly chosen data vectors from the training set $\{\mathbf{x}^{(p)}\}$. The matrices $\{\mathbf{W}_i\}$ are initialized to random orthonormal systems, the eigenvalues in $\{\mathbf{\Lambda}_i\}$ to a constant $\lambda(0)$. Möller and Hoffmann (2004) state that the choice of $\lambda(0)$ is not critical.

3.4.2 Recall

After training, the tupel $\{\mathbf{c}_i, \mathbf{W}_i, \mathbf{\Lambda}_i, \lambda_i^*\}$ has been adapted for every unit i so that the overall model represents the data manifold from which the training vectors $\{\mathbf{x}\}$ have been drawn. Each unit is a hyperellipsoid which describes an iso-potential surface of the normalized Mahalanobis distance plus reconstruction error (Hoffmann and Möller, 2003; Hoffmann, 2004):

$$\begin{aligned} d_i(\mathbf{x}) &= \boldsymbol{\xi}_i^T \mathbf{W}_i \mathbf{\Lambda}_i^{-1} \mathbf{W}_i^T \boldsymbol{\xi}_i + \frac{1}{\lambda_i^*} \left(\boldsymbol{\xi}_i^T \boldsymbol{\xi}_i - \boldsymbol{\xi}_i^T \mathbf{W}_i \mathbf{W}_i^T \boldsymbol{\xi}_i \right) \\ &+ \ln \det \mathbf{\Lambda}_i + (n-m) \ln \lambda_i^* \end{aligned} \tag{3.10}$$

This equation is similar to Eqn. (3.9) but takes the actual volume of the ellipsoid into account (with $\boldsymbol{\xi}_i = \mathbf{x} - \mathbf{c}_i$).

For pattern recall, some dimensions of the data space have to be defined as input and the others as output. The vector $\mathbf{p}$ (of dimension n) contains the input values at the input positions and zero elsewhere. It defines the offset of a constraint space $\mathbf{z}(\boldsymbol{\eta})$ spanning the space of all possible output values:

$$\mathbf{z}(\boldsymbol{\eta}) = \mathbf{M}\boldsymbol{\eta} + \mathbf{p} \tag{3.11}$$

$\boldsymbol{\eta}$ is a parameter vector of size n_{out} (number of output dimensions). M is a matrix of size $n \times n_{out}$. In all applications of NGPCA in this thesis, M is defined in a way that the constraint space is aligned with the coordinate system.

The goal of recall is to generate a pattern $\widehat{\mathbf{z}}$ containing the inputs defined in $\mathbf{p}$ and the outputs generated by the NGPCA model. For this purpose, $\mathbf{z}(\boldsymbol{\eta})$ is inserted into Eqn. (3.10) instead of $\mathbf{x}$:

$$\begin{aligned} d_i(\mathbf{z}(\boldsymbol{\eta})) &= (\mathbf{M}\boldsymbol{\eta} + \boldsymbol{\pi}_i)^T \left(\mathbf{W}_i \mathbf{\Lambda}_i^{-1} \mathbf{W}_i^T + \frac{1}{\lambda_i^*} \left(\mathbf{I} - \mathbf{W}_i \mathbf{W}_i^T \right) \right) (\mathbf{M}\boldsymbol{\eta} + \boldsymbol{\pi}_i) \\ &+ \ln \det \mathbf{\Lambda}_i + (n-m) \ln \lambda_i^* \end{aligned}$$

with $\boldsymbol{\pi}_i = \mathbf{p} - \mathbf{c}_i$.

For each unit i, the parameter vector $\widehat{\boldsymbol{\eta}}_i$ resulting in the smallest distance measure $d_i(\mathbf{z}(\boldsymbol{\eta}))$ has to be determined. Hoffmann and Möller (2003) show that setting $\partial d_i / \partial \boldsymbol{\eta} = 0$ yields

$$\widehat{\boldsymbol{\eta}}_i = \mathbf{A}_i \boldsymbol{\pi}_i$$

with

$$\mathbf{A}_i = -\left(\mathbf{M}^T \mathbf{D}_i \mathbf{M}\right)^{-1} \mathbf{M}^T \mathbf{D}_i$$

and

$$\mathbf{D}_i = \mathbf{W}_i \boldsymbol{\Lambda}_i^{-1} \mathbf{W}_i^T + \frac{1}{\lambda_i^*}\left(\mathbf{I} - \mathbf{W}_i \mathbf{W}_i^T\right) .$$

Since $d_i(\mathbf{z}(\boldsymbol{\eta}))$ is convex, $\widehat{\boldsymbol{\eta}}_i$ actually points to the only minimum. The unit $\hat{i}$ with the smallest distance measure $d_i(\mathbf{z}(\widehat{\boldsymbol{\eta}}_i))$ is selected, yielding the corresponding parameter vector $\widehat{\boldsymbol{\eta}}_{\hat{i}}$. The final pattern $\widehat{\mathbf{z}}$ is determined by inserting $\widehat{\boldsymbol{\eta}}_{\hat{i}}$ into Eqn. (3.11). For each input, a unique output is generated without the danger of getting trapped in a local minimum (Hoffmann and Möller, 2003; Hoffmann, 2004).

Parameter	Description
T	Number of training steps
T_{ortho}	Orthogonalization enforcement cycles
$\epsilon(0)$	Initial value for parameter ϵ (learning rate of NG)
$\epsilon(T)$	Final value for parameter ϵ
$\rho(0)$	Initial value for parameter ρ (neighborhood range of NG)
$\rho(T)$	Final value for parameter ρ
$\sigma^2(0)$	Iteration start value for the residual variance
$\lambda(0)$	Initial eigenvalues

Table 3.3 — Learning parameters for NGPCA.

3.5 Differential Evolution

Differential evolution (DE) has been developed by Storn and Price (1997). It is a method for global optimization over continuous spaces which shares some of the general techniques of evolutionary algorithms. Storn and Price (1997) showed in their initial work that it is superior to well-known methods like adaptive simulated annealing (Ingber, 1993) on a test suite of objective functions.

Meanwhile, DE has been used in various technical applications, e.g. in chemical engineering (Babu and Sastry, 1999) or in the design of digital filters (Storn, 1999). In the following, the algorithm is presented in the specific form in which it is used throughout the studies in this thesis (see Chapts. 4 and 7). Storn and Price (1997) provide the reader with a more general approach.

The goal of the optimization process is to find the global minimum of an analytically unknown function f_{opt}:

$$f_{\text{opt}} : \mathrm{IR}^n \longrightarrow \mathrm{IR}\,, \quad \mathbf{x} \mapsto y = f_{\text{opt}}(\mathbf{x})$$

DE is a parallel search method: In each optimization step g, N_{DE} different parameter vectors $\mathbf{x}_{i,g}$ ($i = 0, ..., N_{\text{DE}} - 1$) are generated and evaluated. From this population, the best parameter vector $\mathbf{x}_g^{\text{best}} = \mathbf{x}_{i_{\text{min}},g}$ is determined with

$$i_{\text{min}} = \underset{i}{\operatorname{argmin}}\{f_{\text{opt}}(\mathbf{x}_{i,g})\}\ .$$

To generate the population of the next generation $g + 1$, the following steps are carried out for each of the parameter vectors $\mathbf{x}_{i,g}$: First, a trial vector $\mathbf{v}$ is computed. This can be done in various different ways. In our studies, we used the version DE2 (from Storn and Price, 1997) as it proved to be the most successful:

$$\mathbf{v} = \mathbf{x}_{i,g} + \gamma\left(\mathbf{x}_g^{\text{best}} - \mathbf{x}_{i,g}\right) + \lambda\left(\mathbf{x}_{r_1,g} - \mathbf{x}_{r_2,g}\right)$$

In our studies on motor learning (Chapt. 4), it was advantageous to set $\gamma = 1$, resulting in the simplified rule

$$\mathbf{v} = \mathbf{x}_g^{\text{best}} + \lambda\left(\mathbf{x}_{r_1,g} - \mathbf{x}_{r_2,g}\right)\ .$$

The indices r_1 and r_2 are drawn at random from the range $[0, N_{\text{DE}} - 1]$ with the restriction that they do not equal neither the index i nor each other. γ and λ are free parameters set by the user. The computation of the trial vector by adding population vector differences to the candidate vector $\mathbf{x}_{i,g}$ is the most important innovation of DE. It enables DE to move the vector population rather fast through narrow and long valleys of the objective function.

In the second step, the trial vector is combined with the original vector $\mathbf{x}_{i,g}$ by a crossover method to determine the vector $\mathbf{x}_{i,g+1}$. Each element $\left(x_{i,g+1}\right)_j$ of $\mathbf{x}_{i,g+1}$ (with $j = 1, ..., n$) is determined by

$$\left(x_{i,g+1}\right)_j = \begin{cases} v_j & : r \leq p_{\text{CR}} \\ \left(x_{i,g}\right)_j & : \text{otherwise} \end{cases}\ .$$

p_{CR} is the crossover probability (also a free parameter set by the user). r is a random number drawn from a uniform distribution in the range $[0; 1]$. Moreover, the implementation of DE we used (Godwin, 1998) ensures that at least one element from $\mathbf{v}$ is carried over to $\mathbf{x}_{i,g+1}$. The crossover method described in Storn and Price (1997) is more restrictive but should yield very similar results in practical usage.

By repeating these two steps for all vectors $\mathbf{x}_{i,g+1}$ of generation g, the population of the next generation $g+1$ is constructed, consisting of vectors $\mathbf{x}_{i,g+1}$. This process is repeated until the maximum number of generations $G_{\max}$ is exceeded or the best parameter vector $\mathbf{x}_g^{\text{best}}$ of generation g falls below an "energy" threshold $E_{\min}$ with $f_{\text{opt}}(\mathbf{x}_g^{\text{best}}) < E_{\min}$. For the studies in this thesis, DE proved to be a reliable optimization method.

Parameter	Description
N_{DE}	Population size
λ	Scaling factor for the computation of the trial vector
γ	Scaling factor for the computation of the trial vector
p_{CR}	Crossover probability
$G_{\max}$	Maximum number of generations
$E_{\min}$	"Energy" threshold

Table 3.4 — Parameters for DE.

Chapter 4
Experimental Studies on Kinematic Motor Learning

Section 2.2 outlines various controller learning strategies, among them staged and continuous learning by averaging which have been developed by the author. The performance of these learning strategies is compared in different (simulated) motor control tasks: First, for saccade-like fixation movements with a robot camera head, and second, for the kinematic control of a planar arm. Both motor control tasks are explored for different task conditions, e.g. the number of links of the planar arm is varied.

The performance measure is always the number of exploration trials needed to achieve a certain controller quality Q^*. The term "exploration trial" is defined as motor command carried out to collect the plant's response. Because exploration trials are the most expensive part of learning, their number is used as performance measure: The fewer trials are needed the better. The quality level Q^* is determined by training a controller network with a set of perfect learning examples and assessing its quality afterwards. These examples were obtained by the optimization method "differential evolution" (see Sect. 3.5). The size of the perfect training set and the quality level Q^* vary from task to task. Important research questions addressed in this chapter cover the following topics:

- How do the learning strategies perform for plants with approximately linear input-output relationships (saccade learning)?
- How do the learning strategies perform for plants with highly non-linear input-output relationships (planar arm)?
- How well do the learning strategies deal with one-to-many mappings in practice (planar arm)?
- How robust are the learning strategies in the presence of noise (saccade learning and planar arm)?

- How well do the learning strategies cope with additional constraints added to the learning task (planar arm)?

The compared learning algorithms are (see Sect. 2.1): feedback error learning (FEL), distal supervised learning (DSL), direct inverse modeling with multilayer perceptrons (DIM), direct inverse modeling with NGPCA (DIM_NGPCA), staged learning by averaging (SLbA), and continuous learning by averaging (CLbA) (only for saccade learning since it is not well suited for the planar arm task). In the following, both learning tasks and all experiments are outlined in close detail. The results are first discussed separately for each learning task. In the final overall discussion, they are reviewed together.

4.1 Comparison Study on Saccade Control

Saccades are fast fixation movements of the eyes. Their purpose is to center interesting targets detected in the visual surroundings on both foveas. With the exception of special experimental settings, saccades are generally assumed to be "ballistic" open-loop movements: Once started, their course cannot be changed. The research on the mechanisms and neural underpinnings of saccade control in humans and primates has gained a lot of interest from psychology, biology, and neurophysiology (for a comprehensive overview, see Leigh and Zee, 1999). In computer science, especially robotics, the field of "active vision" deals with artificial saccades of robot camera heads. While this research is centered to a large extent on the development of technical solutions (e.g., Klarquist and Bovik, 1998), several studies propose models of saccade generation which are closely related to neurophysiological findings (Dean et al., 1994; Gancarz and Grossberg, 1999). In this area between robotics and biology, methods of adaptive saccade learning are of special interest (Bruske et al., 1997; Pagel et al., 1998). In the present study, biological modeling has not been the main purpose, but instead the comparison of the learning strategies FEL, DSL, DIM, DIM_NGPCA, SLbA, and CLbA on a saccade learning task for a robot camera head. The experimental studies have been carried out in four different task conditions:

- Fixation targets restricted to a 2D table surface (without noise)
- Fixation targets restricted to a 2D table surface (with retinal noise)
- Fixation targets located in a 3D cube (without noise)
- Fixation targets located in a 3D cube (with retinal noise)

In the following, first the underlying setup and controller structure is described. Afterwards, the parameter settings chosen for the simulation study are explained. Finally, the results are presented and discussed.

4.1.1 Saccade controller

4.1.1.1 Setup

Figure B.1 shows the robot camera head used for the saccade control task. This setup is thoroughly described in Appendix B. The used camera model is equipped with an $1/3$" CCD (4.8×3.6 mm). Only a central quadratic region with a size of 3.2×3.2 mm is used for target identification (in the following, the term "camera image" refers to this cropped region). The lenses have a focal length of 4 mm, resulting in a diagonal angle of view of 59 degrees. Each camera is mounted on its own pan-tilt unit, providing two degrees of freedom (horizontal pan, vertical tilt). In the 2D version of the fixation task, fixation targets are placed on the surface of the white table in Fig. A.1.

Due to the large overall amount of needed exploration trials for the present study, the comparison results were generated with a simulated geometrical model of this setup. It covers all transformations which are necessary to describe the projection from the target objects on the table surface to coordinates in the camera images. The geometrical model is described in the appendix in Sect. B.1.

4.1.1.2 Control scheme

The sensory context[1] $\mathbf{x}$ the controller receives as input (see Fig. 2.13 in Sect. 2.2.1) consists of a kinesthetic and a visual part. The kinesthetic input comprises the current position of the cameras represented by a conjoint pan-tilt direction and a horizontal and a vertical vergence[2] value: *pan*, *tilt*, $verg_{\text{hor}}$, $verg_{\text{vert}}$. These values are scaled to the range $[-1; +1]$. The corresponding pan and tilt angles of each camera ($\widetilde{pan}_{\text{left}}$, $\widetilde{tilt}_{\text{left}}$, $\widetilde{pan}_{\text{right}}$, and $\widetilde{tilt}_{\text{right}}$) are computed

[1] In accordance with our reasoning on kinematic control in Sect. 2.2.1, we designate $\mathbf{x}$ as sensory context information and not as system state throughout this chapter.

[2] Horizontal vergence indicates the difference between the individual pan directions of each camera, vertical vergence between the individual tilt directions.

by the following equations:

$$\widetilde{pan}_{\text{right/left}} = \frac{pan \pm \lambda_{\text{hor}}(\frac{1}{2}verg_{\text{hor}} + \frac{1}{2})}{1 + \lambda_{\text{hor}}}$$

$$\widetilde{tilt}_{\text{right/left}} = \frac{tilt \pm \lambda_{\text{vert}}verg_{\text{vert}}}{1 + \lambda_{\text{vert}}}$$

λ_{hor} and λ_{vert} are the horizontal and vertical vergence factor with values of 0.5 and 0.2, respectively. They restrict the vergence operating range to avoid postures where both cameras point into completely different directions without any overlapping part of their fields of view. Moreover, the term $\frac{1}{2}verg_{\text{hor}} + \frac{1}{2}$ in the first equation ensures that the optical axes of the cameras never point horizontally into diverging directions. After this conversion, $\widetilde{pan}_{\text{left}}$, $\widetilde{tilt}_{\text{left}}$, $\widetilde{pan}_{\text{right}}$, and $\widetilde{tilt}_{\text{right}}$ are scaled back to the selected operating range of the pan-tilt units (in degrees). The operating range is chosen in a way that at any (*pan*, *tilt*) setting with zero vergence at least a small part of the table surface is visible in at least one camera image.

The visual part of the controller input $\mathbf{x}$ represents the target position in the left and right camera image: $x_{\text{left}}, y_{\text{left}}, x_{\text{right}}, y_{\text{right}}$. These image coordinates are scaled to the range $[-1; +1]$ as well, the image center is at the origin. The desired sensory state $\mathbf{y}^*$ as additional controller input is defined via the sensory variables $x_{\text{left}}, y_{\text{left}}, x_{\text{right}}$ and y_{right}. Successful fixation implies that all of these variables amount to zero. Thus, $\mathbf{y}^*$ is constant for the saccade control task; for this reason, it can be omitted as controller input. Only DIM and DIM_NGPCA require that $\mathbf{y}^* = \left(x^*_{\text{left}}, y^*_{\text{left}}, x^*_{\text{right}}, y^*_{\text{right}}\right)^T$ is provided as controller input because $\mathbf{y}^*$ is an essential element of these learning strategies (see Sect. 2.2.5).

The motor output $\mathbf{u}$ of the controller is defined as change of the motor position. It consists of four values: Δpan, $\Delta tilt$, $\Delta verg_{\text{hor}}$, and $\Delta verg_{\text{vert}}$ (range: $[-2; +2]$). The plant adds the delta values to the current motor position to arrive at the new position. Whenever the valid range $[-1; +1]$ for the new position is exceeded, it is corrected so that no range transgression takes place. Moreover, the plant generates the new target position in the left and right camera image after the camera movement. To summarize the controller input and output:

- Inputs:
  - Sensory context $\mathbf{x}$:
    - Kinesthetic: *pan*, *tilt*, $verg_{\text{hor}}$, $verg_{\text{vert}}$
    - Visual: $x_{\text{left}}, y_{\text{left}}, x_{\text{right}}, y_{\text{right}}$
  - Desired sensory state $\mathbf{y}^*$ (only for DIM and DIM_NGPCA): $x^*_{\text{left}}, y^*_{\text{left}}, x^*_{\text{right}}, y^*_{\text{right}}$

- Motor output $\mathbf{u}$: Δpan, $\Delta tilt$, $\Delta verg_{\text{hor}}$, $\Delta verg_{\text{vert}}$

4.1.2 Experimental procedure

4.1.2.1 Quality measure

The different learning strategies are compared with regard to the number of exploration trials N_{EX} which are required to arrive at a certain controller quality Q^*. The quality Q_C of a controller is computed as average quality of 250 motor outputs $\mathbf{u}$ to random inputs $\mathbf{x}$.[3] The quality Q of a single motor output is computed by the following quality function which is also used for SLbA and CLbA (with $\mathbf{y}^* = \mathbf{0}$; $Q(\mathbf{y}) = Q(P(\mathbf{x}, \mathbf{u})) := Q(P(\mathbf{x}, \mathbf{u}), \mathbf{0})$):

$$\begin{aligned} Q(\mathbf{y}) &= 1 - \frac{r_L + r_R}{2} \\ r_{L/R} &= \frac{1}{\sqrt{2}} \sqrt{x^2_{\text{left/right}} + y^2_{\text{left/right}}} \end{aligned} \quad (4.1)$$

r_L and r_R represent the left and right radial target distance: the Euclidean distance between the image center and the current image coordinates of the fixation target in the left and right camera, respectively. If the target object is not visible in both camera images simultaneously, Q is set to -1.

The quality level Q^* is determined by training the standard controller network (see Fig. 4.2, left) with a set of 500 perfect learning examples over 500 epochs and assessing its quality afterwards. These examples were obtained by an optimization method called "differential evolution" (DE, see Sect. 3.5) without applying retinal noise. The objective function is $1 - Q(P(\mathbf{x}, \mathbf{u}))$ with variable parameters $\mathbf{u}$. The parameters of DE are $N_{\text{DE}} = 40$, $\lambda = 0.4$, $\gamma = 1$, $p_{\text{CR}} = 0.5$, $G_{\text{max}} = 500$, and $E_{\text{min}} = 0.001$ corresponding to a value of $Q(...) = 0.999$; for the parameter definitions see Table 3.4. This procedure was repeated 20 times for both the 2D and 3D task. In the task conditions without retinal noise, the value for Q^* was chosen slightly below the average quality of the resulting controller networks. For the task conditions with noise, the average quality of these controller networks was assessed while applying retinal noise. The value for Q^* was chosen slightly below the resulting quality as well.

The results of the DE controller networks are shown in Table 4.1. For each task condition, two quality values are given. The left one is the average controller quality Q^{DE}, the right one the value chosen for Q^* for the respective

[3] Whenever random sensory input is generated for the saccade task, first a random position for the target object is determined (from either the 2D table surface or from the 3D cube). Afterwards, the initial motor position of the cameras is repeatedly generated at random until the target object is "visible" in both camera images.

Saccade task	Q^{DE}	Q^*	Exploration trials (SD)
2D	0.989	0.985	591179 (67778)
2D with noise	0.98	0.975	
3D	0.985	0.98	516721 (79917)
3D with noise	0.977	0.972	

Table 4.1 — Results of the DE controller networks for the saccade control task. Q^{DE} is the average controller quality for the respective task condition, Q^* the desired quality level.

task condition for the subsequent studies. The average quality of these controller networks (trained with virtually perfect learning examples) marks the upper performance limit of these networks for the saccade control task. For this reason, Q^* is chosen just slightly below their average quality. Moreover, the average number of exploration trials required to collect the training set by DE is reported in Table 4.1 as well. It is rather large for both the 2D and 3D saccade control task in comparison to the results obtained with the different learning strategies (see Tables 4.3 and 4.4).

4.1.2.2 Parameter variation

All learning strategies differ in the way they can be adjusted by implementing certain constraints or by setting specific parameters. To arrive at a fair comparison of their performance, less important parameters were carefully set to assure a good basic performance of each learning strategy ("fixed parameters" in Tables C.1 to C.4 in App. C). Afterwards, one to three parameters with significant impact on the performance were varied systematically for each strategy ("variable parameters" in Tables C.1 to C.4). For each parameter combination, 20 learning passes were carried out. Q_C was measured during the course of learning. Whenever Q_C equaled or exceeded Q^*, the learning pass was halted. The needed number of exploration trials for a certain parameter combination is the average of all 20 learning passes. Furthermore, a learning pass was stopped *without success* whenever the number of learning cycles became larger than an upper limit $T_{\max}$ (for the online methods FEL, DSL, and CLbA, a learning cycle is identical to the generation of a single learning example; for the batch methods DIM and DIM_NGPCA, a learning cycle is identical to a training epoch (DIM) or a single training step (DIM_NGPCA); for SLbA, a learning cycle is identical to a stage; these different definitions result from the different structure of the compared learning strategies). Values for $T_{\max}$ are also reported in Tables C.1 to C.4. A combination of variable parameter values is only designated as "suc-

Term	Explanation
Exploration trial	Execution of a single motor command during the course of learning
Learning pass	A single attempt to acquire the desired quality level through learning with a specific setting of the variable parameters
Learning cycle	Each learning pass consists of a sequence of learning cycles; the exact meaning varies between learning strategies
Training step	An update of the network parameters by a single learning example
Training epoch	An update of the network parameters by an iteration through all of the learning examples in the training set

Table 4.2 — Definition of important terms.

cessful" if *all* 20 learning passes succeeded. For clarity, Table 4.2 provides an explanation of the terms "exploration trial", "learning pass", "learning cycle", "training step", and "training epoch".

4.1.2.3 Task conditions

In the 2D version of the fixation task, the table surface (size: 800×800 mm) with fixation targets was defined by the corners $(250, -700, -300)$ and $(1050, -700, 500)$ (in mm) in the world coordinate system as depicted in Fig. A.1. From the viewpoint of the cameras, most part of the table surface extends rightwards. In the 3D version, the cube had a size of $550 \times 550 \times 550$ mm and was defined by the corners $(275, -925, -275)$ and $(825, -375, 275)$ (in mm). The cube extends above and below the surface of the white table in Fig. A.1. For both the 2D and the 3D version, Fig. 4.1 illustrates that the saccade plant is close to linear. For the center and the corners of the square (2D) and cube (3D), the relationship between the *pan* and *tilt* setting and the values for x_{left} and y_{left} is shown. At least in the visible range ($x_{\text{left}}, y_{\text{left}} \in [-1; 1]$), the resulting graphs are approximately linear.

In the task conditions with noise, the controller input from the visual modality (x_{left}, y_{left}, x_{right}, y_{right}) was disturbed by Gaussian noise with a standard deviation of 0.015 (approx. 1% of the camera images' diagonal).

4.1.3 Learning strategies

4.1.3.1 Networks

The adaptive controllers (expect of DIM_NGPCA) were implemented by multi-layer perceptrons (MLPs; see Sect. 3.1). Figure 4.2 (left) shows the general

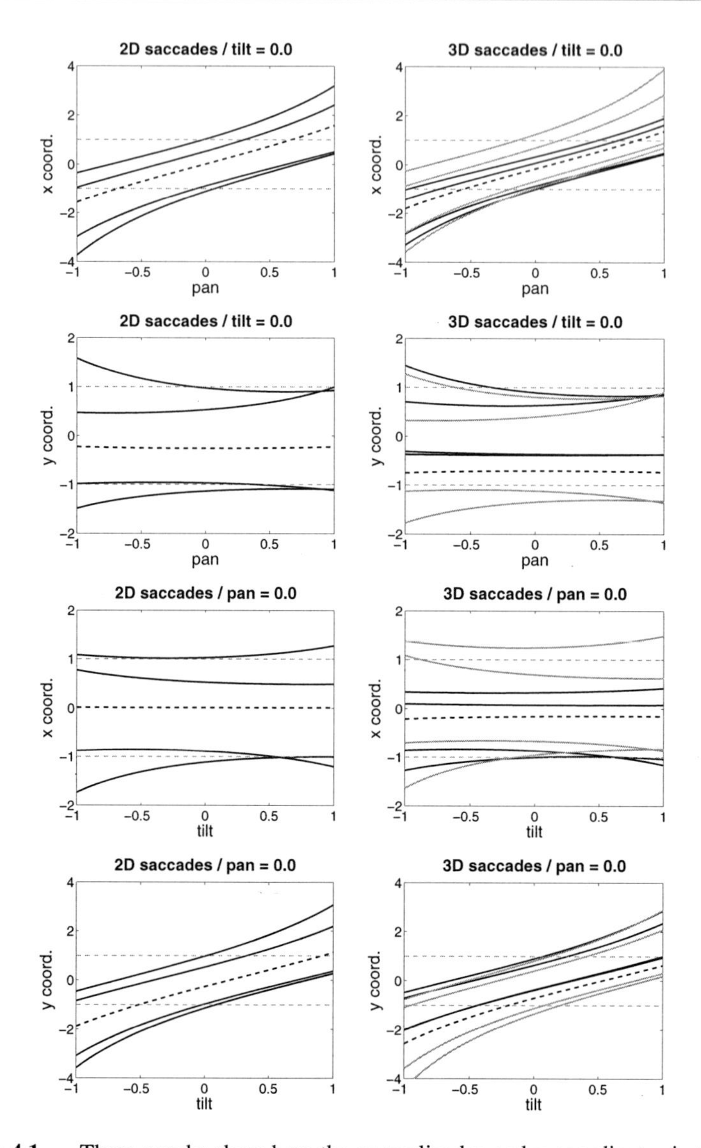

Figure 4.1 — These graphs show how the normalized x and y coordinates in the left camera image for the projection of different points in the world change depending on the *pan* and *tilt* setting ($verg_{\mathrm{hor}}$ and $verg_{\mathrm{vert}}$ are set to 0.0). Bold dashed line: Center of the square (2D) or the cube (3D) with fixation targets. Bold black lines: Corners of the square (2D) or lowermost corners of the cube (3D) with $w_y = -925\,\mathrm{mm}$. Bold gray lines: Uppermost corners of the cube (3D) with $w_y = -375\,\mathrm{mm}$. Horizontal dashed lines: Upper and lower limit of visibility in the camera image.

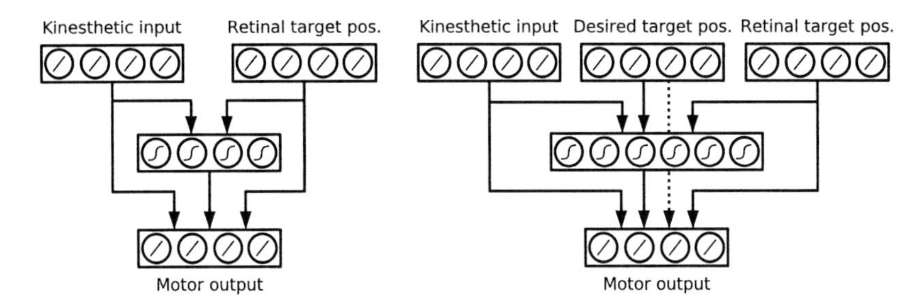

Figure 4.2 — Left: General controller network for the saccade control task. Right: Controller network for DIM (with additional input units for $\mathbf{y}^*$, the desired target position; the dashed arrow line indicates a shortcut connection from the $\mathbf{y}^*$ input layer directly to the output layer).

controller network with linear input units, four hidden sigmoid units (hyperbolic tangent as activation function), and linear output units. A shortcut connection projects directly from the input to the output layer. This facilitates learning of the linear part of the inverse plant. Only the network for DIM (Fig. 4.2, right) is larger (with six hidden sigmoid units) because of the additional input for $\mathbf{y}^*$. Figure 4.3 depicts the combined network used for DSL consisting of the forward model (FM) and the controller. As pointed out in section 2.2.4, first the network weights belonging to the FM are learned. Afterwards, these connections are frozen and the controller part of the network is trained. For all learning strategies, stochastic gradient descent (online backpropagation, see Sect. 3.1.2) was applied for network training. To keep things as straightforward as possible, we did not apply any modifications to the standard backpropagation algorithm such as using an additional momentum term (Zell, 1997). Weights were initialized to random values from the range $[-0.1; 0.1]$. Thus, regarding network training, there is only one free parameter, the learning rate η.

The NGPCA network for DIM_NGPCA is specified as described in Sect. 3.4. The learning parameters are $T = 100000$ (alias $T_{\max}$), $T_{ortho} = 10000$, $\epsilon(0) = 0.5$, $\epsilon(T) = 0.05$, $\rho(0) = 1.0$, $\rho(T) = 0.01$, $\sigma^2(0) = 0.0$, and $\lambda(0) = 10.0$. The number of ellipsoids N and the number of eigenvectors m are varied.

4.1.3.2 Parameter settings

Parameters for FEL In our implementation of FEL, the only free parameter is the gain factor η. It is equivalent to the learning rate of stochastic gradient

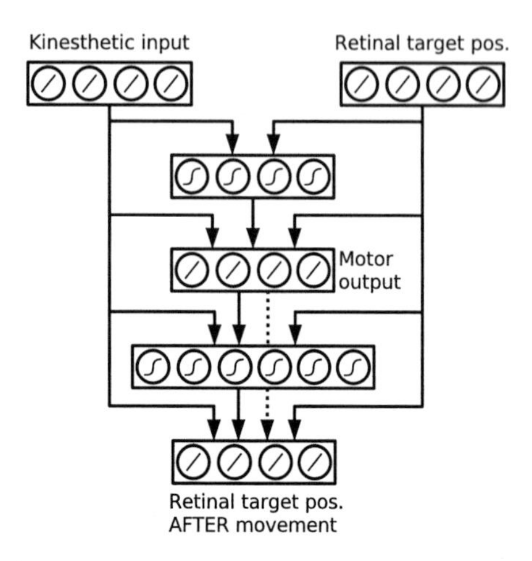

Figure 4.3 — Combined network for DSL for the saccade control task, consisting of the controller and the forward model. The dashed arrow line indicates a shortcut connection.

descent. The feedback controller equation $\Delta\mathbf{u} = \mathbf{G}_{\mathrm{u,x}}\Delta\mathbf{y}$ takes the follwing form for the saccade control task:

$$\Delta \begin{pmatrix} \Delta pan \\ \Delta tilt \\ \Delta verg_{\mathrm{hor}} \\ \Delta verg_{\mathrm{vert}} \end{pmatrix} = \begin{pmatrix} \frac{1}{2} & 0 & \frac{1}{2} & 0 \\ 0 & \frac{1}{2} & 0 & \frac{1}{2} \\ -\frac{1}{2} & 0 & \frac{1}{2} & 0 \\ 0 & -\frac{1}{2} & 0 & \frac{1}{2} \end{pmatrix} \begin{pmatrix} -x_{\mathrm{left}} \\ -y_{\mathrm{left}} \\ -x_{\mathrm{right}} \\ -y_{\mathrm{right}} \end{pmatrix}$$

This is a heuristic form which does not take the kinesthetic input into account. As shown in Sect. B.3 in the appendix, the kinesthetic input plays only a subordinate role for the plant characteristics. Therefore, this omission is justified.

Parameters for DSL DSL has four parameters: the number of learning examples for the FM N_{FM}, the number of epochs used to train the FM, and the learning rates η_{FM} for FM training and η for controller training. The number of epochs used to train the forward model was set equal to N_{FM}. This proved to result in proper learning of the FM without overfitting. η_{FM} was set to a fixed value, N_{FM} and η were varied systematically.

Parameters for DIM The most important parameter for DIM is the number of learning examples N_{CON} in the training set. The learning rate η was set to a fixed value.

Parameters for DIM_NGPCA In addition to the number of ellipsoids N and the number of eigenvectors m, a third variable parameter for DIM_NGPCA is the number of learning examples N_{CON} in the training set.

Parameters for SLbA For SLbA, one needs a strategy how to increase the number of learning examples and training epochs in each stage. It proved to be a favorable approach to increase the number of learning examples from stage to stage, starting with 10 examples in stage 1 and increasing this number by 10 every stage. This corresponds to learning the coarse structure of the problem first with only a small number of learning examples and refining it in later stages. For proper learning without overfitting, the number of training epochs in each stage was chosen to equal the number of learning examples. The applied quality function Q is stated in Eqn. (4.1). The quality threshold for the generation of a single learning example with random sensory context input $\mathbf{x}$ is computed as $\widetilde{Q}_k = Q(P(\mathbf{x}, \mathbf{u}_0))$. In the first stage ($k = 1$), $\mathbf{u}_0$ is a random motor command. For $k > 1$, $\mathbf{u}_0 = C_{k-1}(\mathbf{x})$. The noise which is added to $\mathbf{u}_0$ in the search for a better motor output is drawn from a multivariate Gaussian distribution with zero mean and standard deviation $\sigma = \sigma_0 \left[1 - Q(P(\mathbf{x}, \mathbf{u}_0))\right]$ for all dimensions. Computing σ this way ensures that the better the saccade $\mathbf{u}_0$, the smaller its variation. This is reasonable since large variations of a good saccade are more likely to result in worse than better fixation. The parameter σ_0 was varied systematically. It has a significant impact on the performance of SLbA.

Parameters for CLbA Similar to SLbA, also CLbA relies on the quality function Q in Eqn. (4.1). The quality threshold is computed as $\widetilde{Q}_k = Q(P(\mathbf{x}, \mathbf{u}_0))$ with $\mathbf{u}_0 = C(\mathbf{x})$ for all learning cyles expect for the first where $\mathbf{u}_0$ is a random motor command. The motor noise distribution is defined as stated for SLbA. The two adjustable learning parameters σ_0 and η were varied systematically.

The parameter settings for all learning strategies and task conditions are reported in Tables C.1 to C.4 in App. C.

4.1.4 Results

General remarks The results for the 2D learning task are presented in Table 4.3, for the 3D learning task in Table 4.4. The number of required exploration trials N_{EX} for the best successful combination of variable parameter values and the settings of these parameters are reported there (for DSL, N_{EX} is computed as the sum of the number of exploration trials for the generation of the training set of the FM $N_{\text{EX}}^{\text{FM}}$ and for the subsequent controller training $N_{\text{EX}}^{\text{CON}}$). A more detailed presentation is provided in Appendix D.1: Histogram plots for each learning strategy (Figs. D.2, D.4, D.6, and D.8) show the number of exploration trials for every combination of variable parameters. Parameter combinations for which at least one of the 20 learning passes failed are omitted since their number of exploration trials is not comparable any more with the fully successful combinations in a meaningful way, at least for the learning strategies for which the number of learning cycles is tied to the number of exploration trials (FEL, DSL, SLbA, CLbA). These histograms are mainly provided to prove that the range of variable parameter values was carefully chosen. Mostly, the best parameter value (combination) is at the minimum of an approximately u-shaped distribution. Only for DIM and DIM_NGPCA, the number of exploration trials N_{EX} increases linearly with the size of the training set N_{CON}. The ratio $N_{\text{EX}}/N_{\text{CON}}$ is larger than 1.0 since a single random saccade is not necessarily usable as learning example for DIM or DIM_NGPCA. Instead, the saccadic target is most often not longer visible in both camera images after a random saccade. Learning examples like this are useless. For the same reason, the collection of a single learning example for the training set of the FM in DSL requires multiple exploration trials. The ratios $N_{\text{EX}}/N_{\text{CON}}$ (for DIM) and $N_{\text{EX}}^{\text{FM}}/N_{\text{FM}}$ (for DSL) amount roughly to 10 for all task conditions.

In addition, for DIM_NGPCA a special difficulty arises: The number of exploration trials depends only on the size of the training set N_{CON}. Thus, a multitude of combinations of the number of ellipsoids N and the number of eigenvectors m can be successful for a certain value of N_{CON}. In Tables 4.3 and 4.4, only one of these combinations is reported, but in Figs. D.1, D.3, D.5, and D.7, a grayscale plot shows for each parameter combination how many of the 20 learning passes were successful.

General performance For the 2D task without noise, FEL is clearly the fastest learning strategy with only 184 required exploration trials ($N_{\text{EX}} = 184$), followed by DIM_NGPCA ($N_{\text{EX}} = 400$), DIM ($N_{\text{EX}} = 415$), DSL ($N_{\text{EX}} = 658$), CLbA ($N_{\text{EX}} = 3120$), and SLbA ($N_{\text{EX}} = 8535$). The same ranking order with similar numbers of exploration trials applies to the 2D task with noise; only CLbA shows a considerably worse performance with $N_{\text{EX}} = 5756$. The performance of DIM and DIM_NGPCA is barely distinguishable. Without noise, both require a training set size of $N_{\text{CON}} = 50$, with noise $N_{\text{CON}} = 40$ for DIM_NGPCA and $N_{\text{CON}} = 50$ for DIM.

In the 3D task, the ranking order is different. Without noise, DIM_NGPCA

2D saccade learning task without retinal noise		
Learning strategy	Exploration trials (SD)	Variable parameters
FEL	184 (49)	$\eta = 0.28$
DIM_NGPCA	400 (43)	$N_{\text{CON}} = 50, N = 1, m = 16$
DIM	415 (51)	$N_{\text{CON}} = 50$
DSL	658 (90)	$N_{\text{FM}} = 20, \eta = 0.14$
CLBA	3120 (497)	$\sigma_0 = 0.7, \eta = 0.2$
SLBA	8535 (1162)	$\sigma_0 = 2.0, k_{\max} = 8.0(0.45)$

2D saccade learning task with retinal noise		
Learning strategy	Exploration trials (SD)	Variable parameters
FEL	181 (26)	$\eta = 0.18$
DIM_NGPCA	319 (34)	$N_{\text{CON}} = 40, N = 2, m = 10$
DIM	418 (53)	$N_{\text{CON}} = 50$
DSL	604 (71)	$N_{\text{FM}} = 20, \eta = 0.14$
CLBA	5756 (1681)	$\sigma_0 = 1.3, \eta = 0.12$
SLBA	8855 (1790)	$\sigma_0 = 2.0, k_{\max} = 7.7(0.56)$

Table 4.3 — Results for the 2D saccade learning tasks. Learning strategies are sorted in ascending order with regard to the required number of exploration trials. The corresponding best settings for the variable parameters are shown in the right column (for SLbA, the average required number of stages $k_{\max}$ and its standard deviation (in brackets) are shown as well).

takes the lead ($N_{\text{EX}} = 410, N_{\text{CON}} = 40$), closely followed by DIM ($N_{\text{EX}} = 538$, $N_{\text{CON}} = 50$). The remaining results are: FEL ($N_{\text{EX}} = 1904$), DSL ($N_{\text{EX}} = 4137$), SLbA ($N_{\text{EX}} = 19005$), CLbA ($N_{\text{EX}} = 33737$). Virtually the same ranking order results from the 3D task with noise, also the number of explorations trials stays in the same order of magnitude for all strategies except for CLbA with $N_{\text{EX}} = 58185$. With noise, DIM_NGPCA and DIM share the first place, both with the same size of the training set ($N_{\text{CON}} = 60$).

Statistical tests Since we started this study without any in-advance hypotheses, the only purpose of statistical tests is to support the post-hoc analysis of the data by indicating how likely measured differences are caused by random variations or by real differences between the underlying populations. We restricted the statistical analysis to the most important comparison, namely between learning strategies within each task condition. For this comparison, every learning strategy was matched with every other strategy. The compared measure was the

3D saccade learning task without retinal noise		
Learning strategy	Exploration trials (SD)	Variable parameters
DIM_NGPCA	410 (57)	$N_{\mathrm{CON}} = 40$, $N = 1$, $m = 16$
DIM	538 (63)	$N_{\mathrm{CON}} = 50$
FEL	1904 (322)	$\eta = 0.26$
DSL	4137 (849)	$N_{\mathrm{FM}} = 20$, $\eta = 0.18$
SLBA	19005 (1069)	$\sigma_0 = 0.8$, $k_{\max} = 22.9(0.74)$
CLBA	33737 (5808)	$\sigma_0 = 0.9$, $\eta = 0.2$

3D saccade learning task with retinal noise		
Learning strategy	Exploration trials (SD)	Variable parameters
DIM_NGPCA	648 (71)	$N_{\mathrm{CON}} = 60$, $N = 1$, $m = 16$
DIM	652 (64)	$N_{\mathrm{CON}} = 60$
FEL	1680 (366)	$\eta = 0.2$
DSL	3197 (499)	$N_{\mathrm{FM}} = 20$, $\eta = 0.11$
SLBA	26498 (5308)	$\sigma_0 = 1.4$, $k_{\max} = 16.6(1.35)$
CLBA	58185 (12989)	$\sigma_0 = 1.7$, $\eta = 0.08$

Table 4.4 — Results for the 3D saccade learning tasks. See caption of Table 4.3 for further explanation.

mean number of required exploration trials (as reported in Tables 4.3 and 4.4). The computed pairwise t-tests (two-sided, for independent samples[4]) yielded highly significant results ($p < 0.001$) in all pairwise comparisons except for the ones in which DIM and DIM_NGPCA were compared with N_{CON} being equal (2D task without noise, 3D task with noise). Since the process of training data generation is identical for DIM and DIM_NGPCA, the latter result is inevitable. Overall, these statistical results support the reliability of the reported data.

Additional observations In examining the grayscale plots for DIM_NGPCA (Figs. D.1, D.3, D.5, and D.7), two interesting observations can be made which partly reveal the structure of the underlying learning task. First, the lower bound for the number of eigenvectors m amounts to 10 for successful parameter combinations in the 2D task and to 11 in the 3D task. Second, in every task condition, a single ellipsoid is sufficient for successful learning as long as N_{CON} is large enough. For the 2D task without noise and the 3D task with and without

[4] The degrees of freedom were corrected to compensate for the unequal estimated population variances (Bortz, 1993).

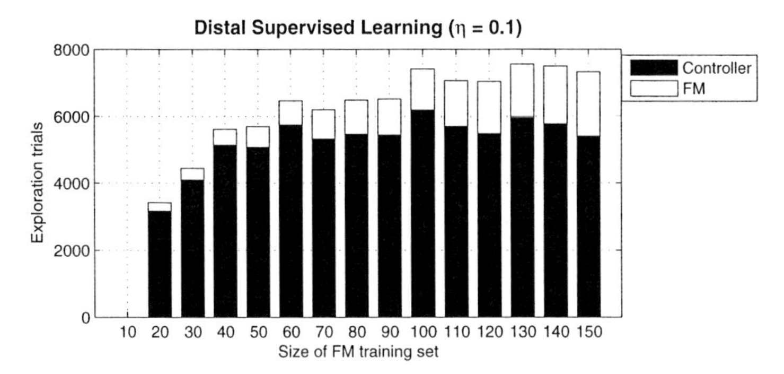

Figure 4.4 — Required exploration trials for DSL with learning rate $\eta = 0.1$ for the **3D saccade learning task with retinal noise**, split into trials needed for forward model (FM) and controller training. Bars are completely omitted whenever at least one of the 20 learning passes failed.

noise, especially the combination of one ellipsoid with 16 eigenvectors belongs to the parameter combinations which succeed with the smallest number of training examples N_{CON}. Taken together, these findings support the initial claim that the sensorimotor data manifold in the saccade learning task is not curved but mainly linear, and moreover that the intrinsic dimensionality of this manifold amounts to 10 or 11, respectively. Thus, the 3D task is more difficult to learn than the 2D task because of the increased dimensionality of the sensorimotor data manifold.

Regarding DSL, the best performance of DSL is obtained with $N_{\mathrm{FM}} = 20$ for all task conditions. For $N_{\mathrm{FM}} = 10$, DSL is rarely successful. For larger values ($N_{\mathrm{FM}} \geq 30$), learning succeeds but an increasing number of exploration trials during controller learning $N_{\mathrm{EX}}^{\mathrm{CON}}$ is required. Only including the successful parameter combinations, the correlations between N_{FM} and $N_{\mathrm{EX}}^{\mathrm{CON}}$ amount to $r = 0.17$ for the 2D task without noise, $r = 0.22$ for the 2D task with noise, $r = 0.45$ for the 3D task without noise, and to $r = 0.56$ for the 3D task with noise. Figure 4.4 illustrates the relationship between N_{FM} and $N_{\mathrm{EX}}^{\mathrm{CON}}$ for the 3D task with noise. Actually, N_{FM} is an indirect measure for FM precision: The correlations between the mean squared error of the FM (per pattern and output unit) on a test set and N_{FM} are negative ($r = -0.70$ for the 2D task without noise, $r = -0.59$ for the other task conditions). Directly correlating the mean squared test error and $N_{\mathrm{EX}}^{\mathrm{CON}}$ yields the following results: $r = 0.01$ for the 2D task without noise, $r = -0.22$ for the 2D task with noise, $r = -0.65$ for the 3D task without noise, and $r = -0.69$ for the 3D task with noise.

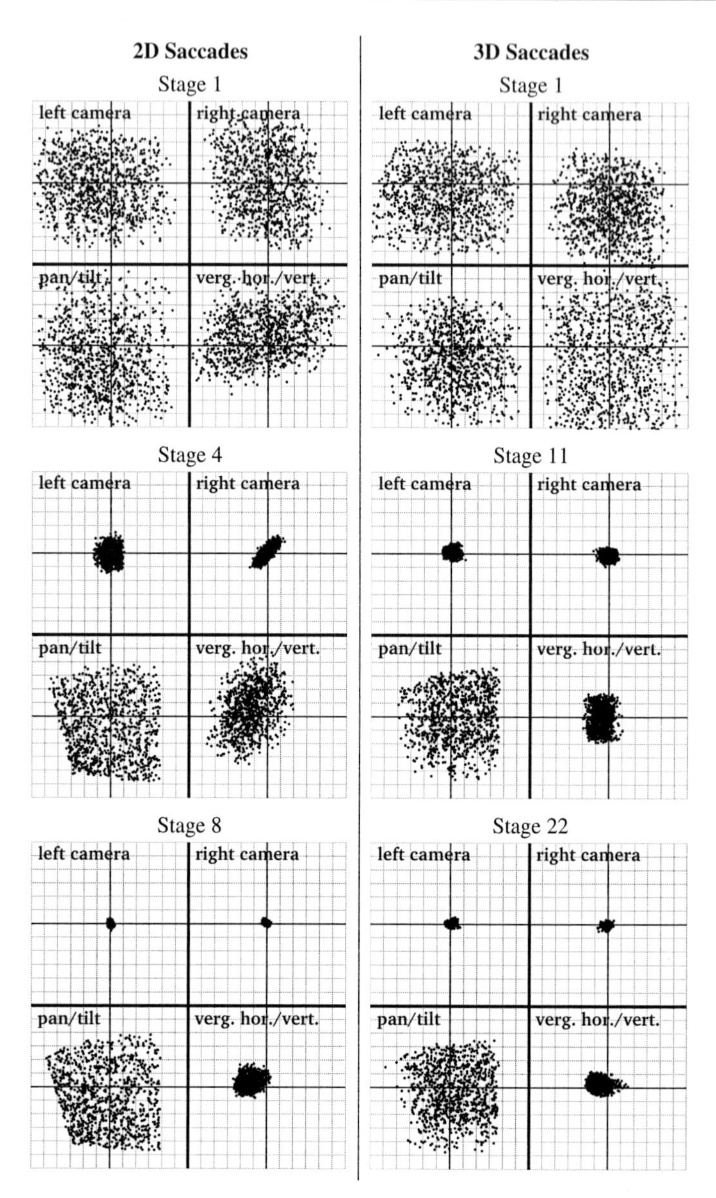

Figure 4.5 — Graphical results of SLbA for the 2D and 3D saccade learning tasks without noise at different stages. The sensory consequences of 1000 controller outputs in response to random controller inputs are marked with black dots in the different panels. In the final stage (last row), the controller output results in precise fixation (minimum scatter around the center of the left and right camera image). The graphical display of the final performance of the DE controller networks looks virtually the same.

	Without noise			**With noise**		
Stage k	Q_{PS}		Q_C	Q_{PS}		Q_C
1	0.729	$>$	0.666	0.567	$<$	0.620
2	0.766	$<$	0.795	0.784	$<$	0.791
3	0.843	$<$	0.905	0.842	$<$	0.884
4	0.919	$<$	0.938	0.910	$<$	0.927
5	0.947	$<$	0.961	0.939	$<$	0.957
6	0.970	$<$	0.975	0.972	$>$	0.969
7	0.981	$<$	0.983	0.976	$>$	0.971
8	0.985	$\approx$	0.985	0.980	$>$	0.977

Table 4.5 — Comparing the average quality Q_{PS} of the learning examples in the training set with the average controller quality Q_C on a test set for the different stages of a single learning pass of SLbA for the **2D saccade learning task** without and with retinal noise.

With regard to SLbA, the required number of stages $k_{\max}$ is of special interest. For the 2D task, it amounts to 8.0 (without noise) or 7.7 (with noise). The standard deviations are smaller than one, indicating small variability. For the 3D task, the values for $k_{\max}$ amount to 22.9 (without noise) or 16.6 (with noise), also with rather small variability. Fig. 4.5 illustrates the results of two single learning passes with optimum parameter settings, one for the 2D task without noise with $k_{\max} = 8$, and one for the 3D task without noise with $k_{\max} = 22$. For the first and last stage and a stage in between, a fourfold panel is shown, depicting the coordinate space of the left and right camera image, the $(pan, tilt)$ subspace, and the $(verg_{\mathrm{hor}}, verg_{\mathrm{vert}})$ subspace. The sensory consequences $\mathbf{x}_{t+1}$ of 1000 controller outputs $\mathbf{u}_t$ in response to random controller inputs $\mathbf{x}_t$ are marked with black dots in the different panels. After the final stage, the controllers are very good performers; this can be concluded from the small scatter around the image centers, indicating precise fixation. In contrast, after the first stage this scatter is still very large. In the 2D task, the shape of the table surface on which the fixation targets are positioned emerges in the $(pan, tilt)$ panel after the last stage. Moreover, the $(verg_{\mathrm{hor}}, verg_{\mathrm{vert}})$ panels of the last stage indicate for both the 2D and the 3D task how small the variation of these variables is for proper fixation movements.

To point out the aspect of quality increase due to the averaging which takes place in controller training, Table 4.5 presents for the 2D task without and with noise the average quality Q_{PS} of the learning examples in the training set side by side with the average controller quality Q_C on a test set for the different stages of a single learning pass. For the 2D task without noise, from the second stage on the controller is always better than the learning examples it gets for training. Finally, in the last stage both quality values are nearly equal. This shows that learning by averaging actually works as supposed. For the 2D task with noise, the positive effect of averaging is lost because of the noise from the sixth stage on. Nevertheless, controller quality still increases from stage to stage because the underlying training set continues to improve.

4.1.5 Discussion

The saccade learning task requires that the controller adapts to an approximately linear input-output relationship. Learning strategies which work by local linear approximation like FEL and DSL, or which use linear models to represent the training data manifold like DIM_NGPCA, seem to profit from this task characteristic. For the simpler version, the 2D task, FEL is clearly the best. As soon as the sensorimotor relationship gets more complex (3D task), DIM_NGPCA and DIM take the lead. The performance of the variants of learning by averaging (LbA) is considerably worse. Obviously, LbA cannot exploit the linear task characteristics as efficient as the other learning strategies. The ranking order between SLbA and CLbA changes depending on the task: In the 2D task conditions, CLbA is first, otherwise SLbA. This indicates that CLbA has difficulties to cope with complex sensorimotor relationships. The influence of noise on the overall ranking order is basically non-existent, only the performance of CLbA drops significantly.

Since the performance of FEL is considerably better than the performance of DSL, it is fair to say that the heuristically determined gain matrix for FEL is better suitable for the local linear adjustment than the estimate $\widehat{\mathbf{J}}^t_{\mathrm{u,x}}$ which is provided by the FM in DSL. A puzzling finding with regard to DSL are the negative correlations between FM precision and the required number of explorations trials for controller training. Intuitively, one would expect a better FM to facilitate controller training and not vice versa. Only if the FM training set is too small, controller training is in the danger to fail, most likely because the small number of training examples is often not sufficient to cover all regions of the sensorimotor space, forcing the FM to extrapolate. As soon as the FM training set is large enough, the negative correlation takes effect. This phenomenon needs to be explored in forthcoming studies.

In summary, for mainly linear tasks like 2D and 3D saccade learning, the learning strategies FEL, DIM_NGPCA, and DIM seem to be best suited. Only if the ratio $N_{\mathrm{EX}}/N_{\mathrm{CON}}$ gets too large because of the task characteristics, this recommendation has to be restricted to FEL. However, every learning strategy is the superior choice for saccade learning compared to optimization methods like DE which require a huge number of exploration trials (see Table 4.1 vs. Tables 4.3 and 4.4). The good performance of FEL supports biologically inspired models of saccade learning which rely on FEL. In these models, the feedback controller is mostly identified as (inherited and pre-wired) "brainstem mechanism" (Dean et al., 1994; Gancarz and Grossberg, 1999).

4.2 Comparison Study on the Control of a Planar Arm

In this section, the performance of the different learning strategies is compared for the kinematic control of a simulated planar arm. This task poses several special difficulties. First, the plant is strongly non-linear. Second, it implies a one-to-many mapping. Since some of the explored learning strategies are not suited for one-to-many mappings, different constraints are imposed on the learning task.

Three different planar arms are explored in this study. They differ with regard to the number of links L which varies between two and four. For the 3- and 4-link arm, two different learning constraints are explored in addition to the no-constraint condition. Moreover, for each planar arm, there is an additional experimental condition with sensor noise. Altogether, this yields ten different comparisons. Not all learning strategies are suited for each combination of planar arm and constraint, thus the list of learning strategies in each comparison varies (see Table 4.7).

4.2.1 Arm controller

4.2.1.1 Setup

The planar arm consists of L links, each of them with unit length (see Fig. 4.6). The basis joint is at the origin of the two-dimensional coordinate system of the working space. The joint angles of the planar arm form the vector $\boldsymbol{\theta} = (\theta_1, ..., \theta_L)'$. The relevant sensory information in this system is the position of the tip of the last link $\mathbf{y} = (y_1, y_2)'$. It is used to define the desired sensory state $\mathbf{y}^*$ as well. The plant P is defined by the following equations:

$$\begin{pmatrix} y_1 \\ y_2 \end{pmatrix} = P(\boldsymbol{\theta}) = \begin{pmatrix} \sum_{i=1}^{L} \cos\left(\sum_{j=1}^{i} \theta_j\right) \\ \sum_{i=1}^{L} \sin\left(\sum_{j=1}^{i} \theta_j\right) \end{pmatrix}$$

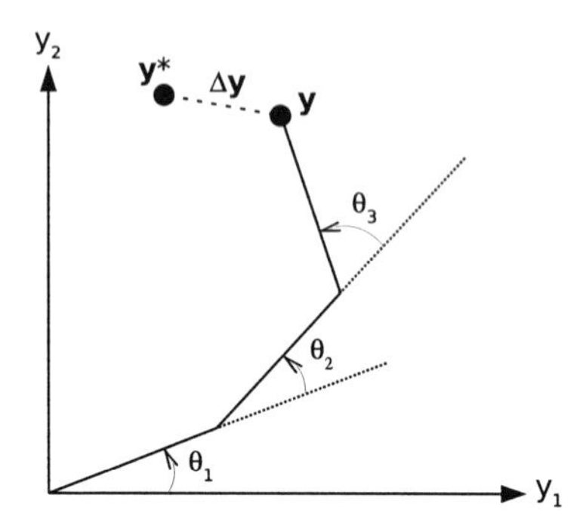

Figure 4.6 — Planar arm with L links. In this illustration, the tip position y deviates from the desired tip position $\mathbf{y}^*$.

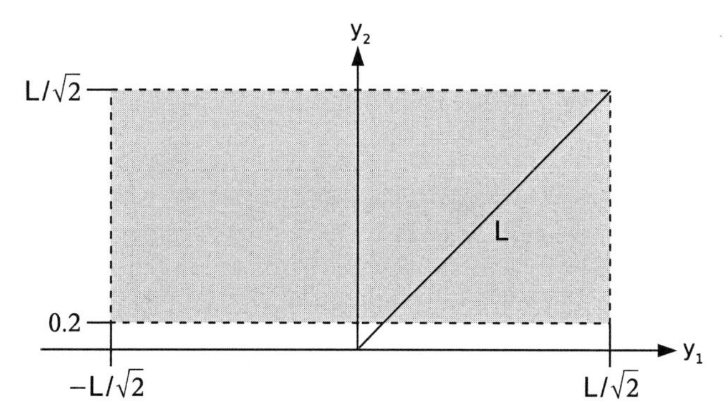

Figure 4.7 — The operating range of the planar arm (gray area). The diagonal line depicts the fully stretched planar arm with L links.

The operating range in which the desired sensory states $\mathbf{y}^*$ are positioned is restricted to an area with $y_1^* \in [-L/\sqrt{2}; L/\sqrt{2}]$ and $y_2^* \in [0.2; L/\sqrt{2}]$ (see Fig. 4.7). The task of the controller is to position the planar arm by a motor command $\mathbf{u} = \boldsymbol{\theta}$ resulting in $\mathbf{y}^* = \mathbf{y}$. In contrast to the saccade controller of Sect. 4.1, $\mathbf{y}^*$ is variable and there is no sensory context input $\mathbf{x}$.

Except for the marginal case of a completely outstretched arm, the inverse kinematics of such a planar arm yields two solutions (for $L = 2$) or infinitely many solutions (for $L \geq 3$). Figure 4.8 illustrates for a 3-link arm that the set of solutions is non-convex: The average of two solutions is no solution in itself. This is clearly visible, when an elbow-down and an elbow-up solution are combined (Fig. 4.8, left), but holds as well for two solutions from the elbow-

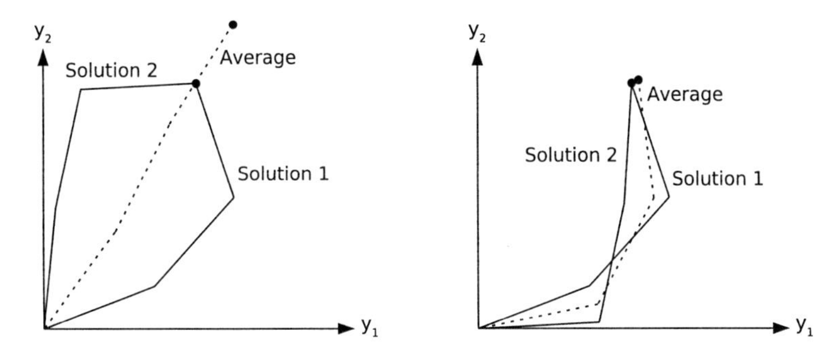

Figure 4.8 — These figures illustrate the non-convexity of the sets of solutions for the inverse kinematics of the 3-link planar arm. Left: The average of an elbow-down and an elbow-up solution deviates considerably from the desired tip position. Right: Even within the set of elbow-down solutions, averaging is not possible.

down class (Fig. 4.8, right). Thus, the planar arm suffers from the one-to-many problem of motor learning (see Sect. 2.2.2). Jordan and Rumelhart (1992) used a 3-link planar arm to illustrate exactly this point and to argue that DSL is superior to DIM because the latter cannot cope with one-to-many problems (at least when a function approximator is used as adaptive controller).

Since SLbA works also on the basis of function approximation, the mixture of elbow-down and elbow-up postures during training resulted in a severe performance drop for this learning strategy during initial preliminary tests. For this reason, the learning task was slightly facilitated by restricting the initialization range of the joint angles for the random generation of motor commands (whenever required in the course of each learning strategy): The angle θ_1 of the basic joint was drawn from the range between $-90°$ and $180°$, all other angles from the range between $0°$ and $180°$. This prevented the mixture of elbow-down and elbow-up solutions during the random generation of motor commands while preserving part of the one-to-many relationship. Beyond this, the operating range of the joint angles was not restricted at all; depending on the learning strategy, new motor commands outside the initialization range could arise.

4.2.1.2 Control scheme

The task of the controller is to generate motor output $\mathbf{u} = \boldsymbol{\theta}$ so that the tip of the last link is placed at the desired position in the working space. As controller input, there is no sensory context $\mathbf{x}$ (see Fig. 2.13) in this control task, only the desired tip position $\mathbf{y}^*$. The basic quality Q of an arm posture $\boldsymbol{\theta}$ in conjunction with a desired tip position $\mathbf{y}^*$ is computed as:

$$Q_{\text{basic}}(P(\boldsymbol{\theta}), \mathbf{y}^*) = Q_{\text{basic}}(\mathbf{y}, \mathbf{y}^*) = 1 - \|\mathbf{y} - \mathbf{y}^*\|$$

As mentioned before, there are different constraints imposed on the learning task. These constraints are reflected by different quality functions for the evaluation of the arm controllers (all computations involving θ angles are in radiant):

- No additional constraint: $Q_0(\mathbf{y}, \mathbf{y}^*) = Q_{\text{basic}}(\mathbf{y}, \mathbf{y}^*)$
- First constraint ("maximum symmetry"): All joint angles θ_i with $i > 1$ should be equal.

$$Q_1(\mathbf{y}, \mathbf{y}^*) = \frac{1}{2} Q_{\text{basic}}(\mathbf{y}, \mathbf{y}^*) + \frac{1}{2}\left(1 - \sqrt{\frac{1}{L-1}\sum_{i=2}^{L}\left(\bar{\theta} - \theta_i\right)^2}\right)$$

$$\text{with} \quad \bar{\theta} = \frac{1}{L-1}\sum_{i=2}^{L}\theta_i$$

Whenever the arm collides with itself, $Q_1(\mathbf{y})$ is set to a penalty value of -100.

- Second constraint ("minimum energy"): The arm should move as little as possible from the zero position (where all joint angles amount to zero).

$$Q_2(\mathbf{y}, \mathbf{y}^*) = \frac{2}{3} Q_{\text{basic}}(\mathbf{y}, \mathbf{y}^*) + \frac{1}{3}\left(1 - \frac{1}{2L}\sqrt{\sum_{i=1}^{L}\theta_i^2}\right)$$

The first constraint completely disambiguates the learning problem. To achieve maximum quality Q_1, only one elbow-down and one elbow-up solution are applicable. Together with the restricted initialization range, controller learning is reduced to a functional relationship. It was expected in advance that such a constraint would work in favor of SLbA. On the contrary, the second constraint poses additional difficulties to SLbA since the maximum achievable quality Q_2 varies depending on the desired tip position. This requires changes of the SLbA learning strategy as outlined in Sect. 4.2.3.1. Since constraints like the first one ("maximum symmetry") and second one ("minimum energy") are not unusual for complex learning tasks, it was considered worthwhile to explore how they influence the performance of the different learning strategies.

4.2.2 Experimental procedure

4.2.2.1 Quality measure

Like in the saccade control task, the different learning strategies are compared with regard to the number of exploration trials which are required to arrive at a certain controller quality Q^*. The quality Q_C of a controller is computed as average quality of 250 motor outputs $\mathbf{u}$ in response to random desired sensory states $\mathbf{y}^*$. The quality Q of a single motor output is computed by the quality function which corresponds to the current learning constraint.

The quality level Q^* is determined by training the standard controller network (see Fig. 4.9, left) with a set of 1000/1500 perfect learning examples over 2000/3500 epochs (first value for the 2-link arm, second value for the 3-link and 4-link arm) and assessing its quality afterwards. These examples were obtained by the optimization method "differential evolution" (DE, see Sect. 3.5) under application of the second constraint: The objective function is $1 - Q_1(P(\boldsymbol{\theta}), \mathbf{y}^*)$ with variable parameters $\boldsymbol{\theta}$. The parameters of DE are $N_{\mathrm{DE}} = 40$, $\lambda = 0.4$, $\gamma = 1$, $p_{\mathrm{CR}} = 0.5$, $G_{\max} = 500$, and $E_{\min} = 0.001$ corresponding to a value of $Q_1(...) = 0.999$; for details, see Table 3.4. This procedure was repeated 20 times. The value of Q^* for each constraint was chosen close to the average quality Q^{DE} accomplished by the resulting controller networks if evaluated with the respective quality function (and with application of additional sensory noise in the condition with noise).

The results of the DE controller networks are shown in Table 4.6. For each task condition, two quality values are given. The left one is the average controller quality Q^{DE}, the right one the value chosen for Q^* for the respective task condition for the subsequent studies. The average quality of these controller networks (trained with virtually perfect learning examples) marks the upper performance limit of these networks for the planar arm task. For this reason, Q^* is generally chosen close to Q^{DE} with the exception of the 4-link arm combined with Q_2. Because this task condition proved to be very hard and time-consuming, Q^* was chosen considerably lower than Q^{DE} in this case. Moreover, the average number of exploration trials required to collect the training set by DE is reported in Table 4.6 as well. It increases considerably with the number of links and demonstrates that motor learning by DE is not a good idea for real world applications.

4.2.2.2 Parameter variation

Like in the saccade learning task, less important parameters of the learning strategies were carefully set to assure a good basic performance of each strat-

Links	Q_0^{DE}	Q_0^*	$Q_{0\mathrm{N}}^{\mathrm{DE}}$	$Q_{0\mathrm{N}}^*$	Q_1^{DE}	Q_1^*	Q_2^{DE}	Q_2^*	Explor. trials (SD)
2	0.979	0.97	0.957	0.945					763413 (61724)
3	0.962	0.96	0.926	0.93	0.981	0.97	0.867	0.86	5508549 (310954)
4	0.947	0.94	0.904	0.9	0.973	0.97	0.889	0.86	10533129 (485243)

Table 4.6 — Results of the DE controller networks for the planar arm control task, evaluated with the different quality functions. Q^{DE} values are the average controller qualities for the respective task conditions, Q^* values are the corresponding desired quality levels. $Q_{0\mathrm{N}}^{\mathrm{DE}}$ and $Q_{0\mathrm{N}}^*$ indicate the values for the task condition with additional noise and no constraint.

egy ("fixed parameters" in Tables C.5 to C.14 in App. C). Afterwards, one to three parameters with significant impact on the performance were varied systematically for each strategy ("variable parameters" in Tables C.5 to C.14). For each parameter combination, 20 learning passes were carried out.[5] Q_C was measured during the course of learning. Whenever Q_C equaled or exceeded Q^*, the learning pass was halted. The needed number of exploration trials for a certain parameter combination is the average of all 20 (or 5) learning passes. Furthermore, a learning pass was stopped without success whenever the number of learning cycles became larger than an upper limit $T_{\max}$ (see Sect. 4.1.2.2 for a definition of learning cycle for the different learning strategies). Values for $T_{\max}$ are reported in Tables C.5 to C.14. A combination of variable parameter values is only designated as "successful" if all 20 learning passes succeeded (see Table 4.2 in the section on saccade learning for an overview of the terms "exploration trial", "learning pass", "learning cycle", "training step", and "training epoch").

4.2.2.3 Task conditions

The task variation for the arm control task has two components as explained before. First, the number of links is varied between two and four, and second, two different constraints are applied to the learning task (in addition to the standard no-constraint condition). Moreover, for each arm, there is one condition with additional sensory noise, added to the measurement of the tip position $\mathbf{y}$. This influences both the accuracy of the sensory error signal and the quality measurement. The noise is generated from a Gaussian distribution with variance $\sigma_{\text{noise}} = 0.015L$ (1.5% of the length of the fully stretched arm). Altogether, this yields ten task conditions as depicted in Table 4.7, in which the learning con-

[5] Because of restrictions in the available computing time, some of the comparisons for the 4-link arm which turned out to be completely without success had to be restricted to 5 learning passes.

	Number of links / quality function									
	2		3				4			
Learning strategy	Q_0	Q_{0N}	Q_0	Q_{0N}	Q_1	Q_2	Q_0	Q_{0N}	Q_1	Q_2
FEL/$\mathbf{J}^+$	√	√	√	√			√	√		
FEL/$\mathbf{J}^t$	√	√	√	√			√	√		
DIM	√	√								
DIM_NGPCA	√	√	√	√	√	√	√	√	√	√
DSL	√	√	√	√	√	√	√	√	√	√
SLbA/a	√	√	√	√	√	√	√	√	√	√
SLbA/b	√	√	√	√	√	√	√	√	√	√

Table 4.7 — Overview of the experimental conditions. Combinations of learning strategy and constraint without tick are excluded from the study.

straints are designated by their respective quality function: $Q_0 \rightarrow$ no constraint, $Q_{0N} \rightarrow$ no constraint with noise, $Q_1 \rightarrow$ "maximum symmetry" constraint, $Q_2 \rightarrow$ "minimum energy" constraint.

4.2.3 Learning strategies

4.2.3.1 Special considerations for the planar arm task

In the following, it is explained which learning strategies are employed in the comparison study on the planar arm in which task conditions (see Table 4.7) and which particular modifications are applied.

FEL The straightforward approach to FEL is to chose the gain matrix as $\mathbf{G}_{\mathrm{u}} = \mathbf{J}^{+}_{\mathrm{u,x}}$ (see Sect. 2.2.3). In addition to this standard approach, a variation of FEL with $\mathbf{G}_{\mathrm{u}} = \mathbf{J}^{t}_{\mathrm{u,x}}$ is explored as well. As we pointed out in Sect. 2.2.4, DSL works by approximating $\mathbf{J}^{t}_{\mathrm{u,x}}$ as gain matrix. Thus, it is interesting to find out how the performance of FEL with a precise gain matrix $\mathbf{J}^{t}_{\mathrm{u,x}}$ differs from the performance of DSL.

FEL is only incorporated into the comparisons without additional constraint. It is possible to define a plant whose output reflects how well the constraints are fulfilled (see DSL), but the analytical identification of the matrices $\mathbf{J}^{+}_{\mathrm{u,x}}$ and $\mathbf{J}^{t}_{\mathrm{u,x}}$ for this extended plant involves so many operations which are not plausible at all for a truly adaptive system that FEL was only included in the comparisons without learning constraint.

DSL DSL is included in every task condition. In the conditions with constraint, the FM of the DSL learning scheme has to learn the output of an extended plant which not only provides the tip position as output but also information relevant for the fulfillment of the constraint. For the first constraint ("maximum symmetry"), the following $L-1$ additional outputs are provided by the plant:

$$y_{i+2} = \theta_{i+1} - \theta_{i+2} \quad \text{with} \quad i = 1, ..., L-2 \quad \text{for} \quad L \geq 3$$

Moreover, y_{L+1} is set to 1 for a posture with collision, and to -1 for a collision-free posture. The desired outputs y_i^* amount to 0 for $i = 3, ..., L$, and to -1 for $i = L+1$.

For the second constraint ("minimum energy"), the following additional output is provided by the plant:

$$y_3 = \frac{1}{2L}\sqrt{\sum_{i=1}^{L} \theta_i^2}$$

The corresponding desired output y_3^* amounts to 0.

DIM Since DIM on the basis of function approximation cannot work for one-to-many problems (Jordan and Rumelhart, 1992), it is only included in the comparisons with the 2-link arm. In searching learning examples for the training set of DIM (and also of DIM_NGPCA), we decided to include only motor commands which result in sensory states $\mathbf{y}$ such that y_1 and y_2 are positioned in the operating range of the desired sensory states $\mathbf{y}^*$. For this reason, the number of exploration trials for DIM and DIM_NGPCA is larger than the number of collected learning examples. By this means, the lacking goal-directedness of DIM is directly reflected by the number of exploration trials and not only by the additional effort of network training (network size, number of learning epochs, etc.).

DIM_NGPCA The combination of DIM with NGPCA should be able to tackle one-to-many problems. For this reason, DIM_NGPCA is included in every comparison. In the conditions with constraint, the desired sensory state as controller input and the output of the plant are extended in the same way as described for DSL. This leads to the additional difficulty that the portion of randomly generated learning examples in the training set which are useful for the control task is considerably reduced (only the ones close to fulfilling the respective constraint by chance). Moreover, the second constraint, which enforces minimum deviation from the zero position of the arm, implies a tradeoff between reaching precision and constraint fulfillment. During controller usage after training, the desired sensory state y_3^* as controller input for this constraint is specified as 0 to indicate that it should be as small as possible. However, none of the learning examples contains an input $y_3^* = 0$ since y_3 is always larger. Thus, the NGPCA network has to extrapolate while generating motor commands during controller usage. The results will show how well DIM_NGPCA can cope with these added difficulties.

SLbA The simple standard strategy of learning by averaging as employed for the saccade learning task did not yield satisfying results for the strongly non-linear plant of the planar arm which includes non-convex solution sets for the controller output. Therefore, three different enhancements are implemented, all of them concerning the generation of a single learning example $[\mathbf{x}, \mathbf{y}^* \longrightarrow \mathbf{u}]$. They are explained in the following with the notation and on the basis of Sect 2.2.7.1, thus including the sensory context $\mathbf{x}$.

First, a simple evolutionary strategy is used in the search for better learning examples. The new motor command $\mathbf{u}$ is not only determined by the random variation of $\mathbf{u}_0$ until $Q(P(\mathbf{x}, \mathbf{u}), \mathbf{y}^*) > \widetilde{Q}_k$. Instead, for each generated motor command $\mathbf{u}$, an additional check is performed: Whenever $Q(P(\mathbf{x}, \mathbf{u}), \mathbf{y}^*) > Q(P(\mathbf{x}, \mathbf{u}_0), \mathbf{y}^*)$, $\mathbf{u}_0$ is substituted by $\mathbf{u}$: $\mathbf{u}_0 := \mathbf{u}$. Afterwards, the new $\mathbf{u}_0$ is the basis for the generation of motor commands $\mathbf{u}$. This process is repeated until $Q(P(\mathbf{x}, \mathbf{u}), \mathbf{y}^*) > \widetilde{Q}_k$ (as usual). This evolutionary strategy speeds up the search for better learning examples.

Second, during each search for a better learning example, the quality threshold $\widetilde{Q}_k$ is lowered with every non-successful attempt to find a motor command $\mathbf{u}$ which exceeds this threshold. Let $\widetilde{Q}_k^{\text{init}}$ be the inital quality threshold, $\mathbf{u}_0^{\text{init}}$ the initial $\mathbf{u}_0$ before the evolutionary process kicks in, and t the number of attempts to find a suitable $\mathbf{u}$ so far. $\widetilde{Q}_k$ is determined as

$$\begin{aligned} \widetilde{Q}_k &= (1 - \lambda^*_{\text{SLbA}})\, Q(P(\mathbf{x}, \mathbf{u}_0^{\text{init}}), \mathbf{y}^*) + \lambda^*_{\text{SLbA}} \widetilde{Q}_k^{\text{init}} \\ \lambda^*_{\text{SLbA}} &= e^{-\lambda_{\text{SLbA}} t} \end{aligned}$$

with $\lambda_{\text{SLbA}} \in]0; 1]$. To interpret these equations: $\widetilde{Q}_k$ decreases exponentially towards the quality of $\mathbf{u}_0^{\text{init}}$, starting from $\widetilde{Q}_k^{\text{init}}$. λ_{SLbA} determines the speed of decay and has to be set by the user as free learning parameter. This enhancement helps to avoid that SLbA gets stuck with the generation of better motor commands in a region where this is very difficult.

Third, an additional learning parameter $\lambda_\sigma \in]0; 1]$ is introduced for the random variation of $\mathbf{u}_0$. The noise which is added to $\mathbf{u}_0$ in the search for a better motor command $\mathbf{u}$ is drawn from a multivariate Gaussian distribution with zero mean and standard deviation $\sigma = \sigma_0 \lambda_\sigma^t [1 - Q(P(\mathbf{x}, \mathbf{u}_0))]$ for all dimensions. t is again the number of attempts to find a suitable $\mathbf{u}$ so far, σ_0 is another free learning parameter also found in the saccade learning task. By this means, the search region for motor commands is reduced step by step, with the expectation that it is easier to find better motor commands in the close vicinity of $\mathbf{u}_0$.

For the arm control task, the quality threshold $\widetilde{Q}_k$ is determined in the following way: In the first stage ($k = 1$), it is set to 0.3. Afterwards, two different algorithms are employed. In version (a), $\widetilde{Q}_k$ is either set to the average quality of the controller of the preceding stage $k - 1$ on its training set, or computed by

$$\widetilde{Q}_k = (1 - Q(P(\mathbf{x}, \mathbf{u}_0), \mathbf{y}^*))\gamma_{\widetilde{Q}} + Q(P(\mathbf{x}, \mathbf{u}_0), \mathbf{y}^*) \tag{4.2}$$

(depending on which value is larger). $\gamma_{\widetilde{Q}}$ is a free learning parameter from the range $[0; 1]$ (in all simulation runs on the planar arm, it is set to $\gamma_{\widetilde{Q}} = 0.5$). In version (b), only Eqn. (4.2) is used. Thus, the quality threshold in version (b) depends solely on the quality of the controller output $\mathbf{u}_0$ in the specific context, while version (a) uses the average controller quality as lower bound for this threshold. This facilitates that the controller quality increases evenly in the overall input space from stage to stage. On the other hand, version (b) is computationally less expensive and has a simpler algorithmic structure which makes it more attractive for both practical usage and biological modeling. SLbA/a and /b are included in all comparisons.

CLbA Several attempts were made to find parameter configurations for CLbA which allow at least an acceptable performance. But this proved to be impossible; learning got stuck on a very low quality level. For this reason, further attempts with CLbA were abandoned, and it is excluded from the comparison study. The failure of CLbA is further discussed in Sect. 4.2.5.

4.2.3.2 Networks

Like in the saccade control task, the adaptive controllers (expect of DIM_NGPCA) were implemented by MLPs. Figure 4.9 (left) shows the general controller network with linear input units, ten hidden sigmoid units (hyperbolic tangent as activation function), and linear output units. The only network input is the desired tip position, no sensory context is provided. Figure 4.9 (right) depicts the combined network used for DSL consisting of the FM and the controller. The hidden layer of the FM has 30 sigmoid units. This rather large

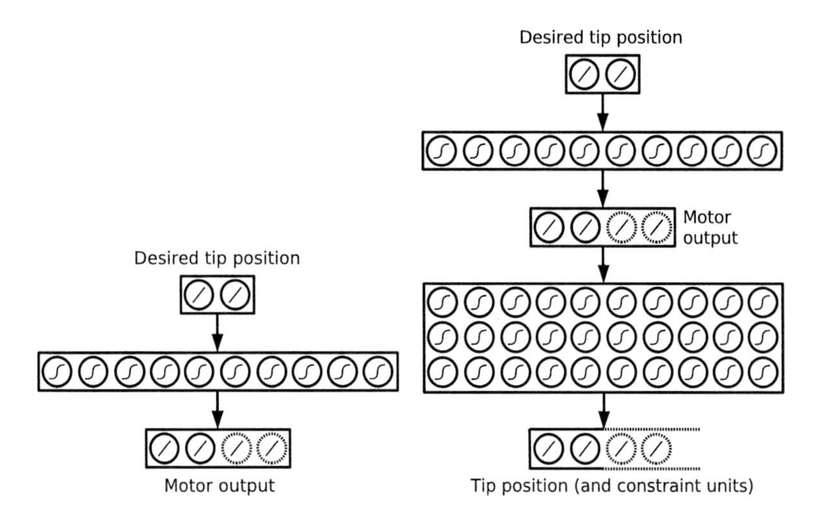

Figure 4.9 — Left: General controller network for the planar arm control task. The motor output consists of two to four joint angles, depending on the number of links of the planar arm. Right: Combined network for DSL for the planar arm control task, consisting of the controller (top three layers) and the forward model (bottom three layers). The output of the forward model contains both the tip position and (optionally) additional constraint units.

number was chosen to ensure that the network is complex enough to acquire a precise FM even for the 4-link arm. As explained in section 2.2.4, first the network weights belonging to the FM are learned. Afterwards, these connections are frozen and the controller part of the network is trained. For all learning strategies, stochastic gradient descent (online backpropagation, see Sect. 3.1.2) was applied for network training. To keep things as straightforward as possible, we did not apply any additional tuning methods. Weights were initialized to random values from the range $[-0.1; 0.1]$. Thus, regarding network training, there is only one free parameter, the learning rate η.

The NGPCA network for DIM_NGPCA is specified as described in Sect. 3.4. The learning parameters are $T_{ortho} = 10000$, $\epsilon(0) = 0.5$, $\epsilon(T) = 0.05$, $\rho(0) = 1.0$, $\rho(T) = 0.01$, $\sigma^2(0) = 0.0$, and $\lambda(0) = 10.0$. The maximum number of training steps $T_{\max}$ differs between task conditions, the number of ellipsoids N and the number of eigenvectors m are variable parameters.

4.2.3.3 Parameter settings

Parameters for FEL The only free parameter is the gain factor η which is equivalent to the learning rate of stochastic gradient descent.

Parameters for DSL DSL has four parameters: the number of learning examples for the FM N_{FM}, the number of epochs used to train the FM, and the learning rates η_{FM} for FM training and η for controller training. The number of epochs used to train the FM was set to a quarter of N_{FM}. This proved to result in proper learning of the FM without overfitting. η_{FM} was set to a fixed value, N_{FM} and η were varied systematically.

Parameters for DIM The most important parameter for DIM is the number of learning examples N_{CON} in the training set. The learning rate η was set to a fixed value.

Parameters for DIM_NGPCA In addition to the number of ellipsoids N and the number of eigenvectors m, a third variable parameter for DIM_NGPCA is the number of learning examples N_{CON} in the training set.

Parameters for SLbA For SLbA, one needs a strategy how to increase the number of learning examples and training epochs in each stage. This strategy was varied depending on the number of links and the selected constraint. It is reported in Tables C.5 to C.14 in the format *LE: a-b-c / EP: a-b-c* with *a* being the start value, *b* the increase from stage to stage, and *c* the maximum value. *LE* indicates the number of learning examples, *EP* the number of epochs. The other learning parameters are λ_{SLbA} (fixed), λ_σ (fixed), and σ_0 (varied systematically) as described in Sect. 4.2.3.1. The quality function is either Q_0, Q_1, or Q_2, depending on the applied constraint.

The parameter settings for all learning strategies and task conditions are reported in Tables C.5 to C.14.

4.2.4 Results

General remarks The results for the 2-link arm are presented in Table 4.8, for the 3-link arm in Table 4.9, and for the 4-link arm in Table 4.10. The number of required exploration trials N_{EX} for the best successful combination of variable parameter values and the settings of these parameters are reported there (for

DSL, N_{EX} is computed as the sum of the number of exploration trials for the generation of the training set of the FM $N_{\mathrm{EX}}^{\mathrm{FM}}$ and for the subsequent controller training $N_{\mathrm{EX}}^{\mathrm{CON}}$; $N_{\mathrm{EX}}^{\mathrm{FM}} = N_{\mathrm{FM}}$ for the planar arm task). A more detailed presentation is provided in Appendix D.2: Histogram plots for each learning strategy (Figs. D.10, D.12, D.14, D.16, D.18, D.20, D.22, D.24, D.26, and D.28) show the number of exploration trials for every combination of variable parameters for all task conditions. Parameter combinations for which at least one of the 20 learning passes failed are omitted since their number of exploration trials is not comparable any more with the fully successful combinations in a meaningful way, at least for the learning strategies for which the number of learning cycles is tied to the number of exploration trials (FEL, DSL, SLbA). Histograms for learning strategies without any success under the given task conditions are completely omitted. These histograms are mainly provided to prove that the range of variable parameter values was carefully chosen. The best parameter value (combination) is ideally at the minimum of an approximately u-shaped distribution although this is often not clearly visible because of the omitted bars. Only for DIM and DIM_NGPCA, the number of exploration trials N_{EX} increases linearly with the size of the training set N_{CON}.

In addition, for DIM_NGPCA a special difficulty arises: The number of exploration trials depends only on the size of the training set N_{CON}. Thus, a multitude of combinations of the number of ellipsoids N and the number of eigenvectors m can be successful for a certain value of N_{CON}. In Tables 4.8 to 4.10, only one of these combinations is reported, but in Figs. D.9, D.11, D.13, D.15, D.17, D.19, D.21, D.23, D.25, and D.27, a grayscale plot shows for each parameter combination how many of the 20 (or 5) learning passes were successful.

General performance Instead of reiterating the numbers given in Tables 4.8 to 4.10, we will only point out the most interesting results here. Moreover, the task conditions are abbreviated in the following: "LxQy" is the task condition with x links and quality function Q_y. The task conditions with noise are L2Q0N, L3Q0N, and L4Q0N. A learning strategy is designated as "successful" in a certain task condition if it is able to exceed the desired quality level Q^* in all learning passes at least with one parameter combination; otherwise, it has failed in this task condition.

Generally, DIM_NGPCA is the fastest learning strategy. The only exceptions are conditions L3Q2 (DSL is best) and L4Q1 (SLbA/a is best). In condition L2Q0, DIM and DIM_NGPCA share the first place with $N_{\mathrm{CON}} = 125$. Under the application of sensory noise (L2Q0N), DIM performs worse (but still comes in second). For the 2-link and the 3-link arm, the winning margin of

2-link arm with Q_0 ($Q_0^* = 0.97$)		
Learning strategy	Exploration trials (SD)	Variable parameters
DIM_NGPCA	552 (44)	$N_{\text{CON}} = 125$, $N = 19$, $m = 3$
DIM	573 (43)	$N_{\text{CON}} = 125$
SLbA/a	23154 (5231)	$\sigma_0 = 0.7$, $k_{\text{max}} = 3.2(0.54)$
SLbA/b	23559 (5418)	$\sigma_0 = 0.7$, $k_{\text{max}} = 3.4(0.57)$
DSL	25% **failed** ($\bar{Q}_C = 0.969$)	$N_{\text{FM}} = 21000$, $\eta = 0.06$
FEL/J^t	15% **failed** ($\bar{Q}_C = 0.966$)	$\eta = 0.1125$
FEL/J$^+$	100% **failed** ($\bar{Q}_C = 0.91$)	$\eta = 0.025$

2-link arm with Q_0 and additional sensor noise ($Q_{0\text{N}}^* = 0.945$)		
Learning strategy	Exploration trials (SD)	Variable parameters
DIM_NGPCA	901 (59)	$N_{\text{CON}} = 200$, $N = 10$, $m = 4$
DIM	2712 (105)	$N_{\text{CON}} = 600$
SLbA/b	41459 (20288)	$\sigma_0 = 0.9$, $k_{\text{max}} = 3.3(0.64)$
SLbA/a	47904 (14769)	$\sigma_0 = 1.1$, $k_{\text{max}} = 3.7(0.71)$
DSL	57090 (15404)	$N_{\text{FM}} = 7000$, $\eta = 0.07$
FEL/J^t	63335 (27819)	$\eta = 0.075$
FEL/J$^+$	60% **failed** ($\bar{Q}_C = 0.94$)	$\eta = 0.025$

Table 4.8 — Results for the 2-link arm. Learning strategies are sorted in ascending order with regard to the required number of exploration trials. The corresponding best settings for the variable parameters are shown in the right column (for SLbA, the average required number of stages k_{max} and its standard deviation (in brackets) are shown as well). For learning strategies which never succeeded in all learning passes the parameter combination with the maximally achieved average final controller quality $\bar{Q}_C$ and the percentage of failed passes with these settings is reported.

DIM_NGPCA compared to the second-best is large (for L2Q0, $N_{\text{EX}} = 552$ for DIM_NGPCA and $N_{\text{EX}} = 23154$ for SLbA/a; for L3Q0, $N_{\text{EX}} = 2696$ for DIM_NGPCA and $N_{\text{EX}} = 62614$ for SLbA/b). For the 4-link arm, this distance is smaller (for L4Q0, $N_{\text{EX}} = 18171$ for DIM_NGPCA and $N_{\text{EX}} = 48650$ for FEL/J^t). In task condition L3Q2, DIM_NGPCA fails completely by a considerable margin ($\bar{Q}_C = 0.80$ compared to $Q_2^* = 0.86$), in condition L4Q1 only by a small amount. In the latter task condition, a slight decrease of the desired quality level would likely have resulted in a success of DIM_NGPCA with at least a few parameter combinations.

The performance of SLbA is mixed with versions (a) and (b) being similar performers. Version (a) has a slight lead since it is faster than (b) in six of

3-link arm with Q_0 ($Q_0^* = 0.96$)		
Learning strategy	Exploration trials (SD)	Variable parameters
DIM_NGPCA	2696 (70)	$N_{\text{CON}} = 1000$, $N = 100$, $m = 4$
SLbA/b	62614 (34174)	$\sigma_0 = 0.4$, $k_{\max} = 6(2)$
SLbA/a	73788 (50860)	$\sigma_0 = 0.4$, $k_{\max} = 6(4)$
FEL/J^t	100615 (36808)	$\eta = 0.03$
DSL	106590 (43683)	$N_{\text{FM}} = 3000$, $\eta = 0.03$
FEL/J$^+$	10% **failed** ($\bar{Q}_C = 0.959$)	$\eta = 0.035$

3-link arm with Q_0 and additional sensor noise ($Q_{0\text{N}}^* = 0.93$)		
Learning strategy	Exploration trials (SD)	Variable parameters
DIM_NGPCA	5439 (52)	$N_{\text{CON}} = 2000$, $N = 40$, $m = 4$
FEL/J^t	90775 (29422)	$\eta = 0.02$
DSL	102530 (41040)	$N_{\text{FM}} = 5000$, $\eta = 0.03$
SLbA/a	179026 (152936)	$\sigma_0 = 0.4$, $k_{\max} = 7(4)$
SLbA/b	232908 (168604)	$\sigma_0 = 0.6$, $k_{\max} = 8(4)$
FEL/J$^+$	15% **failed** ($\bar{Q}_C = 0.929$)	$\eta = 0.04$

3-link arm with Q_1 ($Q_1^* = 0.97$)		
Learning strategy	Exploration trials (SD)	Variable parameters
DIM_NGPCA	8116 (123)	$N_{\text{CON}} = 3000$, $N = 60$, $m = 3$
SLbA/a	34971 (889)	$\sigma_0 = 0.9$, $k_{\max} = 3(0)$
SLbA/b	36325 (5284)	$\sigma_0 = 0.9$, $k_{\max} = 3(0)$
DSL	82765 (30371)	$N_{\text{FM}} = 7000$, $\eta = 0.03$

3-link arm with Q_2 ($Q_2^* = 0.86$)		
Learning strategy	Exploration trials (SD)	Variable parameters
DSL	53950 (13972)	$N_{\text{FM}} = 7000$, $\eta = 0.03$
SLbA/a	65194 (29768)	$\sigma_0 = 0.6$, $k_{\max} = 4(0)$
SLbA/b	70691 (30483)	$\sigma_0 = 0.3$, $k_{\max} = 4(0)$
DIM_NGPCA	100% **failed** ($\bar{Q}_C = 0.80$)	$N_{\text{CON}} = 41000$, $N = 120$, $m = 1$

Table 4.9 — Results for the 3-link arm. See caption of Table 4.8 for further explanation. $k_{\max}$ and its standard deviation are rounded down to integer values.

4-link arm with Q_0 ($Q_0^* = 0.94$)		
Learning strategy	Exploration trials (SD)	Variable parameters
DIM_NGPCA	18171 (175)	$N_{\text{CON}} = 7500, N = 200, m = 6$
FEL/J^t	48650 (15288)	$\eta = 0.015$
DSL	56100 (19144)	$N_{\text{FM}} = 3000, \eta = 0.01$
FEL/J$^+$	158050 (59411)	$\eta = 0.02$
SLbA/b	40% **failed** ($\bar{Q}_C = 0.93$)	$\sigma_0 = 0.55, k_{\max} = 38(11)$
SLbA/a	80% **failed** ($\bar{Q}_C = 0.92$)	$\sigma_0 = 0.1, k_{\max} = 45(8)$

4-link arm with Q_0 and additional sensor noise ($Q_{0\text{N}}^* = 0.9$)		
Learning strategy	Exploration trials (SD)	Variable parameters
DIM_NGPCA	18141 (188)	$N_{\text{CON}} = 7500, N = 140, m = 4$
FEL/J^t	52500 (23972)	$\eta = 0.0125$
DSL	63850 (22477)	$N_{\text{FM}} = 5000, \eta = 0.01$
FEL/J$^+$	142850 (82594)	$\eta = 0.02$
SLbA/b	20% **failed** ($\bar{Q}_C = 0.90$)	$\sigma_0 = 0.3, k_{\max} = 36(8)$
SLbA/a	20% **failed** ($\bar{Q}_C = 0.899$)	$\sigma_0 = 0.2, k_{\max} = 35(11)$

4-link arm with Q_1 ($Q_1^* = 0.97$)		
Learning strategy	Exploration trials (SD)	Variable parameters
SLbA/a	188863 (29839)	$\sigma_0 = 0.7, k_{\max} = 6(0)$
SLbA/b	199904 (45781)	$\sigma_0 = 0.7, k_{\max} = 6(1)$
DSL	20% **failed** ($\bar{Q}_C = 0.97$)	$N_{\text{FM}} = 15000, \eta = 0.01$
DIM_NGPCA	40% **failed** ($\bar{Q}_C = 0.968$)	$N_{\text{CON}} = 90000, N = 360, m = 4$

4-link arm with Q_2 ($Q_2^* = 0.86$)		
Learning strategy	Exploration trials (SD)	Variable parameters
DIM_NGPCA	16976 (133)	$N_{\text{CON}} = 7000, N = 820, m = 1$
DSL	18750 (2826)	$N_{\text{FM}} = 5000, \eta = 0.015$
SLbA/a	306796 (219080)	$\sigma_0 = 0.2, k_{\max} = 16(10)$
SLbA/b	376095 (247611)	$\sigma_0 = 0.5, k_{\max} = 19(11)$

Table 4.10 — Results for the 4-link arm. See caption of Table 4.8 for further explanation. $k_{\max}$ and its standard deviation are rounded down to integer values.

the eight task conditions in which SLbA is successful at all. SLbA gets the first place in task condition L4Q1 and the second place in the task conditions L2Q0, L3Q0, L3Q1, and L3Q2. In the conditions L4Q0 and L4Q0N, none of the parameter combinations for SLbA/a or /b is successful.

Comparing DSL and FEL/J^t, the required number of exploration trials N_{EX} is in the same order of magnitude. In condition L2Q0N, DSL is faster, in conditions L3Q0, L3Q0N, L4Q0, and L4Q0N, FEL/J^t is faster (even if only the number of exploration trials for controller training $N_{\mathrm{EX}}^{\mathrm{CON}}$ in DSL is considered: $N_{\mathrm{EX}}^{\mathrm{CON}} = N_{\mathrm{EX}} - N_{\mathrm{FM}}$ for the planar arm task). First and second places are reached in the following task conditions: L3Q2 (first place for DSL); L3Q0N, L4Q0, and L4Q0N (second place for FEL/J^t); L4Q2 (second place for DSL). FEL/J$^+$ is generally slower or worse than FEL/J^t. Only in task conditions L4Q0 and L4Q0N some of the parameter combinations are successful for FEL/J$^+$; nevertheless, FEL/J$^+$ still requires around three times as many exploration trials N_{EX} as FEL/J^t.

As a general tendency, with an increasing number of links there is an increasing number of required exploration trials N_{EX}. Comparing for example the conditions L2Q0, L3Q0, and L4Q0, the best learning strategy (DIM_NGPCA) requires $N_{\mathrm{EX}} = 552$ for the 2-link arm, $N_{\mathrm{EX}} = 2696$ for the 3-link arm, and $N_{\mathrm{EX}} = 18171$ for the 4-link arm. Thus, the increased complexity and dimensionality of the sensorimotor space for larger link numbers has a direct impact on the required learning effort for DIM_NGPCA (and also for SLbA). On the contrary, FEL/J^t and DSL seem to perform the better the more links are involved. The results for FEL/J^t are: no success at all in condition L2Q0, $N_{\mathrm{EX}} = 100615$ in condition L3Q0, and $N_{\mathrm{EX}} = 48650$ in condition L4Q0. Moreover, FEL/J$^+$ in only successful for the 4-link arm and not for smaller link numbers.

The application of sensory noise has only a small impact on the performance of the different learning strategies. For the 2-link arm, the ranking order does not change between conditions L2Q0 and L2Q0N; without noise, DSL and FEL/J^t are not successful at all, with noise, at least some parameter combinations allow successful learning for these strategies. Comparing conditions L3Q0 and L3Q0N for the 3-link arm, DSL and FEL/J^t profit again from the application of noise which causes a slight reduction in the number of required exploration trials N_{EX}. On the contrary, SLbA/a and /b suffer from the noise through a triplication of N_{EX}; this changes the ranking order as well. For the 4-link arm, the application of noise neither changes the ranking order nor the number of required exploration trials considerably between conditions L4Q0 and L4Q0N.

The influence of the constraints on the learning performance is rather inconsistent. Generally, SLbA profits from the first constraint in conditions L3Q1 and L4Q1. For SLbA/a, $N_{\mathrm{EX}} = 73788$ in condition L3Q0 compared to $N_{\mathrm{EX}} = 34971$ in condition L3Q1. A similar relationship holds for the 4-link arm: no success in condition L4Q0 vs. $N_{\mathrm{EX}} = 188863$ in condition L4Q1. In the latter condition, SLbA/a and /b are the only successful learning strategies although DIM_NGPCA and DSL only fail by a small margin. A slight decrease of the desired quality threshold Q_1^* would likely have changed the picture. In condition L3Q1, DIM_NGPCA has still the lead, but with a triplication of N_{EX} compared to condition L3Q0. For the second constraint, the results are very different between the 3-link and the 4-link arm. Comparing conditions L3Q0 and L3Q2, the ranking order of learning strategies reverses. SLbA/a and /b stay on the same performance level, while DSL improves from $N_{\mathrm{EX}} = 106590$ (L3Q0) to $N_{\mathrm{EX}} = 53950$ (L3Q2) and DIM_NGPCA fails completely in condition L3Q2 by a considerable margin. For the 4-link arm, the ranking order does not change between conditions L4Q0 and L4Q2. DIM_NGPCA stays first with roughly the same number of required exploration trials. DSL stays second but with a much better performance ($N_{\mathrm{EX}} = 18750$ for L4Q2 compared to $N_{\mathrm{EX}} = 56100$ for L4Q0); SLbA improves as well from no success at all (L4Q0) to $N_{\mathrm{EX}} = 306796$ (version (a) in condition L4Q2).

Statistical tests We restricted the post-hoc statistical analysis to the most important comparison, namely between learning strategies within each task condition. For this comparison, every learning strategy was matched with every other strategy (excluding the failed ones). The compared measure was the mean number of required exploration trials (as reported in Tables 4.8 to 4.10). The computed pairwise t-tests (two-sided, for independent samples[6]) yielded significant results ($p < 0.05$) in all pairwise comparisons except for the following ones (p values are reported if smaller than 0.2):

L2Q0 : DIM vs. DIM_NGPCA (N_{CON} equal, thus inevitable);
SLbA/a vs. SLbA/b
L2Q0N: SLbA/b vs. SLbA/a; SLbA/a vs. DSL ($p = 0.07$); DSL vs. FEL/J^t
L3Q0 : SLbA/b vs. SLbA/a; SLbA/a vs. FEL/J^t ($p = 0.07$); FEL/J^t vs. DSL
L3Q0N: FEL/J^t vs. DSL; SLbA/a vs. SLbA/b
L3Q1 : SLbA/a vs. SLbA/b
L3Q2 : DSL vs. SLbA/a ($p = 0.15$); SLbA/a vs. SLbA/b
L4Q0 : FEL/J^t vs. DSL ($p = 0.19$)
L4Q0N: FEL/J^t vs. DSL ($p = 0.14$)

[6] The degrees of freedom were corrected to compensate for the unequal estimated population variances (Bortz, 1993).

L4Q1 : SLbA/a vs. SLbA/b
L4Q2 : SLbA/a vs. SLbA/b

These results suggest that it is not possible to draw any firm conclusions from the direct comparisons between SLbA/a and SLbA/b and between FEL/J^t and DSL. Otherwise, the statistical analysis supports the reliability of the reported data.

Additional observations The grayscale plots for DIM_NGPCA (Figs. D.9, D.11, D.13, D.15, D.17, D.19, D.21, D.23, D.25, and D.27) reveal interesting observations regarding the interaction of the task characteristics and the behavior of NGPCA. First of all, for the conditions without constraint (L2Q0, L2Q0N, L3Q0, L3Q0N, L4Q0, and L4Q0N), the lower limit for the number of eigenvectors m in successful learning passes seems to be equal to L, the number links. Only if a huge number of ellipsoids N is available to the network (in conditions L4Q0 and L4Q0N for $N > 200$), parameter combinations with $m < L$ succeed consistently. The upper limit for m for successful performance seems to be $L + 2$, the overall number of dimensions of the sensorimotor space. Only if additional noise is applied and N is large in relation to the size of the training set N_{CON} (as in conditions L3Q0N and L4Q0N), parameter combinations with $m > L$ can be unfavorable. These results show that there is at least a partial tradeoff between m and N, and that network performance can drop if the data manifold is too crowded with too many ellipsoids with too many eigenvectors.

In the conditions with the first constraint (L3Q1 and L4Q1), DIM_NGPCA fails completely for the 4-link arm, thus there is only evaluable data for condition L3Q1. Here, the lower limit for m is still $L = 3$; moreover, there seems to be a strict upper limit of $m = L + 1 = 4$. A closer inspection of the experimental data reveals that the average final controller quality $\bar{Q}_C$ of DIM_NGPCA is not just slightly below the desired quality level Q_1 for $m > 4$, but instead really abysmal (partly, $\bar{Q}_C$ is even negative, indicating the occurrence of collisions). Thus, the first constraint seems to impose a structure on the training data manifold which is difficult for NCPCA to cope with when the number of eigenvectors is too large. This result was barely foreseeable and illustrates the need to test the parameter settings for NGPCA carefully.

Applying the second constraint, DIM_NGPCA fails completely in condition L3Q2, thus only condition L4Q2 remains for closer inspection. Here, success is only possible for $m \leq 2$ eigenvectors. The smallest number of learning examples with $N_{\mathrm{CON}} = 7000$ is required for the combination of $m = 1$ and $N = 820$. For $m = 2$, at least $N_{\mathrm{CON}} = 13000$ learning examples are necessary (with $N = 520$). Overall, at least $N = 420$ ellipsoids are necessary (starting

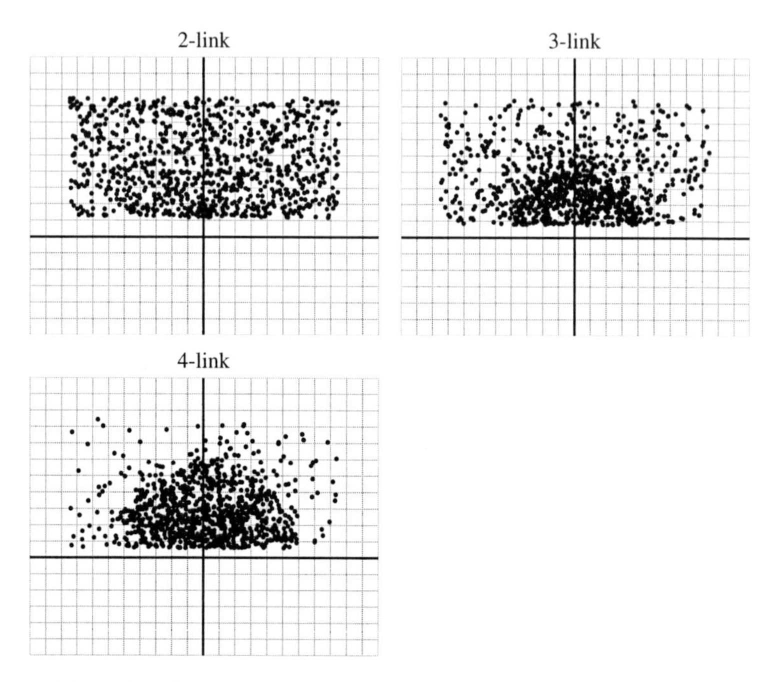

Figure 4.10 — Distribution of 1000 learning examples for DIM_NGPCA in the (y_1, y_2) working space for the different planar arms. Only the position of the tip y is shown as black dot.

with $N_{\text{CON}} = 22000$ and $m = 2$). This strange behavior of NGPCA can be explained by the special characteristic of the second constraint which has already been discussed before in Sect. 4.2.3.1 for DIM_NGPCA. Basically, the second constraint forces the NGPCA network to extrapolate for the generation of motor commands. Obviously, this extrapolation only yields the desired results if the training data manifold is represented by a densely packed large number of units with only one or two directions for linear interpolation. This network structure reminds of a lookup table where interpolation plays a subordinate role. Thus, no far-reaching extrapolation along the direction of the eigenvectors takes place, but instead the best-fitting motor output is recalled from the PCA unit closest to the desired sensory state. Obviously, this "lookup table approach" of DIM_NGPCA does not work as well in condition L3Q2 for the 3-link arm.

Figure 4.10 shows the distribution of 1000 learning examples in the working space which are collected for DIM_NGPCA for the different planar arms (the tip position y is shown as black dot). These learning examples are collected

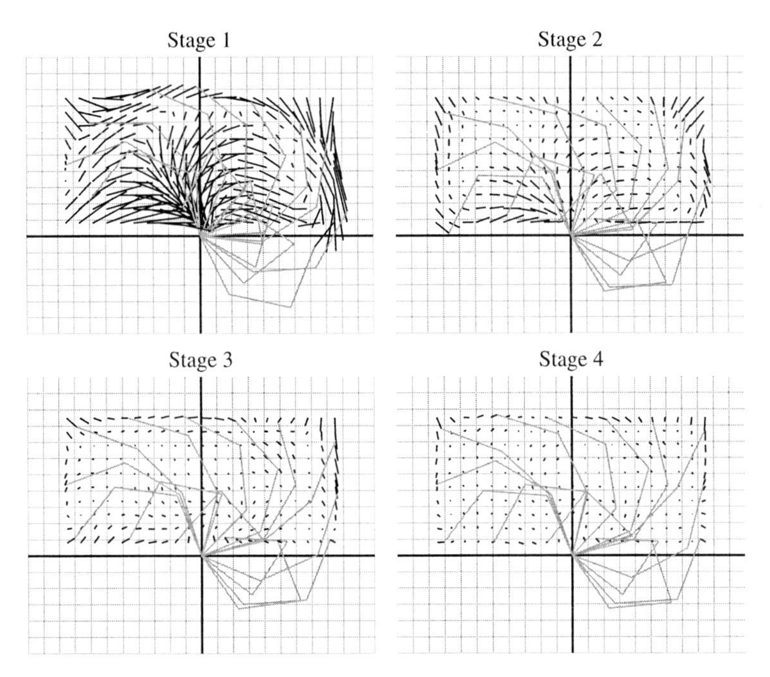

Figure 4.11 — Graphical results of SLbA/b for Q_0 for the 3-link arm, depicted in the (y_1, y_2) working space. The controller performance after different stages is illustrated by black bars which indicate the distance between the desired tip position $\mathbf{y}^*$ and the tip position $\mathbf{y}$ resulting from the arm posture which is generated by the controller output $\mathbf{u}$. These error bars are shown for a regularly spaced grid of desired tip positions covering the whole operating range. Moreover, for 14 desired tip positions at the outer border of the operating range the corresponding controller-generated arm posture is shown in gray color.

by generating random motor commands $\mathbf{u} = \boldsymbol{\theta}$ and assessing the plant output $\mathbf{y} = P(\boldsymbol{\theta})$ afterwards. Only if $\mathbf{y}$ is within the operating range of the desired sensory states $\mathbf{y}^*$, the learning example $[\mathbf{y} \longrightarrow \mathbf{u}]$ is added to the training set. Therefore, the ratio between the number of required exploration trials N_{EX} and the size of the training set N_{CON} for DIM/DIM_NGPCA is larger than one. It amounts to 4.5 for the 2-link arm, to 2.7 for the 3-link arm, and to 2.4 for the 4-link arm. Although this ratio works in favor for DIM_NGPCA for larger link numbers, Fig. 4.10 shows that larger link numbers are actually worse: The distribution of learning examples becomes more and more unbalanced, lumping around the origin of the working space while the outer corners of the operating range are only sparsely populated with learning examples. Therefore, the

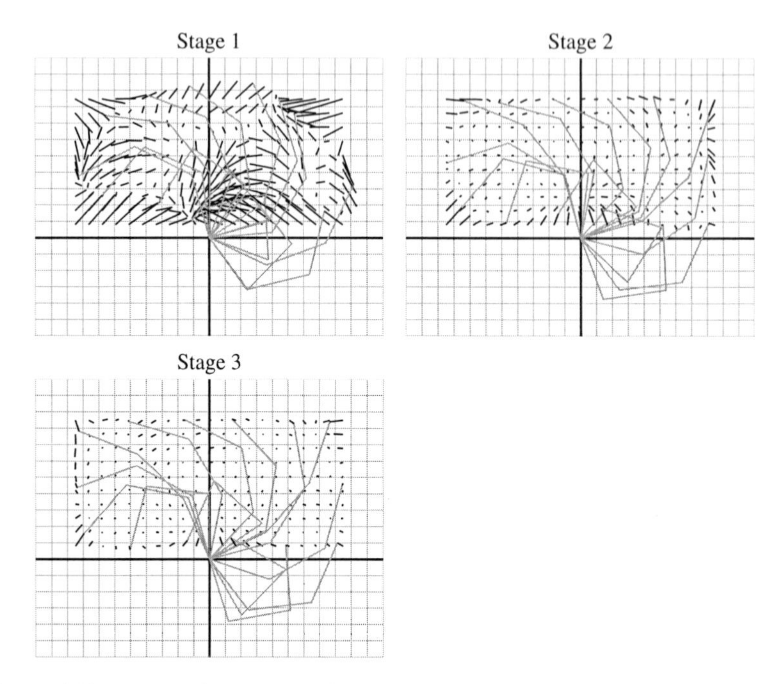

Figure 4.12 — Graphical results of SLbA/b for Q_1 for the 3-link arm. For further explanation see the caption of Fig. 4.11.

overall number of learning examples has to increase to guarantee that controller training is successful in the periphery as well. For even longer planar arms, this effect will work strongly against DIM_NGPCA while the ratio $N_{\mathrm{EX}}/N_{\mathrm{CON}}$ will converge to a value around 2 (since the operating range takes roughly half of the working space in which the main part of the learning examples is generated). This uneven distribution of learning examples in the training set for DIM_NGPCA is an indirect consequence of the lacking goal-directedness of DIM.

As in the saccade learning task, we explore for DSL if the correlations between the mean squared error of the FM (per pattern and output unit after training) on a test set and the number of exploration trials during controller learning $N_{\mathrm{EX}}^{\mathrm{CON}}$ are positive or negative. In contrast to saccade control, the results for the planar arm are quite mixed in this respect: $r = 0.20$ for L2Q0N with $n_s = 29$ (n_s is the number of successful parameter combinations; only these are considered for the computation of r); $r = 0.07$ for L3Q0 with $n_s = 5$; $r = -0.17$ for L3Q0N with $n_s = 5$; $r = -0.24$ for L3Q1 with $n_s = 16$; $r = 0.05$ for L3Q2

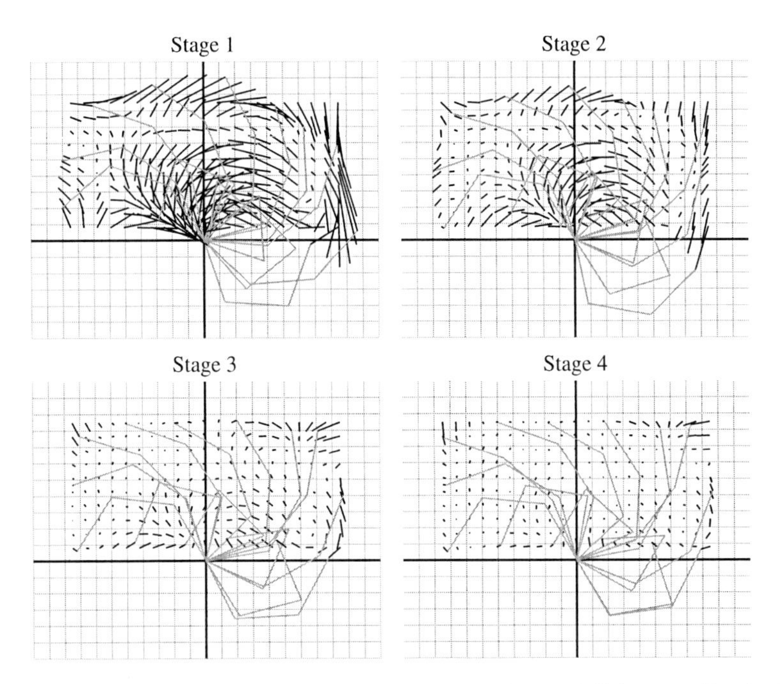

Figure 4.13 — Graphical results of SLbA/b for Q_2 for the 3-link arm. For further explanation see the caption of Fig. 4.11.

with $n_s = 14$; $r = 0.13$ for L4Q0 with $n_s = 19$; $r = 0.39$ for L4Q0N with $n_s = 14$; $r = -0.01$ for L4Q2 with $n_s = 19$. These results are inconclusive.

With regard to SLbA, it is interesting to note that the number of required stages increases strongly with the number of links. For example, in condition L2Q0 $k_{\max}$ amounts to 3.2 for SLbA/a, in condition L3Q0 to 6, and in condition L4Q0 to 25 (only considering the 20% of successful learning passes in this condition). With noise, $k_{\max}$ gets slightly larger, whereas both constraints help to reduce $k_{\max}$ considerably (e.g., $k_{\max} = 6$ in condition L4Q1). As examples for the course of learning with SLbA, Table 4.11 reports the average quality Q_{PS} of the training set side by side with the average controller quality Q_C on a test set for the different stages of a single learning pass of SLbA/b for the task conditions L3Q0, L3Q0N, L3Q1, and L3Q2. The quality values in this table demonstrate that averaging has here the same positive effect (as theoretically supposed) as in the saccade learning task: At least in the first and second stage, the controller quality is larger than the quality of its learning examples. In later stages, especially in condition L3Q0N with sensor noise, learning slows down

	L3Q0			**L3Q0N**			**L3Q1**			**L3Q2**		
Stage k	Q_{PS}		Q_C	Q_{PS}		Q_C	Q_{PS}		Q_C	Q_{PS}		Q_C
1	0.53	$<$	0.59	0.52	$<$	0.57	0.46	$<$	0.85	0.49	$<$	0.60
2	0.84	$<$	0.90	0.84	$<$	0.88	0.94	$<$	0.96	0.76	$\approx$	0.76
3	0.96	$>$	0.94	0.95	$>$	0.92	0.98	$>$	0.97	0.84	$\approx$	0.84
4	0.98	$>$	0.96	0.97	$>$	0.92				0.88	$>$	0.87
5				0.97	$>$	0.93						

Table 4.11 — Comparing the average quality Q_{PS} of the learning examples in the training set with the average controller quality Q_C on a test set for the different stages of a single learning pass of SLbA/b for the task conditions L3Q0, L3Q0N, L3Q1, and L3Q2.

and relies mainly on the improvement of the training set from stage to stage.

Figures 4.11 to 4.13 further illustrate the course of learning for SLbA/b for the 3-link arm with quality functions Q_0 to Q_2 (conditions L3Q0, L3Q1, and L3Q2). In each figure, the controller performance after all stages from the very first to the very last is indicated with black error bars (difference between desired tip position $\mathbf{y}^*$ and the tip position $\mathbf{y}$ resulting from the controller output $\mathbf{u} = C_k(\mathbf{y}^*)$). Moreover, the corresponding arm posture $\mathbf{u}$ is depicted as well for 14 positions $\mathbf{y}^*$ at the border of the operating range. The presented learning passes have been carried out with the optimal parameter settings from Table 4.9. Independent of the quality function, learning progresses noticeably from stage to stage with shorter and shorter error bars. Depending on the region within the operating range, learning takes place with different speed, the outer corners being the most difficult part. The first constraint (Q_1) facilitates learning in the second stage compared to the no-constraint condition (Q_0), while the second constraint (Q_2) has the contrary effect. For Q_1, the finally generated arm postures show a clear symmetry (which is enforced by the first constraint — obviously successfully); this is not the case for Q_0 and Q_2.

Figure 4.14 shows the final results for the other learning strategies in condition L3Q0; controllers are trained with the optimal parameter settings from Table 4.9. It is noticeable that all learning strategies which rely on local linear approximation (DSL and FEL) converge to very similar final controller output (comparing the arm postures of DSL, FEL/$\mathbf{J}^t$, and FEL/$\mathbf{J}^+$ on the one hand and of DIM_NGPCA and SLbA/b (in Fig. 4.11) on the other hand). In the analogous comparison for the 4-link arm in Fig. 4.15, this similarity remains only between FEL/$\mathbf{J}^t$ and DSL (which works basically with an approximation of $\mathbf{J}^t$) while the final arm postures for FEL/$\mathbf{J}^+$ show a very uneven characteristic. SLbA/a and /b generate also similar arm postures. However, these figures present only

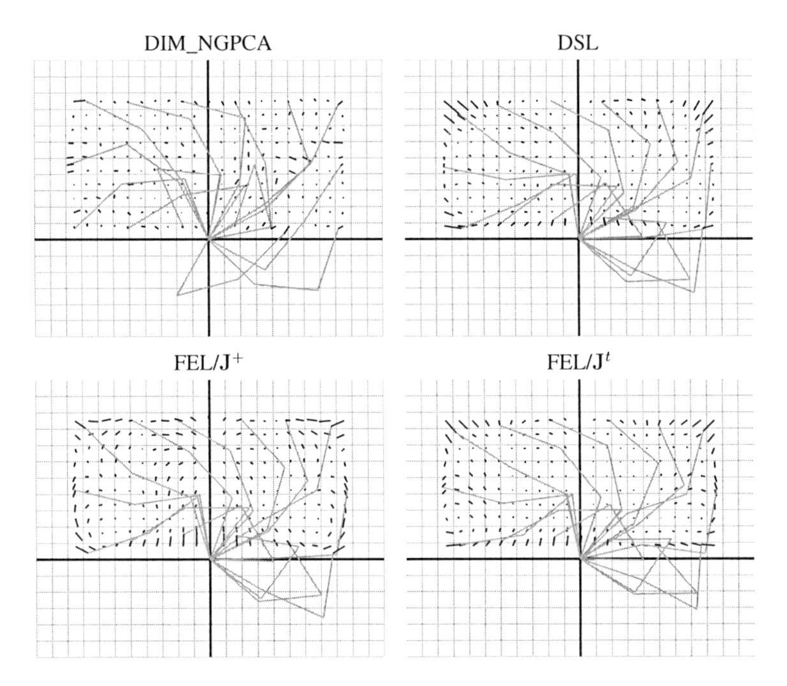

Figure 4.14 — Comparison of various learning strategies for Q_0 for the 3-link arm, depicted in the (y_1, y_2) working space. The controller performance after training is illustrated by black error bars. Moreover, for certain desired tip positions the corresponding controller-generated arm posture is shown in gray color. For a more detailed explanation, see the caption of Fig. 4.11. Since FEL/J$^+$ is not capable to reach the desired quality level in this task condition, the error bars are longer on average for FEL/J$^+$ than for the other learning strategies.

the results of a single learning pass, and especially for the 4-link arm the final arm postures generated by the controller networks do vary from pass to pass. Nevertheless, Figs. 4.14 and 4.15 illustrate that the one-to-many problem of the no-constraint conditions is solved by all learning strategies by converging to distinct solutions in the different regions of the operating range. These solutions vary between learning strategies and also between learning passes.

In Fig. 4.16, the influence of the different constraints on the final controller output is compared for SLbA/b for the 3-link and the 4-link arm. For Q_1 (the "maximum symmetry" constraint), it is clearly visible from the controller-generated arm postures that this symmetry has been successfully achieved. Without any constraint (Q_0), the arm postures show less symmetry (3-link arm) or

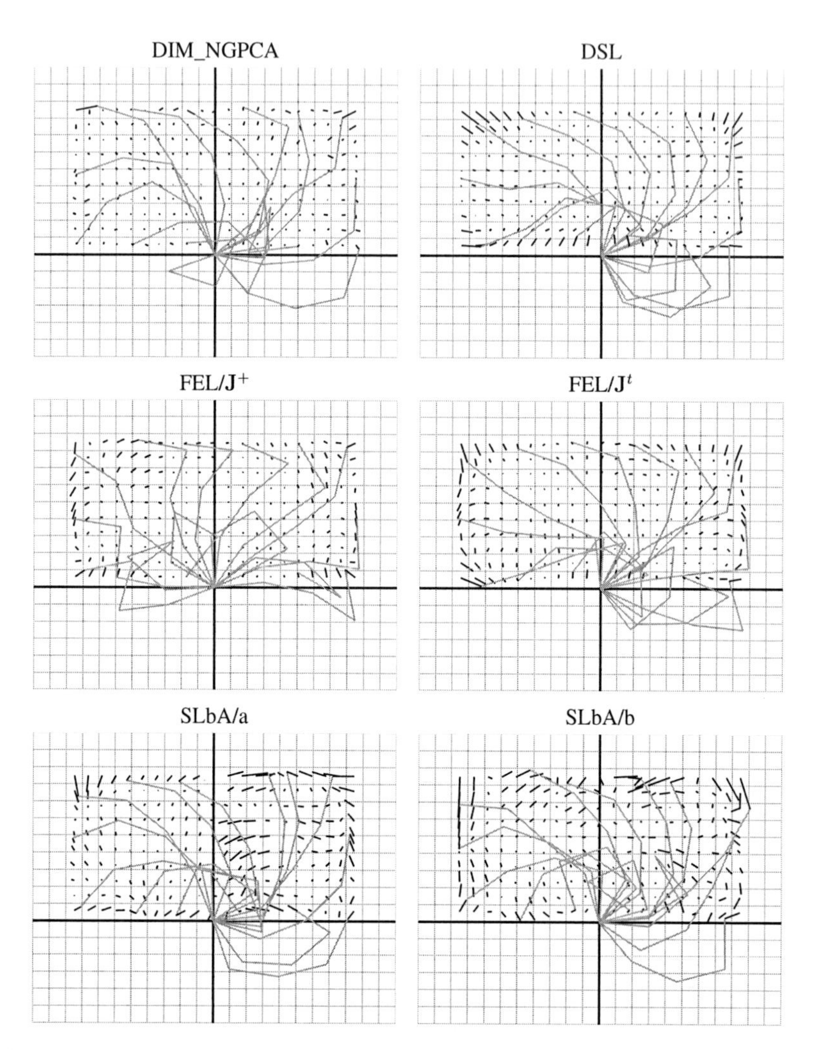

Figure 4.15 — Comparison of various learning strategies for Q_0 for the 4-link arm. For further explanation see the caption of Fig. 4.14. Since SLbA/a and /b are not capable to reach the desired quality level in this task condition, the error bars are longer for SLbA/a and /b on average than for the other learning strategies.

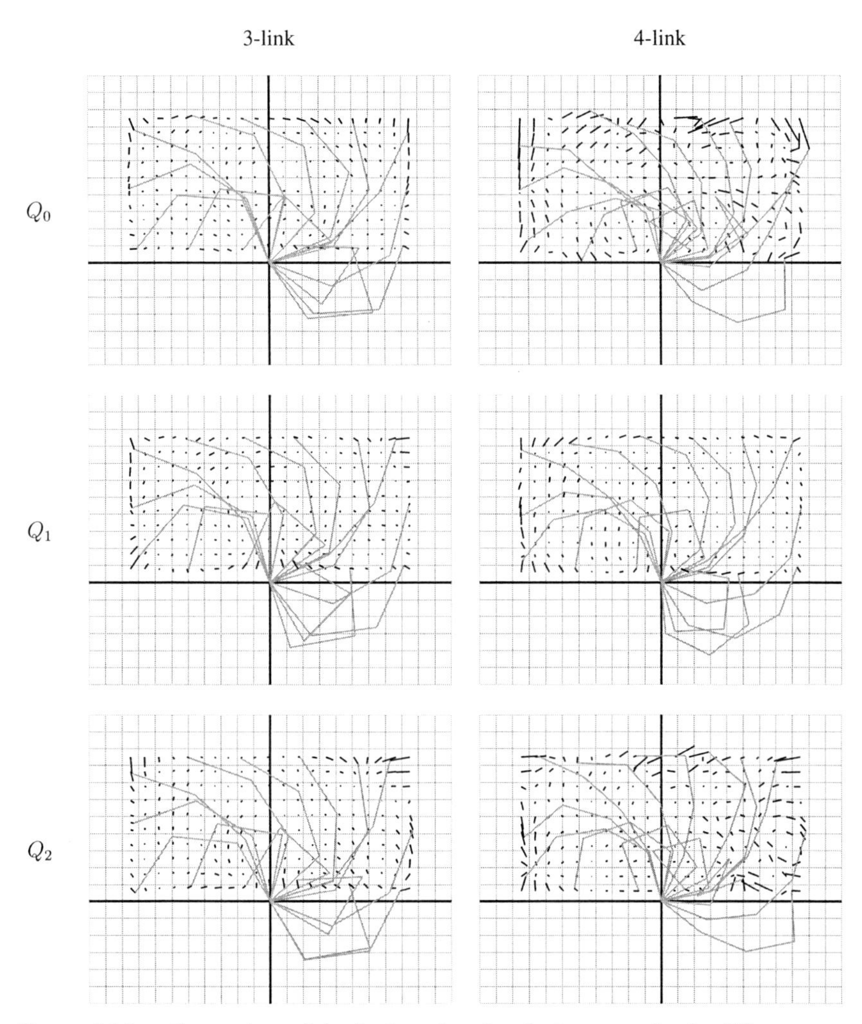

Figure 4.16 — Comparison of the final results after the last stage for the different contraints with SLbA/b for the 3-link (left) and the 4-link (right) arm. For further explanation see the caption of Fig. 4.14.

barely any symmetry (4-link arm). The arm postures for Q_2 and Q_0 are very similar. With regard to the 4-link arm, the constraint Q_2 seems to help in reducing the length of the error bars. Although this constraint introduces a tradeoff between reaching precision and "energy minimization", the positive effect of reducing the ambiguity of the one-to-many mapping seems to prevail for SLbA.

4.2.5 Discussion

In contrast to the saccade learning task, the plant in the planar arm task is non-linear. As result, DIM_NGPCA is the clear performance leader while the local linear approximation techniques like FEL and DSL are not really competitive any longer. These linear techniques seem to be less well suited to non-linear tasks. Nevertheless, successful NGPCA networks are also composed of a multitude of ellipsoids to cope with the non-linearity. SLbA is a fair performer, but decreasingly less so with an increasing number of links of the planar arm.

The strength of DIM_NGPCA is that it combines the simple learning strategy of DIM with the capability of NGPCA to learn one-to-many mappings. DIM_NGPCA offers a unique approach to the one-to-many problem by storing multiple solutions of the inverse kinematics simultaneously if encountered during learning. Here, we use a recall mechanism which only generates the solution with the smallest distance measure (see Sect. 3.4.2). This solves one of the two main problems of DIM. The second problem, its lacking goal-directedness, remains but is not that important for planar arms with up to 4 links. However, with an increasing number of links the distribution of learning examples in the operating range becomes more and more unfavorable as shown in the results section. Thus, at some point the missing goal-directedness will become an issue. Another implication of the missing goal-directedness shows up for the first constraint (Q_1), for which only a small subset of the learning examples (the ones close to "maximum symmetry") represents the desired sensory outcome. However, the first constraint has only a mild negative impact on the performance of DIM_NGPCA (at least for the 3-link arm), which demonstrates that the lacking goal-directedness does not always pose a problem in practice. Overall the performance of DIM_NGPCA becomes less reliable when learning constraints are enforced by additional units for the desired sensory input. Especially for the "energy minimization" constraint (Q_2), the successful NGPCA networks require at least 420 ellipsoids with exactly 2 eigenvectors and remind more of a lookup table than of a "healthy" approximation with local PCAs. Such an NGPCA network has 10080 parameters compared to the 74 weights of the MLP controller networks which perform successfully with DSL or SLbA. One can state in favor of DIM_NGPCA that it even manages to learn a task which requires extrapolation, but on the downside this is very expensive in terms of network complexity, storage requirements, and computation effort during recall. Sensor noise has a negative impact on DIM_NGPCA as well, but not up to a point where the performance leadership of DIM_NGPCA is endangered. To learn noisy data, larger training sets are required. In the 2-link task, where no one-to-many mapping is involved, DIM with MLPs performs just as good as DIM_NGPCA in the condition without noise, but suffers more from noise.

Surprisingly, the performance of the local linear approximation techniques (FEL and DSL) increases for larger link numbers. These learning strategies might surpass DIM_NGPCA at some point. However, this has not been tested in the present study. FEL/$\mathbf{J}^t$ shows the best performance, closely followed by DSL and with a larger margin by FEL/$\mathbf{J}^+$. Thus, the best gain matrix is the exact transpose of the Jacobian $\mathbf{J}^t_{\mathrm{u,x}}$. The estimate $\widehat{\mathbf{J}}^t_{\mathrm{u,x}}$ which is generated by the FM in DSL is slightly worse (but these are not firm conclusions; see the statistical analysis). Since the precision of the estimate $\widehat{\mathbf{J}}^t_{\mathrm{u,x}}$ depends on the precision of the FM, it is as puzzling as in the saccade learning task that the correlations between the mean squared error of the FM on a test set and the number of exploration trials during controller learning in the DSL learning scheme are not clearly positive, although the exact transpose of the Jacobian yields the overall best results (in FEL). The pseudoinverse $\mathbf{J}^+_{\mathrm{u,x}}$ is the worst choice as gain matrix; for the planar arm, one would clearly prefer $\mathbf{J}^t_{\mathrm{u,x}}$. However, based on the available data it is not clear to what extent this ranking order of gain matrices generalizes to other learning tasks. From the local linear approximation techniques, DSL was the only one tested with constraints. These were imposed on the learning task by adding sensory output units to the FM. Each of these output units had a fixed desired value signaling perfect constraint fulfillment. Generally, DSL experienced a speedup of learning through the constraints, only for the 4-link arm the first constraint ("maximum symmetry") caused DSL to fail by a very small margin. Additional sensor noise has only a small impact on FEL and DSL; for the 2-link arm, it even helped DSL and FEL/$\mathbf{J}^t$ to get into the set of successful learning strategies. In addition, FEL and DSL cope very well with the one-to-many mapping of the planar arm task by converging to distinct solutions which vary between learning passes due to the stochastic nature of network initialization and training pattern generation.

The performance of SLbA decreases with an increasing number of links in the no-constraint conditions. The most likely reason is that the one-to-many nature of the learning task becomes the more dominant the more links are involved. For the 2-link arm, there is no ambiguity at all; here, SLbA gets the second place while FEL and DSL fail completely (at least without noise). For the 3-link arm, SLbA still comes in second although the one-to-many problem is present and — even worse — the non-convexity of the solution sets (see Sect. 4.2.1.1) violates one precondition of learning by averaging: It is not guaranteed that the average of two arm postures has a quality which is larger than the quality of the worse of these two postures. Although this precondition is violated, SLbA is fairly successful for the 3-link arm, and averaging over learning examples has the expected positive effect for subsequent controller performance (see Table 4.11). But for the 4-link arm, SLbA finally fails because it cannot cope with the increased ambiguity. Nevertheless, as soon as the ambiguity is removed by the first constraint (there is only one posture for each tip position with "maximum symmetry"), SLbA even becomes the best learning strategy while DSL and DIM_NGPCA fail (at least for the 4-link arm). The second constraint ("minimum energy") does not help to reduce the ambiguity completely, instead it introduces a tradeoff between reaching precision and the distance of the arm posture to the resting position in joint space. However, this constraint helps SLbA as well to become successful for the 4-link arm although only with a huge number of required exploration trials compared to the other learning strategies. The impact of sensor noise on SLbA is rather negative for the 2- and the 3-link arm (duplication respectively triplication of required exploration trials), but for the 4-link arm additional noise lifts SLbA nearly into the group of successful learning strategies: With noise, SLbA fails only by a very tight margin. The performance difference between SLbA/a and /b is very small with a slight edge for version (a) (as outlined in the results section; however, the performance difference between SLbA/a and /b is not statistically significant in any task condition). Thus, it might be favorable to use the overall quality of the controller of the preceding stage as minimum for the quality threshold. This may ensure more even learning speed across the operating range: Problematic areas are lifted earlier on the average performance level.

The complete failure of CLbA on the planar arm task illustrates that learning by averaging is not that well suited for online learning. For non-linear tasks, the averaging between different imperfect learning examples obviously only takes place in the desired way if the whole set of learning examples is presented simultaneously for batch learning as in SLbA. We assume that during online learning the effect of averaging is counteracted by catastrophic interference, this means "forgetting" of the approximated function in one part of the input space while learning takes place in other parts of the input space. In non-linear tasks, the target function which has to be approximated by the MLP changes significantly between different local parts of the input space, while linear tasks are well-tempered in this respect. For this reason, learning by averaging in an online fashion suffers especially from catastrophic interference for non-linear tasks as was observed for the planar arm with CLbA.

In conclusion, for non-linear and redundant task domains like the planar arm, the application of DIM_NGPCA can be recommended as long as the negative impact of the lacking goal-directedness of DIM is not too strong, for example due to an uneven distribution of learning examples in the operating range. Otherwise, DIM_NGPCA is fairly efficient and resistent to noise. If additional constraints have to be incorporated into the learning task, SLbA is a strong contender since the constraints can be specified in a straightforward fashion in the quality function which is used during learning. Moreover, SLbA proved to be the overall most reliable learning strategy if constraints are applied. On the downside, SLbA slows down if sensor noise is present and does not cope well with one-to-many mappings and non-convex solution sets. The local linear approximation techniques (FEL and DSL) show no distinct advantage other than their resistance to noise. They are not that fast and also not that reliable with constraints (speaking of DSL; for FEL, constraints get even more complicated since one needs to determine the Jacobian of the plant which is extended by the constraint output units analytically). However, the available data shows the trend that DIM_NGPCA slows down with an increasing number of links while DSL and FEL speed up. Thus, for motor tasks with high-dimensional sensorimotor spaces DSL and FEL may be faster than DIM_NGPCA since the latter is hampered by the lacking goal-directedness in the generation of learning examples.

4.3 Overall Discussion

In summary, the initial reasearch questions have been answered by the results of the two learning tasks in the following way: For linear plants, local linear approximation techniques like FEL and DSL work very well while DIM_NGPCA is a good allround performer for both linear and non-linear plants. CLbA is the only learning strategy which fails completely on non-linear tasks. DIM_NGPCA deals well with one-to-many mappings as do FEL and DSL while SLbA fails if the redundancy of the learning task is too large. With regard to sensory noise, the most considerable performance drop can be observed for CLbA, SLbA, and DIM. If additional learning constraints are imposed, the most reliable learning strategy is SLbA (closely followed by DSL). Overall, the results are consistent with the theoretical summary on the different learning strategies in Sect. 2.2.8 (see also Table 2.1).

However, it has to be emphasized that the performance of DIM-related techniques like DIM_NGPCA depends heavily on two task characteristics: First, how much sampling effort is required to find learning examples which are close to the desired operating range of the controller in sensory space, and second, how much sampling effort is required to fill all areas within the operating range with at least the minimum of learning examples needed for good interpolation during controller learning. Although these task characteristics are not directly linked to the dimensionality of the combined input/output space of the controller, one can expect that a larger number of dimensions often implies enlarged sampling effort (as it does for the planar arm).

Moreover, the comparison in this study relies on the number of required exploration trials. This performance indicator works in favor of learning strategies which are based on batch learning, this means collecting a set of learning examples first and using it afterwards multiple times for controller adaptation. In contrast, learning strategies like FEL and DSL cannot reuse learning examples, thus every cycle of controller adaptation requires additional exploration trials to generate a new learning example. If one aims at biological plausibility and rejects batch learning for this reason, the number or required exploration trials for DIM/DIM_NGPCA and SLbA (and also for DSL due to FM training) gets much larger. For example, the DIM_NGPCA networks with optimal parameter settings in condition L3Q0 for the planar arm (3-link arm without constraint and without noise) would require around seven times as many exploration trials if they were not allowed to reuse the collected learning examples during controller adaptation. However, in this task condition DIM_NGPCA would remain fastest even with this additional burden.

Other performance indicators besides the required number of exploration trials could be the number of required network adaptation cycles (as suggested in the previous paragraph), the overall computational effort, the minimum required number of adjustable network parameters, etc. In this study, we referred to number of exploration trials since these are connected to "real" movements of the agent. We judge these real movements to be more relevant than indicators of computational effort because the latter are not only linked to the motor learning strategy but also to the applied neural network algorithm. For the same reason, we evaluated batch learning strategies just by the number of required exploration trials without any attempt to make this number equivalent to the online learning strategies as discussed in the previous paragraph. Future neural network algorithms (e.g., an advanced version of NGPCA) may allow much faster learning so that a single cycle through the training set would be sufficient for successful controller learning. The absolute number of learning examples in the training set, on the other hand, has a lower limit to allow precise interpolation between the provided data points for any neural network algorithm, and this number is directly linked to the required number of exploration trials.

Our goal in this study was to compare the effort of the different learning strategies to reach a very high quality level, close to the achievable optimum. Because of this approach, some strategies simply failed in certain task conditions because they were not able to reach this level consistently. Generally, one can expect that the performance comparisons yield different results if the demanded quality levels are varied, depending on the relationship between the number of exploration trials and acquired controller quality for the different learning strategies. One strategy may advance quickly in the beginning but stagnate close to the demanded very high quality level, a second strategy may proceed slower in the beginning but does not stagnate until the demanded very high quality level is already surpassed. It might be interesting to explore these relationships in future studies. Furthermore, one could extend the present study to multiple noise levels instead of just one in each task configuration. Similar to the demanded quality level, one can expect the different learning strategies to break down at different levels of applied sensory noise (or even motor noise). However, all of these additional variations will require a lot of additional computational effort if they are combined with an as thorough exploration of the parameter space for each learning strategy as in the present study.

Additional learning constraints have only been applied in the planar arm task. Depending on the learning strategy, they were implemented differently. SLbA adjusts itself easily to the constraints since they are directly encoded in the quality function. For DIM_NGPCA and DSL, constraints were imposed by additional desired sensory input units of the controller (DIM_NGPCA) or by additional sensory output units of the FM (DSL). For DIM_NGPCA and DSL, this constraint implemention caused less reliable learning success, while SLbA partly even relied on additional constraints for successful learning to reduce the ambiguity of the one-to-many mappings. Without constraints, the performance of SLbA was worse than expected; nevertheless, for learning tasks with constraints SLbA is the most direct approach without the need to modify the controller input or to specify additional plant outputs, and moreover the most reliable approach throughout our studies.

In conclusion, it has to be stated again that DIM in combination with NG-PCA is the overall winner of this comparison study. This contradicts the view that DIM is the least favorable approach among DIM, FEL, and DSL although this view is often expressed in the literature (e.g., Jordan, 1996; Kawato, 1990). It could be shown that the first argument against DIM, its inability to deal with one-to-many mappings, can be overcome by abstract recurrent neural networks. Moreover, the impact of its second weakness, the lacking goal-directedness, depends heavily on the task characteristics, thus one can expect DIM to be fairly efficient for many kinematic control tasks like in the present study. The third criticsm (with regard to biological plausibility), the hypothesized need for neural rewiring, can be counteracted by neural architectures where the input layer of the DIM controller is a neural map which can be activated both from sensory afferences and from memory (to represent desired sensory states) — this could be part of a larger recurrent neural network architecture which represents sensory states and which can reproduce these states when it is triggered by memory traces (this idea extends the neural map approach by Kuperstein, 1988). To resolve the remaining second weakness of DIM more convincingly, future research might explore goal-directed search strategies in motor space to replace the random sampling for the generation of learning examples.

Chapter 5

Visual Forward Models for Camera Movements

5.1 Adaptive Acquisition of a Prediction Mapping[1]

5.1.1 Visuomotor prediction

In Sect. 1.5 and Chapt. 2, we argued how important sensory or state prediction by forward models (FMs) is for motor control, perception, and cognition in general. Since many species rely on the visual sense for movement control, for orientation and navigation, and for the identification of relevant objects and events in the environment, it is of special interest to explore the mechanisms of visual prediction. Accordingly, the focus in this section is on the learning of FMs in the visual domain. In our understanding, visual FMs predict representations of entire visual scenes. In the nervous system, this could be the relatively unprocessed representation in the primary visual cortex or more complex representations generated in higher visual areas. Studies on predictive remapping (see Sect. 1.3.4.2) suggest that visual prediction takes place at various processing levels in the brain (Duhamel et al., 1992; Melcher, 2007; Umeno and Goldberg, 1997; Walker et al., 1995). Regarding robot models, the high-dimensional sensory input and output space of visual FMs poses a tough challenge to any machine learning or neural network algorithm. Moreover, there might be unpredictable regions in the FM output (because parts of the visual surrounding only become visible after execution of the motor command whose consequences are to be predicted). In the following, a learning algorithm is suggested which solves both problems in the context of robot "eye" movements.

[1] This section is a slightly modified and updated version of the publication by Schenck and Möller (2007; © Springer). The permission for republication is kindly granted by Springer.

Figure 5.1 — Left: Visual forward model (FM). Right: Single component of a visual forward model predicting the intensity of a single pixel $(x_{\mathrm{Out}}, y_{\mathrm{Out}})$ of the output image (adapted from Schenck and Möller, 2007, © Springer).

5.1.2 Structure of the visual forward model

The task of the robot model is to predict the visual consequences of eye movements. In the model, the eye is replaced by a camera which is mounted on a pan-tilt unit. Prediction of visual data is carried out on the level of camera images. In analogy to the sensor distribution on the human retina, a retinal mapping is applied which decreases the resolution of the camera images from center to border. This mapping is used to make the prediction task more difficult and more realistic. The input of the visual FM is a "retinal image" at time step t (called "input image" in the following) and a motor command m_t.[2] The output is a prediction of the retinal image at the next time step $t+1$ (called "output image" in the following; see left part of Fig. 5.1).

The question is how such an adaptive visual FM can be implemented and trained by exploration of the environment. A straight-forward approach is the use of function approximators which predict the intensity of single pixels. For every pixel $(x_{\mathrm{Out}}, y_{\mathrm{Out}})$ of the output image, a specific forward model $\mathrm{FM}_{(x_{\mathrm{Out}}, y_{\mathrm{Out}})}$ is acquired which forecasts the intensity of this pixel (see right part of Fig. 5.1). Together, the predictions of these single FMs form the output image as in Fig. 5.1 (left). This simple approach was explored in a diploma thesis (Große, 2005) under the author's supervision: Various neural network topologies and learning algorithms were tested and compared, but unfortunately none of them produced satisfactory learning results. In conclusion, the direct prediction approach suffers from the high dimensionality of the input space (since the retinal image at time step t is part of the input) and is not successfully applicable in practice. In the work of Hoffmann (2007), where images with a size of 40×40 pixels are directly predicted, an additional denoising model is required. This model has to be trained for the specific environment of the mobile robot.

Hence, in this study we pursue a different approach. Instead of forecasting pixel intensities directly, our solution is based on a "back prediction" of where a

[2] In contrast to the preceding chapters, the symbol m is used in this chapter to denote motor commands, and the symbol s to denote sensory states. This change is applied to avoid confusion with the image coordinates x and y.

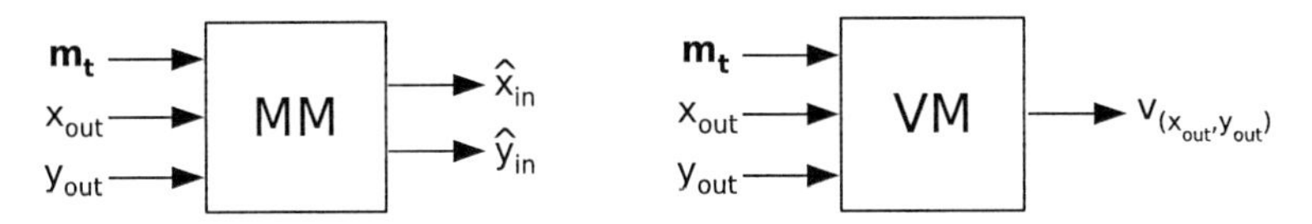

Figure 5.2 — Left: Mapping model (MM). Right: Validator model (VM) (for details see text) (adapted from Schenck and Möller, 2007, © Springer).

pixel of the output image has been in the input image before the camera's movement. The necessary mapping model (MM) is depicted in Fig. 5.2: As input, it receives the motor command $\mathbf{m}_t$ and the location of a single pixel $(x_{\mathrm{Out}}, y_{\mathrm{Out}})$ of the output image; as output it estimates the previous location $(\widehat{x}_{\mathrm{In}}, \widehat{y}_{\mathrm{In}})$ of the corresponding pixel (or region) in the input image. The overall output image is constructed by iterating through all of its pixels and computing each pixel intensity as $\widehat{\mathrm{I}}^{\mathrm{Out}}_{(x_{\mathrm{Out}},y_{\mathrm{Out}})} = \mathrm{I}^{\mathrm{In}}_{(\widehat{x}_{\mathrm{In}},\widehat{y}_{\mathrm{In}})}$ (using bilinear interpolation).[3] Moreover, an additional validator model (VM) generates a signal $v_{(x_{\mathrm{Out}},y_{\mathrm{Out}})}$ indicating whether it is possible at all for the MM to generate a valid output for the current input. This is necessary because even for small camera movements parts of the output image are not present in the input image. In this way, the overall FM (Fig. 5.1, left) is implemented by the combined application of a mapping and a validator model.

The basic idea of the learning algorithm for the MM is outlined in the following for a specific $\mathbf{m}_t$ and $(x_{\mathrm{Out}}, y_{\mathrm{Out}})$. During learning, the motor command is carried out in different environmental settings. Each time, both the actual input and output image are known afterwards, thus the intensity $\mathrm{I}^{\mathrm{Out}}_{(x_{\mathrm{Out}},y_{\mathrm{Out}})}$ is known as well. It is possible to determine which of the pixels of the input image show a similar intensity. These pixels are candidates for the original position $(x_{\mathrm{In}}, y_{\mathrm{In}})$ of the pixel $(x_{\mathrm{Out}}, y_{\mathrm{Out}})$ before the movement. Over many trials, the pixel in the input image which matches most often is the most likely candidate for $(x_{\mathrm{In}}, y_{\mathrm{In}})$ and therefore chosen as MM output $(\widehat{x}_{\mathrm{In}}, \widehat{y}_{\mathrm{In}})$. When none of the pixels matches often enough, the MM output is marked as non-valid (output of the VM).

5.1.3 Method

To acquire such an MM and VM as in Fig. 5.2, the following steps are executed. First, a grid of points is defined in the input space of the MM and VM (com-

[3] In this study, pixel intensities of the retinal input and output images are three-dimensional vectors in RGB color space.

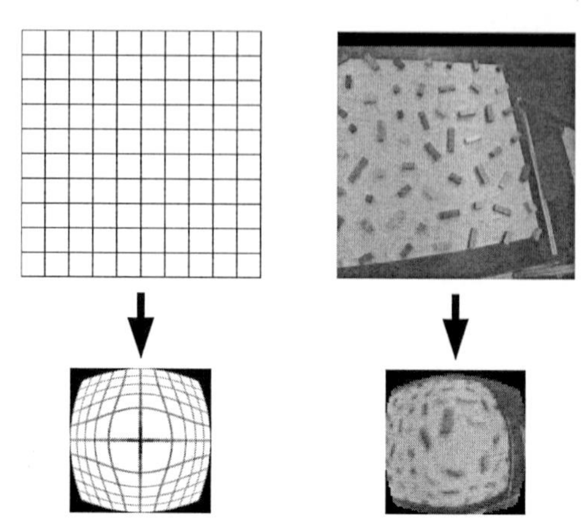

Figure 5.3 — Retinal mapping. Left: For an image depicting a regular grid. Right: For a camera image (right part adapted from Schenck and Möller, 2007, © Springer).

posed of $\mathbf{m}_t$ and $(x_{\text{Out}}, y_{\text{Out}})$), ranging from the minimum to the maximum value in each input dimension. For each grid point, the most likely estimate $(\widehat{x}_{\text{In}}, \widehat{y}_{\text{In}})$ is determined by collecting candidate pixels in many different visual surroundings. Along the way, the VM output $v_{(x_{\text{Out}}, y_{\text{Out}})}$ is determined as well. Thereafter, one radial basis function network (RBFN) is trained to interpolate the MM output between the grid points, and another RBFN to interpolate the VM output. The resulting networks can be applied to image prediction afterwards. In the following, the methods are outlined in more detail.

5.1.3.1 Setup

The robot setup is shown in Fig. A.1 (left) in App. A. Only the right camera is used during training, although the FM can be used afterwards for both the left and right camera since they share the same geometry. A central quadratic region of the original camera image (captured in RGB color) with a resolution of 240×240 pixels is used for further processing (and called "camera image" in the following for simplicity). The horizontal and the vertical angle of view of this region amount to 48.5 degrees. The camera is mounted on a pan-tilt unit with two degrees of freedom. In this study, the valid range for the pan angle is between -60.4 and 23.8 degrees, for the tilt angle between -42.9 and 21.4 degrees (for a more thorough description, see App. B). In this range, the camera image always captures at least a small part of the white table shown in Fig. A.1 (left) below the cameras.

The pan and tilt axes cross in close vicinity to the nodal point of the camera-lens system. For this reason, the effect of changing the pan and tilt position by a certain amount (Δpan,Δtilt) is almost independent of the current camera position (a detailed analysis of the camera geometry which supports this claim is provided in App. B). Accordingly, the motor input $\mathbf{m}_t$ of the FM just consists of Δpan and Δtilt. Both values can vary between -29 and $+29$ degrees. For the same reason, object displacements in the camera images during camera movements are virtually independent from the object distance to the camera. Thus, depth information is irrelevant for the learning task.

For the training of the visual FM, not the real setup was used, but instead "virtual" camera movements were carried out using an image database. This image database contains the camera images for more than $120,000$ different camera positions within the above-mentioned pan-tilt range. Instead of using the camera directly, we retrieved the images from the database. The recorded scene shows the white table with 56 colored wooden blocks on its surface — 14 blocks each of the colors red, green, blue, and yellow.

5.1.3.2 Retinal mapping

As mentioned before, the input and output images of the FM are "retinal" images with decreasing resolution from image center to border. Camera images are converted to such retinal images by a "retinal mapping". The effect of this conversion is depicted in Fig. 5.3. The basic idea of this mapping is best outlined in polar coordinates. The origins of the coordinate systems are located at the image centers. They are scaled in a way that in both images the maximum radius (along the horizontal/vertical direction) amounts to 1.0. r_R is the radius of a point in the retinal image, r_C is the radius of the corresponding point in the camera image, the angle of the polar representation is kept constant. r_C is computed by $r_C = \lambda r_R^\gamma + (1 - \lambda) r_R \ , \ \gamma > 1 \ , \ 0 \leq \lambda \leq 1$. Here we use $\gamma = 2.5$ and $\lambda = 0.8$. The resolution of the final retinal image is 69×69 pixels. To avoid aliasing artifacts in the heavily subsampled outer regions of the original image, adaptive smoothing is applied (with a binomial filter whose mask size is proportional to the local subsampling factor).

While the input image of the FM is an unmodified retinal image, the output image is a center crop with a size of 53×53 pixels. This is necessary to clip the black corners of the retinal image without valid information (see Fig. 5.3) which are just a technical artifact.

5.1.3.3 Grid of cumulator units

The input space of the MM and VM consists of four dimensions: Δpan, Δtilt, x_{Out}, and y_{Out}. In this space, a four-dimensional grid $\mathbf{P}$ of points $\mathbf{p}_{ijkl} = (\Delta\text{pan}^{(i)}, \Delta\text{tilt}^{(j)}, x_{\text{Out}}^{(k)}, y_{\text{Out}}^{(l)})$ is inscribed, with $i, j = 1, .., 11$ and $k, l = 1, .., 13$. $\Delta\text{pan}^{(i)}$ and $\Delta\text{tilt}^{(j)}$ vary from -29 to $+29$ degrees with constant step size (covering the whole valid $\Delta\text{pan}/\Delta\text{tilt}$ range), while $x_{\text{Out}}^{(k)}$ and $y_{\text{Out}}^{(l)}$ form an equally spaced rectangular grid covering the whole output image.

To each point $\mathbf{p}_{ijkl}$, a so-called "cumulator unit" C_{ijkl} is attached. Such a unit is basically a single-band image with the same size as the input image. Each "pixel" of this unit can hold any positive integer value including zero. They are used to collect candidate pixels for the MM output $(\widehat{x}_{\text{In}}, \widehat{y}_{\text{In}})$.

5.1.3.4 Learning process

The goal of the learning process is to accumulate activations in the cumulator units. At the beginning, all pixels of these units are set to zero. In each learning trial, the pan-tilt unit is first moved into a random (pan,tilt) position. The input image for the FM is recorded and processed. Afterwards, the algorithm iterates through all points of the grid $\mathbf{P}$, the corresponding motor command is executed (relative to the initial random position), and the output image is generated from the camera image after the movement.[4] For each point $\mathbf{p}_{ijkl}$, the intensity of the output image at the coordinates $(x_{\text{Out}}^{(k)}, y_{\text{Out}}^{(l)})$ is compared to the intensities of all pixels $(x_{\text{In}}, y_{\text{In}})$ in the current input image. Whenever the intensity difference is below a certain threshold α, the value of pixel $(x_{\text{In}}, y_{\text{In}})$ in cumulator unit C_{ijkl} is increased by one. The intensity difference is computed as Euclidean distance in RGB color space. The threshold α is set to 3.5% of the overall intensity range in a single color channel. In the present study, 100 trials were carried out, each with $11 \times 11 \times 13 \times 13 = 20449$ iteration steps (size of the grid $\mathbf{P}$).[5] In each trial, the initial camera position was varied, resulting in different input images.

Figure 5.4 illustrates four final cumulator units C_{ijkl} in the grid $\mathbf{P}$. Their positions along the Δpan and Δtilt dimensions are marked on the two-dimensional grid on the left (camera movements to the lower right of increasing length, starting at position 1 with zero movement). Their position $(x_{\text{Out}}^{(k)}, y_{\text{Out}}^{(l)})$ in output

[4] To save time and effort, $\Delta\text{pan}^{(i)}$ and $\Delta\text{tilt}^{(j)}$ are iterated in the outer loops so that every distinct motor command in the grid has only to be carried out once.

[5] Because of the symmetric effect of $\pm\Delta$pan and $\pm\Delta$tilt, the effective number of iteration steps could be quadruplicated to 81796 by varying the sign of Δpan and Δtilt systematically while mirroring the input image accordingly — without carrying out any additional actual camera movements.

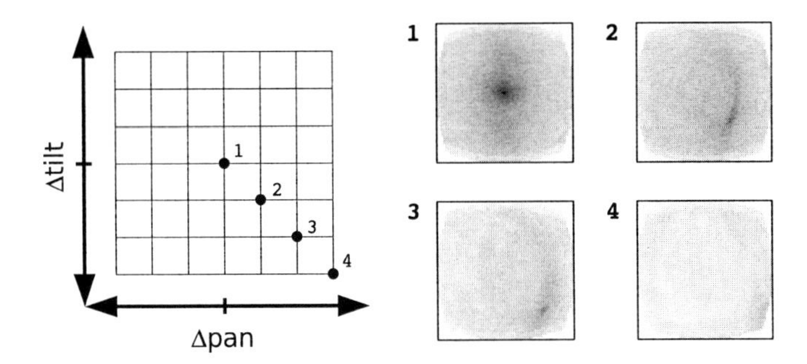

Figure 5.4 — Cumulator units for the center pixel for four different (Δpan,Δtilt) positions. All depicted cumulator units were normalized by the same scaling factor so that a pixel value of zero corresponds to white and the overall maximum pixel value to black (adapted and updated from Schenck and Möller, 2007, © Springer).

image coordinates is the center pixel. The pixel color in the cumulator units reflects the size of the accumulated sum from white (zero) to black (maximum sum). Unit 1 with zero camera movement shows a clear maximum exactly in the center. Thus, the most likely origin of the center pixel in the output image is the center pixel in the input image. This is exactly what is expected when no camera movement takes place. Unit 2 is associated with a small camera movement to the lower right. The intensity maximum is no longer in the center of the unit, but in the lower right corner: When the camera moves into a certain direction, the new image center has its origin in the direction of the movement. Because of the retinal mapping, the intensity maximum moves a large distance towards the border of the cumulator unit although the corresponding camera movement is rather small. Unit 3 with a larger camera movement shows a similar effect. Moreover, its maximum intensity is obviously weaker than in unit 1. This is mainly caused by the retinal mapping with its heavy subsampling in the outer image regions (causing fewer matches with the correct candidate pixel). Finally, unit 4 shows no visible maximum in print at all. Actually, the corresponding camera movement is so large that the center pixel of the output image has no valid counterpart in the input image, therefore it is unpredictable.

5.1.3.5 Generating a raw version of the MM and VM

After the cumulator units have been acquired in the learning process, raw versions of the MM and VM can be created whose output is defined at the grid positions $\mathbf{p}_{ijkl}$ in input space. The output $(\widehat{x}_{\mathrm{In}}, \widehat{y}_{\mathrm{In}})$ of the MM at grid point

$\mathbf{p}_{ijkl}$ are the coordinates of the pixel with maximum intensity in the cumulator unit C_{ijkl}. The outputs $v_{(x_{\text{Out}},y_{\text{Out}})}$ of the VM at point $\mathbf{p}_{ijkl}$ is set to 1 (signaling valid output of the MM at this point) whenever the maximum pixel intensity in unit C_{ijkl} is above a certain threshold. Otherwise, $v_{(x_{\text{Out}},y_{\text{Out}})}$ is set to 0. The threshold is computed as the product of the maximum pixel intensity of all cumulator units and a factor $\beta = 0.41$. This proved to be the value resulting in the most correct separation.

Figure 5.5 shows the output of the MM and VM for 6×6 different motor commands ($\Delta\text{pan}^{(i)}$,$\Delta\text{tilt}^{(j)}$). For each motor command, the pixel coordinate space of the input image is shown in a single panel. The two-dimensional grid in each panel connects points along the $x_{\text{Out}}^{(k)}$ and $y_{\text{Out}}^{(l)}$ directions of $\mathbf{P}$. The position of each grid point corresponds to the output $(\widehat{x}_{\text{In}}, \widehat{y}_{\text{In}})$ of the MM at this point. Only points with valid output are shown (determined by the VM). The lower right panel with no movement shows an identity mapping between $\left(x_{\text{Out}}^{(k)}, y_{\text{Out}}^{(l)}\right)$ and $(\widehat{x}_{\text{In}}, \widehat{y}_{\text{In}})$ (as expected). The other panels reflect the relationship between the camera movement and the pixel shift between input and output image. The strong curvature of the grid is mainly caused by the retinal mapping.

5.1.3.6 Network training

The output of the raw versions of the MM and the VM is only defined at the grid points $\mathbf{p}_{ijkl}$. To get the output in-between, function interpolation is necessary. For this purpose, the raw versions of the MM and the VM were replaced by radial basis function networks (RBFN) (for details, see Sect. 3.2) in the final step of the learning algorithm. These networks have the same input-output structure as the MM and the VM, respectively (see Fig. 5.2). The training data for both networks was generated from the output of the raw versions of the MM and the VM at the grid points $\mathbf{p}_{ijkl}$ (overall, there are $11 \times 11 \times 13 \times 13 = 20449$ grid points). For the MM network, training data was restricted to the 10523 grid points with valid output (as indicated by the raw version of the VM).

Both the MM and the VM network were initialized with the K-means algorithm, the weights between the layer with Gaussians and the output layer were computed by a standard pseudoinverse technique (Moody and Darken, 1989). The variances of the Gaussian units were determined with the parameters $\alpha_\sigma = 1.0$ and $\beta_\sigma = 0.1$ (see Table 3.2 for the parameter definitions). Input and output values were scaled to the range $[-0.6; 0.6]$.

The MM network is an RBFN with 1500 Gaussians for each output unit (x_{Out} and y_{Out}). The training set consisted of the 10523 valid input-output pairs of the raw MM. The mean squared error per pattern per output unit amounted to $6.1 \cdot 10^{-5}$ after network adaptation. The VM network has 1500 Gaussians in the

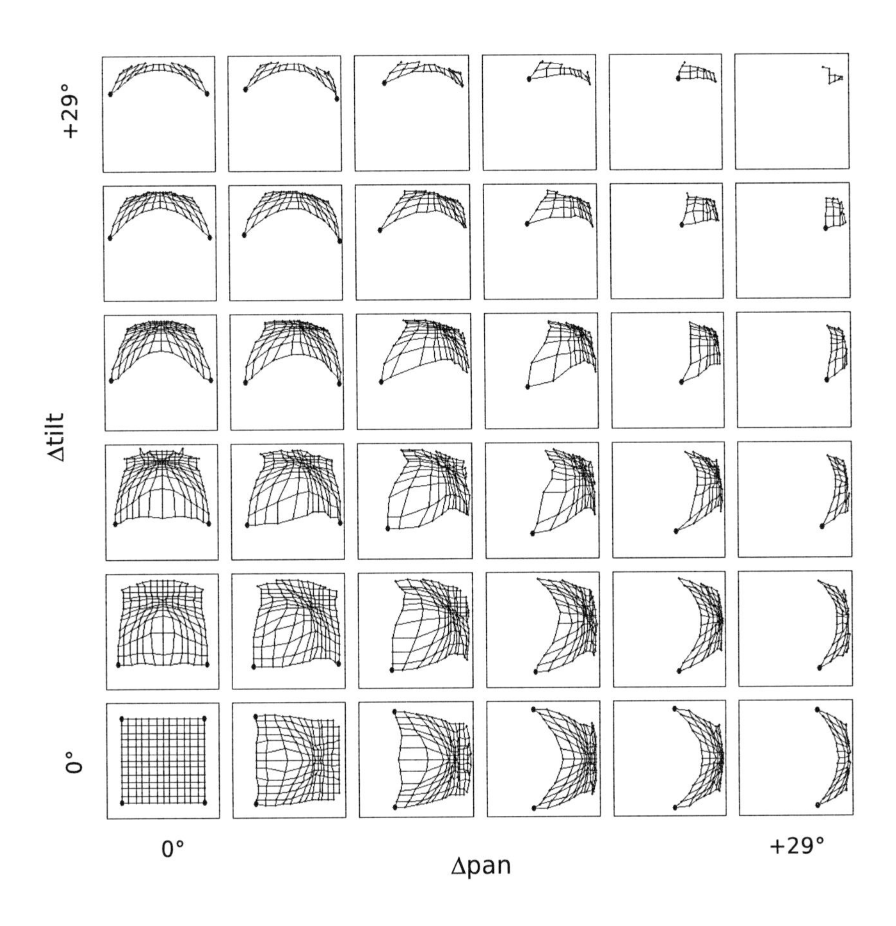

Figure 5.5 — Mapping from pixel coordinates $(x^{(k)}_{\text{Out}}, y^{(l)}_{\text{Out}})$ (grid points) in the output image to pixel coordinates $(\widehat{x}_{\text{In}}, \widehat{y}_{\text{In}})$ in the input image for 6×6 different (Δpan,Δtilt) positions. Only the upper right corner of the grid in the motor subspace is shown since the effects of ±Δpan and ±Δtilt are mirror-symmetric. The overall grid contains 11×11 different (Δpan,Δtilt) positions.

hidden layer for its single output unit. It basically had to learn a classification task with a training set covering all 20449 grid points. While the mean squared error per pattern per output unit still amounted to $5.2 \cdot 10^{-2}$ after network adaptation, only 1.1% of the grid points were misclassified.

It is possible to use alternative methods for function interpolation, e.g., to construct the RBFNs directly from the grid points without learning (even during the acquisition of the cumulator units as a kind of "online" method), or to use other non-linear regression methods.

5.1.4 Results

The MM and VM network are used to implement the overall visual FM for predicting the output image as explained in Sect. 5.1.2. Especially, non-predictable regions of the output image are marked by the VM network. The prediction works rather precise as shown exemplary in Fig. 5.6. The actual and the predicted output image are compared for four different motor commands (Δpan,Δtilt) (camera movements to the lower right of increasing length as in Fig. 5.4). Moreover, the region of each output image which is marked as non-predictable by the VM network is shown in black color in the third row of images. The input image (the same for all four movements) is displayed as well. Movement 1 is a zero movement. The actual and the predicted output image are very similar and show the center crop from the input image. Movements 2 and 3 are of increasing size. The non-predictable regions mask parts of the output images which have no correspondence in the input image. The center of the predicted images is slightly blurred and distorted because the mapping generated by the MM network has to enlarge a region of a few pixels in the input image to a much larger area (especially for movement 3). Movement 4 is so large that the center of the output image is non-predictable. Nevertheless, the small upper left part of the output image which is predicted corresponds closely to the actual output.

This visual inspection of a few exemplary camera movements demonstrates the learning success. At the current stage of development, the additional application of quantitative evaluations is not useful because of the lack of competing learning algorithms for visual FMs. Furthermore, quantitative measures like the Euclidean distance in pixel space are difficult to interpret because the FM has to enlarge parts of the input image while the actual output maintains the optimum resolution in the image center. We pointed out in Sect. 5.1.3.1 that depth information is irrelevant for our learning task because of the camera geometry. Therefore, it is possible to rearrange objects in the field of view of the camera without any harm to the prediction performance of the visual FM.

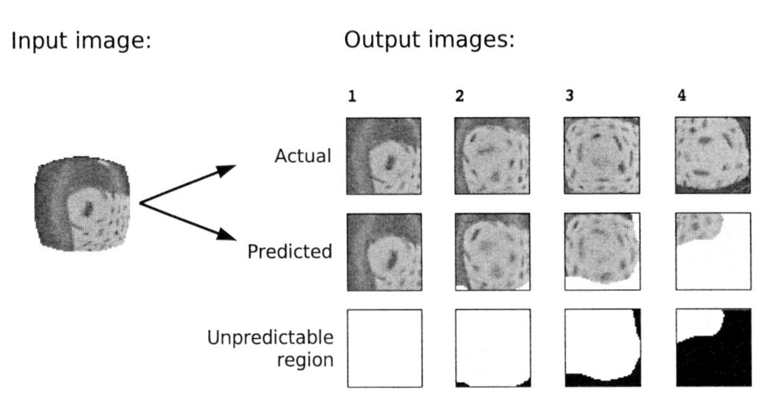

Figure 5.6 — Comparison of actual and predicted output images at four different (Δpan,Δtilt) positions (the same as in Fig. 5.4) (adapted and updated from Schenck and Möller, 2007, © Springer).

5.1.5 Discussion and conclusions

The proposed learning algorithm for visual FMs overcomes the problem that these models have a high-dimensional input and output space due to the size of visual data. Forecasting pixel intensities is replaced by forecasting a mapping between output and input pixel locations. The only restriction regarding image size during the learning process is imposed by the size of the computer memory because it has to hold the cumulator units during the learning process. After learning, the acquired mapping in the MM network can be applied to images of arbitrary size.

The learning process relies on matching pixels between the output and input image. By imposing a retinal mapping, it is demonstrated that this learning principle even works when strong image distortions are involved (including color changes caused by smoothing and subsampling in the outer areas of the camera images). Future research will reveal to which extent the performance of the learning algorithm deteriorates in response to even more ambiguous visual data (e.g., by using monochrome images). The distinction between cumulator units with a large and a small maximum pixel intensity offers a natural solution for the detection of unpredictable image regions. A small maximum indicates that no correct pixel match exists, while an existing correct match accumulates to a large maximum during the learning process.

At the current stage of development, the application of a grid of cumulator units spanned in the input space of the MM and VM only allows for low-dimensional motor commands $\mathbf{m}_t$ because of the storage requirements of these units. To overcome this problem, the next step of development is an online learning scheme to adapt to the maximum (the modal value) of the intensity distribution in each cumulator unit without the need to store the distribution. Preliminary results of this approach are presented in the next section (Sect. 5.2). This would make it possible to extend the scheme towards more dimensions in motor space. Even further, the goal is to replace the fixed grid structure in motor space by random movements (while maintaining the grid in $(x_{\mathrm{Out}}, y_{\mathrm{Out}})$ space with the appropriate spacing for the distortions caused by the imaging system). In addition, the learning algorithm in its current form is limited insofar as it cannot be applied to setups in which depth information is relevant for precise prediction, e.g. a camera mounted on a mobile robot or on a pan-tilt unit whose axes are too far away from the nodal point of the camera-lens system. In setups like this, it is necessary to determine the depth information and to incorporate it into the learning algorithm. Future work will address this issue.

The visual FM of this study belongs to the class of anticipatory mechanisms which generate sensory anticipations (in contrast to the prediction of future system states). The FM works at the lowest level of abstraction by predicting direct sensor output (in the model: continuous pixel intensities in RGB color space of a retinal image). It remains an open question at which level visual FMs work in biological organisms. Studies on predictive remapping (see Sect. 1.3.4.2) discovered neurons which shift their visual receptive fields in anticipation of an upcoming saccade in the superior colliculus (Walker et al., 1995), in the lateral intraparietal area (Duhamel et al., 1992), and in the frontal eye field (Umeno and Goldberg, 1997). Overall, it seems to be very likely that predictive remapping is an important brain mechanism in the context of visual anticipation.

The basic ideas of the proposed learning algorithm might offer an explanation for the acquisition of visual FMs in biological organisms: first, learning the input-output relationship by matching low-level visual features, and second, identifying predictable regions by detecting that a good match emerges during learning. It might be a worthwhile future research project to combine this idea with the concept of shifting visual receptive fields as in predictive remapping.

In robot models of sensorimotor processing, visual FMs can be used to explore the various functions of FMs stated in Sect. 1.5. In the following chapter (Chapt. 6), we will suggest a model of grasping to extrafoveal targets which is built around the visual FM of this study. Moreover, we will propose how the visual FM could be used in the field of saccade learning. These models provide insights about possible functional principles on an abstract modeling level. Moreover, for robotics applications, visual FMs may become an important building block of truly autonomous systems, both for motor control and for perceptual competences.

5.2 Replacing Cumulator Units by Online Learning

In the learning algorithm for the visual FM in the preceding section, the cumulator units basically serve the purpose of determining the modal value of the distribution of the matching pixels. This a kind of batch learning scheme: First, a data distribution is sampled, and all encountered examples are stored, and afterwards all stored data points are used to determine the modal value. However, it would be more elegant to use an online learning scheme for modal values without the need to store the learning examples. For other statistical values like the mean and the median value, this is possible by using an appropriate error function and gradient descent. Unfortunately, such a straightforward approach does not work for modal values. In the following, we will first present the learning of the mean and the median value for illustrative purposes; afterwards, we will develop a novel learning algorithm for modal values.

5.2.1 Learning of the mean and the median value

The starting point of the following considerations is a function approximator whose parameters are adapted by gradient descent along an error function (like the MLP in Sect. 3.1). Usually, a function approximator has to learn a mapping between many different pairs of input and output values. For each distinct input, there is one correct training output. For simplicity, we assume an unspecific single fixed input instead. Moreover, the task of the approximator is to learn either the mean or the median value of a distribution of corresponding output data points. In practice, such a distribution may originate in noise which scatters around the "true" output value. The function approximator has a single output $\widehat{s}$, output data points are denoted as s_i.[6] i is the index of the learning cycle, $\widehat{s}_i$ is the output of the approximator in cycle i.

Mean value The correct error function for the learning of the mean value is

$$E = \frac{1}{2}(\widehat{s}_i - s_i)^2 , \tag{5.1}$$

as is proven in the following. The gradient of this error function with regard to the approximator output amounts to

$$\frac{\partial E}{\partial \widehat{s}_i} = \widehat{s}_i - s_i$$

[6] s is used as symbol for an arbitrary data point within this section to avoid confusion with the image coordinates x and y.

(generally, this gradient is used to compute the gradients of the adaptable parameters of the approximator by the chain rule afterwards). After a sufficient number of learning cycles, the approximator output has converged to an (approximately) stable value $\widehat{s}$, and the expected value of this gradient amounts to zero (otherwise learning would not have converged yet):

$$\begin{aligned}\left\langle \frac{\partial E}{\partial \widehat{s}} \right\rangle &= \lim_{n\to\infty} \frac{1}{n} \sum_{i=1}^{n} (\widehat{s} - s_i) \stackrel{!}{=} 0 \\ \Leftrightarrow \lim_{n\to\infty} \frac{1}{n} \sum_{i=1}^{n} \widehat{s} &= \lim_{n\to\infty} \frac{1}{n} \sum_{i=1}^{n} s_i \\ \Leftrightarrow \widehat{s} &= \bar{s}_i\end{aligned}$$

Thus, the output of the function approximator after learning is the mean value of the output data distribution. If one deals with noisy data and can savely assume that the noise distribution is symmetric, the error function in Eqn. (5.1) yields the desired learning results.

Median value The correct error function for the learning of the median value is

$$E = |\widehat{s}_i - s_i| \, . \tag{5.2}$$

The proof starts again by determining the gradient:

$$\frac{\partial E}{\partial \widehat{s}_i} = \begin{cases} -1 & , \widehat{s}_i < s_i \\ +1 & , \widehat{s}_i > s_i \\ \text{nd.} & , \widehat{s}_i = s_i \end{cases}$$

Again, after the approximator output has converged to an (approximately) stable value $\widehat{s}$, the expected value of this gradient amounts to zero (we ignore the non-defined case here):

$$\left\langle \frac{\partial E}{\partial \widehat{s}} \right\rangle = \lim_{n\to\infty} \frac{1}{n} \sum_{i=1}^{n} \begin{cases} -1 & , \widehat{s} < s_i \\ +1 & , \widehat{s} > s_i \end{cases} \stackrel{!}{=} 0$$

For this equation to be fulfilled, the count of -1 and $+1$ in the sum has to be equal, thus $\widehat{s}$ is as often larger than s_i as it is smaller. In conclusion, $\widehat{s}$ is the median value of the distribution of data points s_i.

For noisy data with asymmetric noise distributions, learning the median value is more appropriate than learning the mean value. Therefore, the error function in Eqn. (5.2) is advantageous for such data. Moreover, the median value is less sensitive to outliers than the mean value. On the downside,

Eqn. (5.2) is at a general disadvantage compared to Eqn. (5.1), because the derived gradient does not take the size of the difference $\widehat{s}_i - s_i$ into account. This could result in slower learning if the approximator output $\widehat{s}_i$ is way off in the beginning of the learning process.

General remark These derivations apply as well to function approximators with multiple outputs since each output dimension can be treated on its own in the same way as shown for a single output.

5.2.2 Learning of the modal value

Algorithm To the best of the author's knowledge, there is no way to impose learning of the modal value by using gradient descent. The learning algorithm which is presented in this section is based on a very different idea: to fit an hyperellipsoid to the data distribution so that the center of this hyperellipsoid is approximately located at the modal value. The hyperellipsoid is defined by its center $\mathbf{c}$ and the covariance matrix $\mathbf{\Sigma}$.[7] It is acquired by an online learning rule which is directly related to the usual "offline way" of computing the mean value and the covariance matrix of a data set:

$$\mathbf{c}_{i+1} \leftarrow (1-\eta_1)\mathbf{c}_i + \eta_1 \mathbf{s}_i \tag{5.3}$$

$$\mathbf{\Sigma}_{i+1} \leftarrow (1-\eta_2)\mathbf{\Sigma}_i + \eta_2(\mathbf{c}_i - \mathbf{s}_i)(\mathbf{c}_i - \mathbf{s}_i)^T \tag{5.4}$$

i is the index of the learning cycle, $\mathbf{s}_i$ is the data point which is sampled in cycle i. η_1 and η_2 are learning parameters, which are determined in the following way in each cycle i:

$$\eta_1 = \widetilde{\eta_1}\, G_{(\mathbf{c}_i;\mathbf{\Sigma}_i)}(\mathbf{s}_i) \tag{5.5}$$

$$\eta_2 = \widetilde{\eta_2}\, G_{(\mathbf{c}_i;\mathbf{\Sigma}_i)}(\mathbf{s}_i) \tag{5.6}$$

$G_{(\mathbf{c}_i;\mathbf{\Sigma}_i)}$ is the Gaussian which corresponds to the hyperellipsoid:

$$G_{(\mathbf{c}_i;\mathbf{\Sigma}_i)}(\mathbf{s}_i) = \exp\left(-\frac{1}{2}(\mathbf{c}_i - \mathbf{s}_i)^T \mathbf{\Sigma}_i^{-1}(\mathbf{c}_i - \mathbf{s}_i)\right)$$

$\widetilde{\eta_1}$ and $\widetilde{\eta_2}$ in Eqns. (5.5-5.6) are fixed learning rates. Computing the actual learning rates η_1 and η_2 by Eqns. (5.5-5.6) has the following effects in combination with Eqns. (5.3-5.4):

[7] The eigenvectors $\mathbf{w}_j$ and eigenvalues λ_j of $\mathbf{\Sigma}$ determine the shape of the hyperellipsoid: The axes of the hyperellipsoid point into the directions of the eigenvectors, and the length of each axis is equal to the corresponding $\sqrt{\lambda_j}$. In the context of principal component analysis (see Sect. 3.4), λ_j is interpreted as the variance of the data distribution in the direction of the corresponding eigenvector $\mathbf{w}_j$.

- The smaller the distance $||\mathbf{c} - \mathbf{s}_i||$, the larger the influence of the data point $\mathbf{s}_i$ on the update of the hyperellipsoid.
- During learning, the hyperellipsoid gets smaller and smaller, because data points which are close to the center have a large impact in Eqn. (5.4) because of η_2.
- From the region close to the modal value of the data distribution, more data points are sampled than from other regions; thus, even if the hyper-ellipsoid is relatively far away from this region, the modal value attracts the center of the hyperellipsoid because of the frequency of updates from this region.
- Furthermore, the hyperellipsoid gets a shape, in which it is elongated towards the region around the modal value, while it shrinks in the other directions; accordingly, the actual learning rates η_1 and η_2 stay large for data points which are sampled from the region around the modal value.
- The closer the center of the hyperellipsoid gets to the modal value, the easier the hyperellipsoid can shrink, which in return reduces the influence of data points which are not close to the modal value.
- Finally, the hyperellipsoid ideally settles at the modal value and shrinks to an extremely small size.

For the initialization of $\mathbf{c}_0$ and Σ_0 it proved to be a good approach in our tests to place $\mathbf{c}_0$ at the center of the likely range of the data distribution; the hyperellipsoid is aligned initially with the coordinate system and is defined by rather large variance values such that it covers a large proportion of the likely range of the data distribution.

Experiments In the following, it is shown how the learning algorithm performs on four different cumulator units which were generated in the previous study (as described in Sects. 5.1.3.3/5.1.3.4). The pixel intensities in these units are interpreted as (unnormalized) density function which defines a two-dimensional data distribution over the input pixel position space. In Fig. 5.7, these units are shown with their respective mean, median, and modal values. The two-dimensional mean and median values are composed from the particular values for each single dimension, which are computed from the respective marginal distributions. The modal values mark the density maximum in the two-dimensional data space. Unit A has a clear modal value in the center which is identical to the mean and the median value. In unit B, the modal value is still

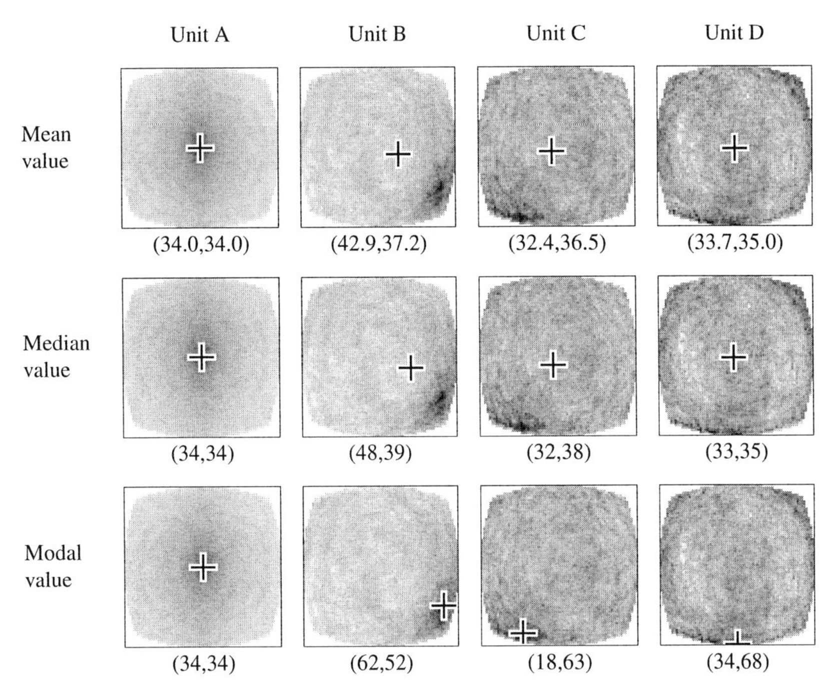

Figure 5.7 — The mean, the median, and the modal value are marked with a cross in four different cumulator units. Below each image, the respective value is reported in pixel coordinates (running from 0 to 68 in each dimension). The density maxima in each cumulator unit are shown in black color, the minima in white color.

clearly identifiable, but shifted towards the lower right corner and no longer identical to the mean and the median value. In unit C, the density distribution is rather uniform with a lot of local density maxima. The overall density maximum is at the lower left corner very close to the border. Unit D is even worse: The overall density maximum is only at a small peak at the lowermost position. The overall density distribution has many other maxima which are nearly as strong.

In the experiments, 3000 data points are drawn at random from the data distribution in each cumulator unit and fed to the online learning algorithm for the modal value. The fixed base learning rates are $\widetilde{\eta}_1 = 0.5$ and $\widetilde{\eta}_2 = 0.2$. The course of learning is illustrated in Fig. 5.8. For each unit, the shape and position of the ellipsoid is shown after different numbers of learning cycles. Moreover, the data points which have been drawn from the distribution so far are depicted. For units A to C, learning is successful — the difference between

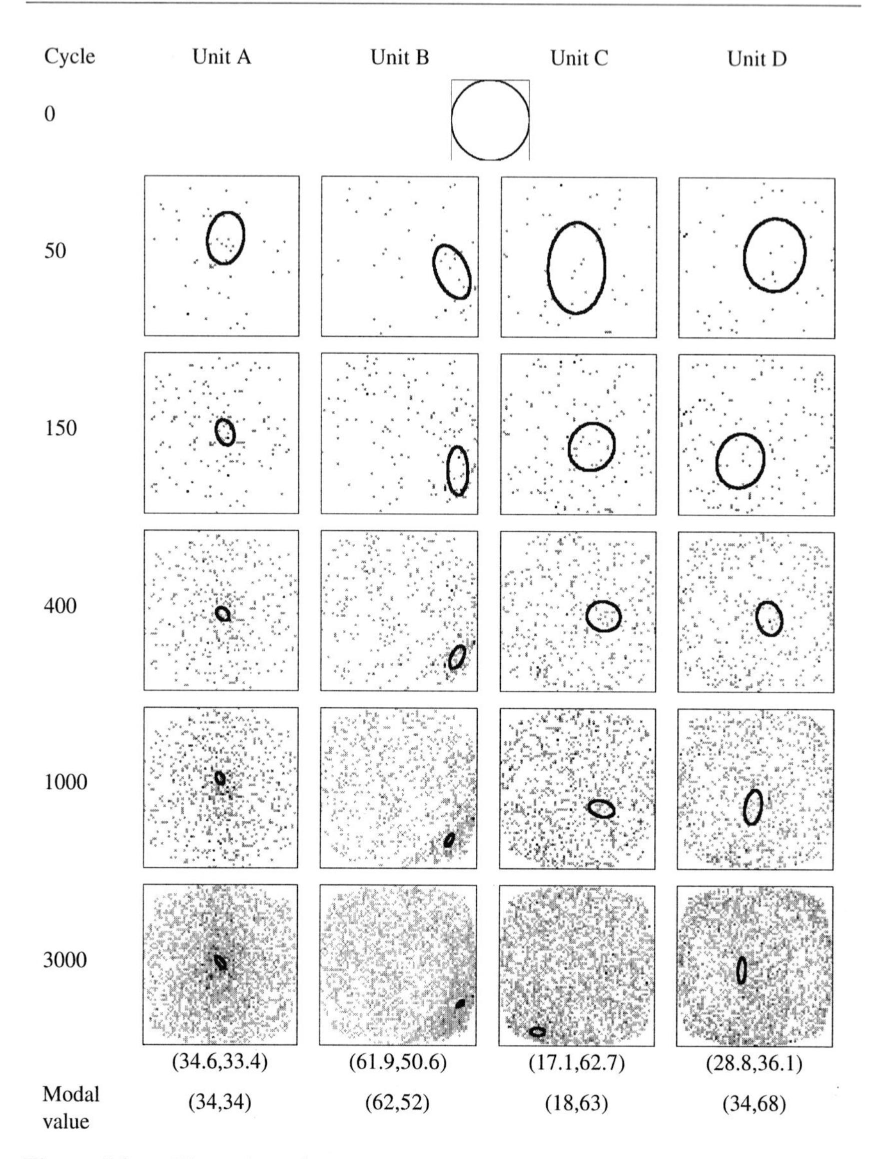

Figure 5.8 — Illustration of the learning process for the four cumulator units. For each reported learning cycle and unit, the respective shape and position of the ellipsoid is depicted (the initial ellipsoid is only shown once for all units and in smaller size). Moreover, the data points which have been sampled so far are shown as well (black indicates data points with the maximum sampling frequency, white with the minimum sampling frequency for the respective cycle and unit). For the last cycle, the position of the center of the ellipsoid is reported (all values in pixel coordinates running from 0 to 68).

the final position of the center and the modal value amounts to less than two pixels in each dimension (in pixel coordinates running from 0 to 68 for the cumulator units). The larger the number of learning cycles, the smaller the ellipsoids become. In unit C, the ellipsoid moves first into the wrong direction and nearly settles down on a local density maximum (in cycle 1000), but finally it gets the "kick" to move on to the global density maximum (in cycle 3000). For unit D with the most disadvantageous data distribution, learning gets stuck before the modal value is reached. However, at least the shape of the ellipsoid is oriented into the right direction, illustrating exemplary this feature of the learning algorithm (as does the ellipsoid for unit B in cycle 50). Overall, the learning examples in Fig. 5.8 demonstrate clearly that the proposed learning scheme converges towards the modal values and not to the mean or median values which are depicted for comparison in Fig. 5.7.

A quantitative analysis yields the following results: Over 10 learning passes, the average Euclidean distance between the center of the ellipsoid and the modal value after 1000 learning cycles amounts to 2.3 (in pixel coordinate space) for unit A, to 2.7 for unit B, to 14.2 for unit C, and to 22.7 for unit D. After 3000 cycles, the distance values for unit A and B improve to 1.5 and for unit C to 7.4, while the distance value for unit D stays at 19.6. The worse performance of unit C compared to unit A and B is mainly caused by one completely failed learning pass in which the ellipsoid settles down far away from the modal value of unit C. Without this failed pass, the average distance value for unit C after 3000 learning cycles amounts to 2.5.

Conclusion The experimental results demonstrate clearly that the proposed learning algorithm can be successfully applied to the identification of modal values, at least for two-dimensional data distributions like the cumulator units. However, it is not guaranteed that the proposed learning algorithm converges to the modal value. On rather uniform data distributions with only a small density peak like in unit D, learning can get stuck with a shrinked hyperellipsoid before the modal value has been reached. Moreover, on data distributions with many different modal values, learning may not end up at the modal value with the highest density.

More thorough tests on data sets of higher dimensionality have to be carried out in the future. Furthermore, in contrast to function approximators which map different input values to different output values, the proposed algorithm works so far without any additional input data, thus it cannot adapt to modal values which change depending on the input. To overcome this limitation, one has to find ways to combine the algorithm with function approximator techniques. A promising candidate for the underlying function approximator is "supervised growing neural gas" (Fritzke, 1998), because it supports single units between which interpolation takes place; these single units could be used to encode the parameters of the hyperellipsoids.

5.2.3 Learning of the raw MM and VM with the new modal value algorithm

The proposed online learning scheme can be directly used to replace the cumulator units in the visual FM which has been presented in the first part of this chapter. The FM is composed from a mapping model (MM) and a validator model (VM). In the following, we will describe a new way to learn the "raw versions" of the MM and the VM (see Sect. 5.1.3.5). The raw MM and the raw VM are only defined at the points of the grid **P** which is inscribed in the input space of the MM and the VM. In Sect. 5.1.3.3, a cumulator unit was attached to each grid point. Now this cumulator unit is replaced by an ellipsoid, and the center of the ellipsoid defines the output of the raw MM at the particular grid point.

During the learning process of the raw MM and the raw VM (see Sect. 5.1.3.4), the matching pixels in the input image are determined in each iteration.[8] These pixels are assigned a value of 1, while the other pixels get a value of 0. In the following, this image consisting of 0 and 1 pixels is called "matching-pixel image". In the previous version of the learning algorithm for the raw MM/VM, the matching-pixel image was added to the cumulator unit at the respective grid point. In the new version of the learning process, in which the cumulator units are replaced by the ellipsoids, up to 10 of the pixel positions to which a 1 is assigned are drawn at random to update the respective ellipsoid in each iteration.[9] In this way, the ellipsoid converges to the modal value (of the hypothetical cumulator unit) at the particular grid point over many iterations.

Learning the raw VM is more difficult without the cumulator units. Before, its output has been determined on the basis of the maximum pixel intensity in each cumulator unit in relation to the overall maximum pixel intensity. Without cumulator units, this straightforward approach is no longer possible. Instead, each ellipsoid gets its own additional "load" parameter ζ^{Cum}. Each time an ellipsoid is updated by the presentation of a new data point, the current matching-pixel image $\mathrm{I}^{\mathrm{Match}}$ is used to change the load parameter in the following way

[8] In this context, the term "iteration" means a single learning step at a single grid point during an overall learning trial.

[9] The update is restricted to 10 pixels per matching-pixel image to avoid that the ellipsoid gets stuck too early at the wrong position because of a non-representative pattern in a single matching-pixel image.

(with $\zeta_0^{\mathrm{Cum}} = 0$):

$$\zeta_{i+1}^{\mathrm{Cum}} \leftarrow \zeta_i^{\mathrm{Cum}} + \sum_{(x,y)\in \mathrm{I}^{\mathrm{Match}}} \left[N_{(\mathbf{c}_i;\mathbf{\Sigma}_{\mathrm{Cum}})}\left(\begin{pmatrix} x & y \end{pmatrix}^T \right) \cdot \mathrm{I}_{(x,y)}^{\mathrm{Match}} \right] \tag{5.7}$$

i is the update cycle of the ellipsoid, $\mathbf{c}_i$ is the current center of the ellipsoid, x and y are the horizontal and vertical pixel coordinates in the matching-pixel image (the sum is computed over all of its pixels), $\mathrm{I}_{(x,y)}^{\mathrm{Match}}$ is the intensity of a single pixel in the matching-pixel image (0 or 1), and $N_{(\mathbf{c}_i;\mathbf{\Sigma}_{\mathrm{Cum}})}$ is a normalized Gaussian distribution:

$$N_{(\mathbf{c}_i;\mathbf{\Sigma}_{\mathrm{Cum}})}(\mathbf{s}) = \frac{1}{2\pi\sqrt{|\mathbf{\Sigma}_{\mathrm{Cum}}|}} \exp\left(-\frac{1}{2}(\mathbf{c}_i - \mathbf{s})^T \mathbf{\Sigma}_{\mathrm{Cum}}^{-1} (\mathbf{c}_i - \mathbf{s}) \right)$$

In Eqn. (5.7), s is the pixel coordinate vector $\begin{pmatrix} x & y \end{pmatrix}^T$. $\mathbf{\Sigma}_{\mathrm{Cum}}$ is a fixed diagonal matrix

$$\mathbf{\Sigma}_{\mathrm{Cum}} = \begin{pmatrix} \sigma_{\mathrm{Cum}}^2 & 0 \\ 0 & \sigma_{\mathrm{Cum}}^2 \end{pmatrix}$$

with σ_{Cum} being set to 0.16 times the width of the (quadratic) matching-pixel image. Basically, these equations define the convolution of a narrow normalized Gaussian, which is positioned at the current center of the ellipsoid, with the matching-pixel image. The result of this computation is cumulated in ζ_i^{Cum} from update cycle to update cycle. Finally, after the last update cycle $i_{\max}$ of the particular ellipsoid, the final ζ value is computed as $\zeta = \zeta_{i_{\max}}^{\mathrm{Cum}} / i_{\max}$.

ζ^{Cum} grows the faster the larger the density maximum around the center of the ellipsoid is. In this way, one can estimate the "strength" of the modal value without storing the cumulator unit. After all learning trials of the raw MM and VM have been finished, the output of the VM is determined in the following way: The output $v_{(x_{\mathrm{Out}},y_{\mathrm{Out}})}$ of the VM at a particular grid point is set to 1 (signaling valid output of the MM at this point) whenever the ζ value of the ellipsoid at this point is above a certain threshold. Otherwise, $v_{(x_{\mathrm{Out}},y_{\mathrm{Out}})}$ is set to 0. The threshold is computed as the product of the maximum ζ value of all ellipsoids and a factor β.

The learning results for the raw MM and VM, which are shown in Fig. 5.9 (in analogy to Fig. 5.5 for the algorithm with the cumulator units), were generated with a grid with 7×7 positions in motor space and 11×11 positions in output pixel position space. 100 learning trials were carried out, the parameter β was chosen as $\beta = 0.012$. Figure 5.9 shows the output of the raw MM and VM for 4×4 different motor commands (Δpan,Δtilt) on the basis of the described online learning scheme for modal values. For each motor command, the pixel coordinate space of the input image is shown in a single panel. The

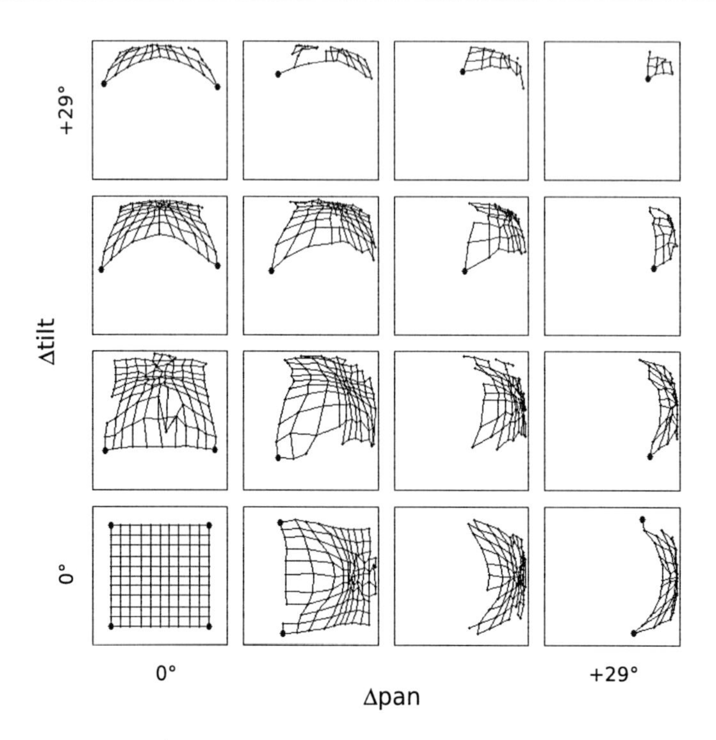

Figure 5.9 — Mapping from pixel coordinates $(x_{\text{Out}}^{(k)}, y_{\text{Out}}^{(l)})$ (grid points) in the output image to pixel coordinates $(\widehat{x}_{\text{In}}, \widehat{y}_{\text{In}})$ in the input image for 4×4 different (Δpan,Δtilt) positions. Only the upper right corner of the grid in the motor subspace is shown. The overall grid contains 7×7 different (Δpan,Δtilt) positions. This figure has been generated on the basis of the online learning scheme for modal values without the storage of cumulator units.

two-dimensional grid in each panel connects points along the $x_{\text{Out}}^{(k)}$ and $y_{\text{Out}}^{(l)}$ directions of the grid $\mathbf{P}$. The position of each grid point corresponds to the output $(\widehat{x}_{\text{In}}, \widehat{y}_{\text{In}})$ of the MM at this point. Only points with valid output are shown (determined by the VM).

Overall, the results appear rather similar to the results which are obtained on the basis of the cumulator units (see Fig. 5.5). However, one has to note that the output of the raw MM is erratic at some grid points (most likely because of ellipsoids which got stuck before they reached the modal value). Moreover, the output of the raw VM generates too many false 0 outputs, marking valid MM outputs as non-valid. We expect that the first problem can be solved by linking the learning of neighboring ellipsoids in the grid in some way (similar to a self-organizing map; Kohonen, 1995). The second problem requires further refinement of the calculation of the ζ values. Nevertheless, these results demonstrate clearly that the online learning algorithm for modal values can be successfully applied to real-world problems. For the visual FM, it replaces the cumulator units by ellipsoids with just eight parameters (center, covariance matrix, load value ζ^{Cum}, number of update cycles), in this way resolving the problem of the large storage requirements of the cumulator units. This in turn allows theoretically for motor commands of higher dimensionality for the visual FM.

Chapter 6

Grasping to Extrafoveal Targets

6.1 Introduction

In everyday live, reaching and grasping movements are mainly carried out under visual control. The most important information about the position and shape of target objects is obtained from accompanying eye movements and retinal activation. A considerable amount of research adresses the question how eye and arm movements are coordinated and which information is used at which stage of motor planning and execution (e.g., Bekkering and Sailer, 2002; Frens and Erkelens, 1991; Horstmann and Hoffmann, 2005; Mather and Fisk, 1985; Neggers and Bekkering, 2000; Prablanc et al., 1979). Experimental studies show that saccades for target fixation usually precede arm movements (Abrams et al., 1990; Neggers and Bekkering, 1999; Vercher et al., 1994). Even when the onset time of eye and arm movements is the same, eye movements are finished more rapidly, providing the eye orientation as input for the completion of the arm movement. Nevertheless, as everyday experience shows, it is possible for humans to reach for and grasp objects while the saccade to the target is suppressed. But this ability comes at a price: Several studies have shown that the accuracy of limb movements suffers in such a setting (Abrams et al., 1990; Mather and Fisk, 1985; Prablanc et al., 1979; Vercher et al., 1994). In conclusion, grasping and reaching to both fixated and to non-fixated target objects is possible, although the former allows for more precise arm and hand movements.

In a previous study (Hoffmann et al., 2005), we explored the necessary coordinate transforms for both settings, and presented a computational model for grasping movements with a robot arm. In this chapter, the focus is on the sensorimotor processing for grasping to non-fixated target objects which are projected on the extrafoveal region of the retina. The starting point is the premotor theory of attention (Rizzolatti et al., 1994) which states that spatial attention is a consequence of the preparation of goal-directed, spatially coded movements. Because the neural mechanisms for foveal vision in primates and humans appear to be highly developed, oculomotor maps coding space for eye movements play a central role in selective attention according to this theory. Experimental evidence for the close coupling of saccade preparation and visual attention has been found in several studies, for example in the work of Deubel and Schneider (1996) and Irwin and Gordon (1998). Moreover, there is a considerable overlap between frontoparietal control structures which are activated during covert shifts of visual attention and during saccade preparation, as functional imaging studies have shown (Beauchamp et al., 2001; Nobre et al., 2000; Perry and Zeki, 2000). Muggleton et al. (2003) were able to modulate attentionally guided performance in visual search tasks by transcranial magnetic stimulation over the frontal eye fields. In summary, there is strong experimental evidence for the link between visual attention and saccade preparation. The link between manual response preparation and shifts of spatial attention has been less convincing, but several studies (Deubel et al., 1998; Eimer et al., 2006; Schiegg et al., 2003) provide support for the claim that covert preparation of manual responses is linked to shifts of spatial attention as well.

In this chapter, a computational model of grasping to extrafoveal targets is proposed, which is implemented on a robot setup. This model is based on the premotor theory of attention and adds one specific hypothesis: Attention shifts caused by saccade planning imply a prediction of the retinal images after the saccade. The foveal region of these predicted retinal images is required to determine movement parameters for the manual interaction with objects at the target location of the attention shift.

Without visual prediction, grasping towards extrafoveal target objects is difficult because of the heavy distortions found in retinal images (with the term "retinal image" we refer here to the activation pattern of receptors in the retina). These distortions have at least three distinct sources. First, the retina has approximately the shape of a half-sphere (Atchinson and Smith, 2000). This brings about that the projection of one and the same object on the retina has a different shape, depending on its position relative to the optical axis of the eye. Second, the lens system of the eye suffers from chromatic and monochromatic aberrations in various forms, causing varying image quality (focus, shape of point spread function) throughout the retina (Atchinson and Smith, 2000). And third, the distribution of light receptors (rods and cones) on the retina is non-uniform. Cones are used for color vision under strong light, rods for monochromatic vision under low light levels. The cones are densely packed in the fovea (around 0 degrees eccentricity) with rapidly decreasing density towards the periphery of the eye. The density of rods decreases much slower towards the periphery but they are completely absent from the fovea (see Fig. 6.12 in McIlwain, 1996, p. 93). Because of the non-uniform sensor distribution, the pattern

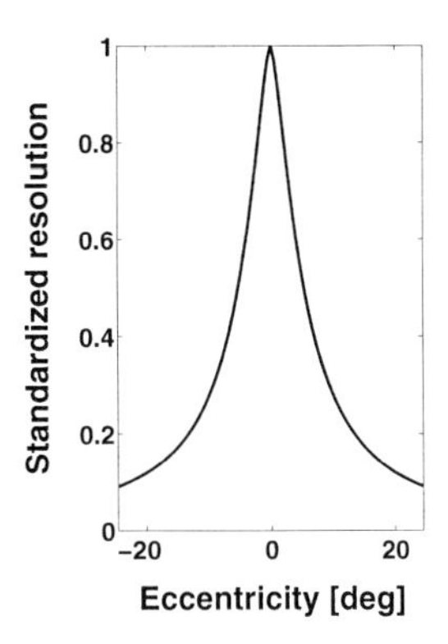

Figure 6.1 — Resolution of the artificial retinal images (as specified in Sect. 5.1.3.2) depending on the distance from the image center. Resolution values are standardized to a maximum of 1.0.

of rod and cone activation caused by the projection of a certain object on the retina varies considerably with the retinal position of this projection. This non-uniformity is also found in the retinotopic maps in the visual cortex (Mallot, 1985).

Considering this background information, a grasping task in which the eyes do not fixate the target object poses a special difficulty because the object-related retinal activation differs depending on the object's position relative to the eyes. Any mechanism which extracts the necessary information for proper grasping (e.g., object orientation) from this activation pattern has be tuned to the exact position on the retina onto which the object is projected. This would cause considerable computational overhead and the need to learn complex input-output relationships between retinal activation and grasping parameters. To avoid this overhead, the system could predict how the foveal representation of the target object would look like after a successful saccade, and use a much simpler sensorimotor model which takes just the predicted foveal activation as input to generate the grasping parameters as output. This model could be the same sensorimotor model as the one which is applied to fixated target objects. Thus, visual prediction would allow to apply one and the same model for the sensorimotor processing for both grasping to foveal and extrafoveal target objects. We hypothesize that such a prediction actually takes place when humans and other primates grasp towards extrafoveal targets. In accordance with the premotor theory of attention, the first step would be that spatial attention is shifted towards the object by preparing the motor command for making a saccade towards this object (but this saccade is never carried out). The second step is to use this saccadic motor command as input for a visual forward model to generate the predicted foveal representation of the target object. In the third step, the planned new eye position and the predicted foveal representation of the target object are provided as input for the sensorimotor model which associates this information with an appropriate motor command for grasping.

For the visual prediction, the visual forward model (FM) from Sect. 5.1 is used. We do not make strong assumptions about the visual representation underlying the prediction. In the robot implementation, the prediction takes place on the level of artificial retinal images (see Sect. 5.1.3.2) which mimic roughly the cone distribution on the human retina (see Fig. 6.1). The important analogy to biological retinal activation patterns is the fact that a target object appears in a different shape depending on its location in the retinal image.

The anticipation of sensory consequences in the nervous system of biological organisms is supposed to be involved in several sensorimotor processes which are outlined in detail in Sect. 1.5 and Chapt. 2. In the model of grasping to extrafoveal targets, the prediction of visual data serves as a replacement for sensory feedback and is used in the planning process for motor control (although it is only a one-step "planning" for the generation of a single movement). The learning of adaptive visual FMs is a rather new field. It is difficult because of the high dimensionality of visual data and because part of the output may be non-predictable. The learning algorithm for visual FMs in Sect. 5.1 overcomes both problems.

In the following, the components of the overall model are explained in detail, and it is described by which learning procedures they are acquired. Afterwards, the final experiments and their results are presented. The purpose of these experiments is to show that a robot implementation of the proposed model is actually capable of grasping to extrafoveal targets. Moreover, we hypothesize that grasping of fixated targets results in slightly better performance than grasping to extrafoveal targets, and that grasping of extrafoveal targets without visual prediction results in low grasping success, illustrating the need for a visual FM. These hypotheses are tested in the experiments.

6.2 Overall System Architecture

6.2.1 Overview

The overall model consists of three parts (see. Fig. 6.2). First, a saccade controller acquired through continuous learning by averaging (CLbA; see

Sects. 2.2.7.2 and 4.1); second, a visual FM predicting retinal images with decreasing image resolution towards the corners in analogy to the sensor distribution on the human retina (see Sect. 5.1); and third, an arm controller for grasping movements which receives the output of the saccade controller and the orientation of the target object as inputs (similar to the controller presented by Hoffmann et al., 2005).

When the model is used for grasping to extrafoveal targets, a single trial starts with the presentation of the grasping target, a red wooden block, at a random location within the working space on a table surface. The cameras are in a random posture. The saccade controller generates the necessary motor command for proper fixation with the cameras, but this movement is not carried out, only the suggested motor command is recorded as input for the visual FM and the arm controller. Afterwards, the visual FM predicts the retinal images after the (hypothetical) saccade. From these predicted images, the orientation of the block is determined. Finally, the arm controller uses both the saccadic motor command and the block orientation in the predicted images as inputs to generate the grasping movement.

In the final experiments, the grasping performance of four different versions of the robot model is compared: (1) for grasping towards target objects which are precisely fixated by a series of saccades, using the actual retinal images instead of the predicted ones; (2) for grasping towards target objects which are fixated just by one saccade, also using the actual retinal images instead of the predicted ones; (3) for grasping towards non-fixated target objects using visual prediction; (4) for grasping towards non-fixated target objects without visual prediction.

6.2.2 Setup

The experimental setup (see Fig. A.1) consists of a robot arm with six rotatory degrees of freedom and two-finger gripper (for details see App. A). Moreover, a robot camera head belongs to the setup as described in Sect. 5.1.3.1 and App. B. For the training of the saccade controller (Sect. 6.3) and the visual FM (Sect. 6.4), not the real setup was used, but instead "virtual" camera movements were carried out using an image database (see Sect. 5.1.3.1).

6.3 Saccade Controller

The saccade controller in this study is very similar to the one in Sect. 4.1. As main difference, it is not trained on the basis of a geometrical model, but in-

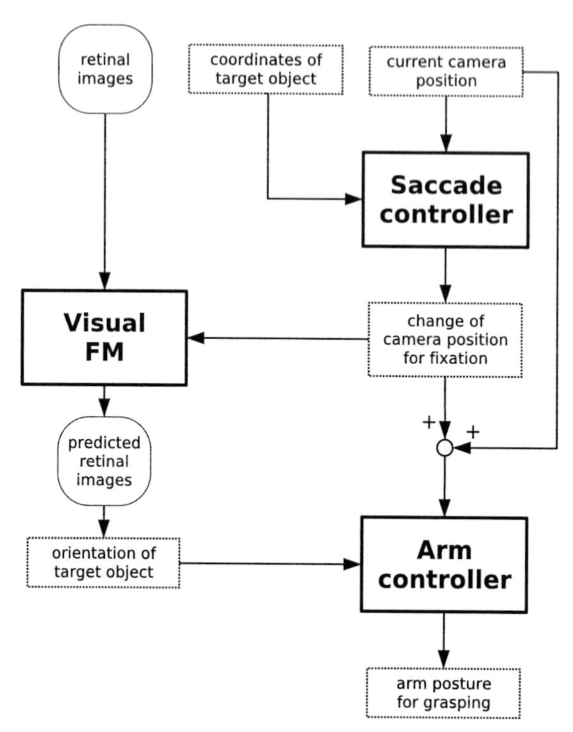

Figure 6.2 — The overall system architecture. The main components of the model are a saccade controller, a visual FM, and an arm controller (for details see text) (adapted from Schenck and Möller, 2007, © Springer).

stead for the real setup. As controller learning strategy, CLbA is applied (see Sect. 2.2.7.2). For convenience, a description of this learning strategy especially for the saccade learning task is provided in the following.

6.3.1 Controller input and output

The task of the saccade controller is to fixate target objects with both cameras so that the target object is projected onto the center of both camera images. In time step t, the saccade controller receives the current sensory state $\mathrm{s}_{\mathrm{SAC}}^{(t)}$ as input[1], composed of a kinesthetic and a visual part (see Fig. 6.3). The

[1] In contrast to Chapts. 2 and 4, the symbol m is used in this chapter to denote motor commands, and the symbol s to denote sensory states. This change is applied to avoid confusion with the image coordinates x and y.

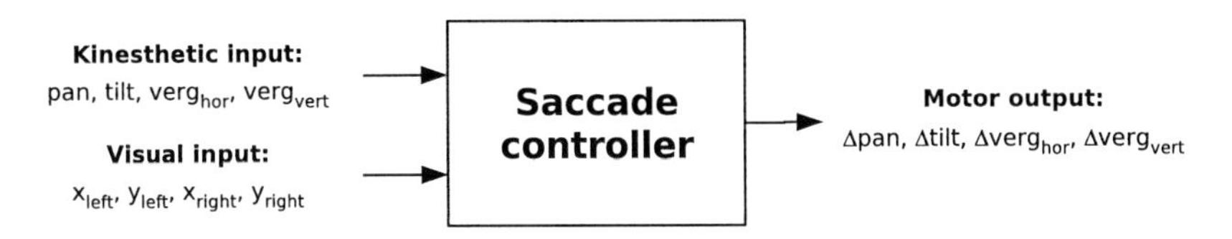

Figure 6.3 — Input and output of the saccade controller (adapted from Schenck and Möller, 2007, © Springer).

kinesthetic input $\mathbf{s}_{\mathrm{KIN}}^{(t)}$ consists of the current position of the cameras, defined by a conjoint pan-tilt direction (pan, tilt), and a horizontal and vertical vergence value ($\mathrm{verg}_{\mathrm{hor}}$, $\mathrm{verg}_{\mathrm{vert}}$). The visual part $\mathbf{s}_{\mathrm{VIS}}^{(t)}$ represents the position of the target object in the left and right camera image relative to the image center: x_{left}, y_{left}, x_{right}, y_{right}. The motor output $\mathbf{m}_{\mathrm{SAC}}^{(t)}$ of the saccade controller is defined as change of the motor position. It consists of four values: Δpan, Δtilt, $\Delta\mathrm{verg}_{\mathrm{hor}}$, and $\Delta\mathrm{verg}_{\mathrm{vert}}$. The new position of the cameras is computed as $\mathbf{s}_{\mathrm{KIN}}^{(t+1)} = \mathbf{s}_{\mathrm{KIN}}^{(t)} + \mathbf{m}_{\mathrm{SAC}}^{(t)}$, and the cameras are moved accordingly. All sensory variables are scaled to the range $[-1; 1]$, the motor output variables to the range $[-2; +2]$.

6.3.2 Image processing

The image processing is restricted to a central area of 213×213 pixels in each camera image. For simplicity, in the following (throughout Sect. 6.3) the term "camera image" refers to this cropped region. The image processing extracts the position of the target object in the left and right camera image. Before any saccade, an appropriate target object has to be selected, after the saccade, it has to be re-identified to evaluate the success of the saccadic movement. In our setup, target objects are colored wooden blocks (see Fig. 6.4). To identify the blocks, the camera images are first converted into the CIELAB color space (Jain, 1989). Afterwards, the resulting a- and b-chroma channels are matched against default intensity values for red, green, blue, and yellow objects. The resulting segments for each color are denoised and smoothed by a median filter. Their center of mass is calculated, finally yielding a list of red, green, blue, and yellow object coordinates.

Before any saccade, one of the detected objects is chosen from either the left or right camera image, depending on the current task. Afterwards, the matching target object in the other camera image is identified by searching for the image region with the highest local pixel intensity correlation. In this

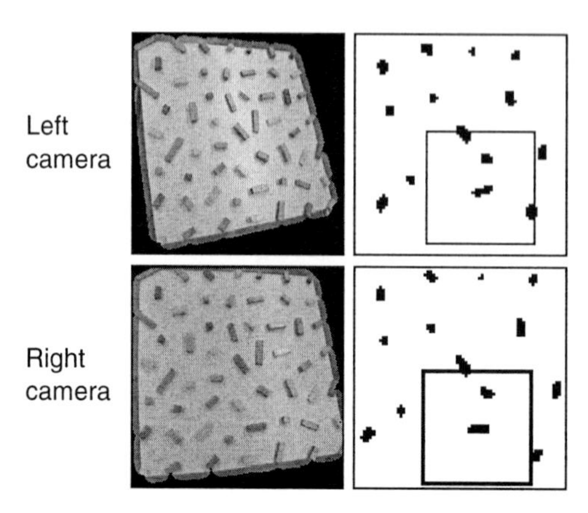

Figure 6.4 — Camera images (left column; the surroundings of the table are already blanked) and salience images for red objects (right column) before a saccade. The selected target is marked with a rectangle in the salience images. It has been identified first in the right camera (bold rectangle); by a correlation approach, the corresponding image region in the left camera has been found.

way, the target object coordinates $\mathrm{s}_{\mathrm{VIS}}^{(t)} = \left(x_{\mathrm{left}}^{(t)}, y_{\mathrm{left}}^{(t)}, x_{\mathrm{right}}^{(t)}, y_{\mathrm{right}}^{(t)}\right)$ are determined. After the saccade, the target object is re-identified in both camera images by the same correlation approach, providing the coordinates $\mathrm{s}_{\mathrm{VIS}}^{(t+1)} = \left(x_{\mathrm{left}}^{(t+t)}, y_{\mathrm{left}}^{(t+t)}, x_{\mathrm{right}}^{(t+t)}, y_{\mathrm{right}}^{(t+t)}\right)$.

6.3.3 Implementation

The saccade controller is implemented by a multi-layer perceptron (MLP; see Sect. 3.1). It has eight inputs and four linear output units (see also Fig. 6.3). The single hidden layer has four units with hyperbolic tangent as activation function; in addition, the inputs are also directly connected to the output layer ("shortcut connections"). In the beginning, the network weights are initialized to random values, resulting in erratic output. The network is trained by providing proper learning examples for weight adjustment as outlined in the following section.

6.3.4 Learning by averaging

Like most motor learning tasks, saccade learning suffers from the problem of the "missing teacher signal". Whenever an incorrect motor command is carried out, the resulting error is only measurable in the sensory domain. The motor error and therefore the correct motor output remains unknown (for a more thorough discussion, see Sect. 2.2.2). In Sect. 2.2.7.2, a new motor learning algorithm called "continous learning by averaging" (CLbA) is suggested to overcome this problem. Expressed for the domain of saccade control, its basic idea is to search at random in the neighborhood of the network output in motor space for saccades which are slightly better than the saccade produced by the controller network and which bring the target object closer to the center in the camera images. These improved saccades are used as learning example for network adaptation. In the process of learning, over- and undershoot saccades cancel each other out, resulting in more precise motor output of the network. This "canceling out" only works because the MLP as function approximator adapts to the average of the over- and undershoot saccades.

For a precise description of the algorithm, we first define the radial target distance as $r = r(\mathbf{s}_{\text{VIS}}) = \frac{1}{2\sqrt{2}}\left(\sqrt{x^2_{\text{left}} + y^2_{\text{left}}} + \sqrt{x^2_{\text{right}} + y^2_{\text{right}}}\right)$, with $r = 0.0$ indicating a perfect saccade where the center of mass of the target object is projected exactly on the center of both camera images, and with $r = 1.0$ being the worst value (as long as the target is not completely lost which is even worse). The basic steps of the algorithm in each learning trial are:

- Generate a sensory state $\mathbf{s}^{(t)}_{\text{SAC}} = \left(\mathbf{s}^{(t)}_{\text{KIN}}, \mathbf{s}^{(t)}_{\text{VIS}}\right)$ by positioning the cameras at random and selecting one of the available target objects at random.

- Assess the output of the controller network in response to this input: $\mathbf{m}^{(t)}_{\text{SAC}} = C(\mathbf{s}^{(t)}_{\text{SAC}})$. Compute $\mathbf{s}^{(t+1)}_{\text{KIN}}$, move the cameras to their new position, and determine $r = r(\mathbf{s}^{(t+1)}_{\text{VIS}})$.

- Generate a new motor command $\widehat{\mathbf{m}}^{(t)}_{\text{SAC}}$ by adding Gaussian noise with a variance of $(r\sigma_0)^2$ to each motor parameter of $\mathbf{m}^{(t)}_{\text{SAC}}$. Compute the corresponding $\widehat{\mathbf{s}}^{(t+1)}_{\text{KIN}} = \mathbf{s}^{(t)}_{\text{KIN}} + \widehat{\mathbf{m}}^{(t)}_{\text{SAC}}$, move the cameras to their new position, and determine $r(\widehat{\mathbf{s}}^{(t+1)}_{\text{VIS}})$. Repeat this step until $r(\widehat{\mathbf{s}}^{(t+1)}_{\text{VIS}}) < r(\mathbf{s}^{(t+1)}_{\text{VIS}})$ and $r(\widehat{\mathbf{s}}^{(t+1)}_{\text{VIS}}) < r(\mathbf{s}^{(t)}_{\text{VIS}})$.

- Use the final motor command $\widehat{\mathbf{m}}^{(t)}_{\text{SAC}}$ as learning example for the controller network. For weight adaptation, we use simple gradient descent with a learning rate η (see Sect. 3.1.2).

The saccade controller network of this study was trained with parameter values $\sigma_0 = 1.3$ and $\eta = 0.12$. After 450 learning trials, the average radial target distance over 50 test saccades amounted to $r < 0.018$. In the course of these 450 learning trials, 4435 saccades were carried out. Compared to the results

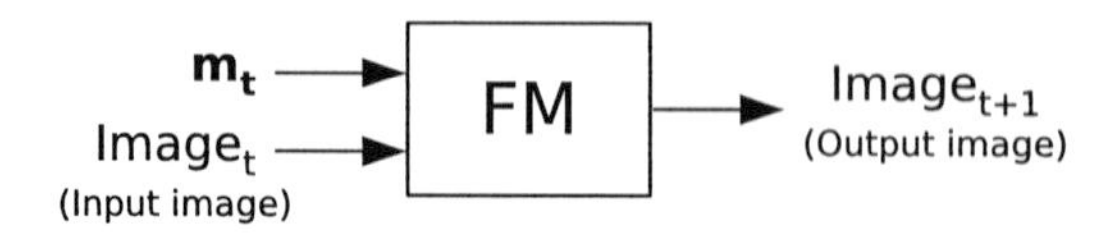

Figure 6.5 — Visual forward model (FM) (adapted from Schenck and Möller, 2007, © Springer).

for the 2D saccade learning task with noise on the simulated camera setup in Sect. 4.1, the value of $r < 0.018$ corresponds to a quality $Q_C > 0.982$ which is slightly above the desired quality level $Q^* = 0.975$ for this task condition. In conclusion, this single result provides evidence for a close match between the real setup and the simulated setup.

6.4 Visual Forward Model

6.4.1 Application in the overall model

The task of a visual FM is to predict future visual sensory states. In the framework of the used robot setup, this means to predict how the retinal images (which are generated from the original camera images) will look like after a movement of the camera head. The input of a visual FM is the current retinal image at time step t and the motor command $\mathrm{m}_{\mathrm{FM}}^{(t)}$, the output is a prediction of the resulting retinal image in time $t + 1$ (see Fig. 6.5). Learning of this input-output relationship is difficult because of the high dimensionality of the image data, and because of the fact that part of the future image may not be predictable at all. The learning algorithm for visual FMs which is presented in Sect. 5.1 overcomes both problems. Exactly the FM from Sect. 5.1 is applied in the overall model on grasping to extrafoveal targets for the left camera. The motor command $\mathrm{m}_{\mathrm{FM}}^{(t)}$, which specifies the relative movement of the left camera, is computed from $\mathrm{m}_{\mathrm{SAC}}^{(t)}$, the overall movement command for the robot camera head, which is generated by the saccade controller (see Fig. 6.2).

6.4.2 Potential application for saccade learning

The saccade controller in Sect. 6.3 could benefit from a visual FM as well. In the following, an integration of the FM and the saccade controller is proposed which has not yet been implemented. In the image processing for the saccade controller, first a suitable target object for fixation has to be determined in either the left or right camera image. Afterwards, the object has to be identified in the other camera image as well. Most importantly, the target object has to be re-identified in both camera images after the saccade for the computation of the sensory error. This re-identification is computationally very expensive since it involves extensive matching of image regions. Moreover, when working with distorted retinal images like the ones used for the visual FM, simple matching algorithms do not work (for this reason, the saccade controller in its current implementation just uses non-distorted camera images).

A straightforward solution for the re-identification of target objects is the application of the visual FM developed in Sect. 5.1. For each camera, it can predict from the executed motor command to which image location the center of mass of the target object has moved (via the internal mapping model), or if it has been lost (via the internal validator model). Afterwards, the exact position of the target object can be determined by a search process which is restricted to the very close neighborhood of the predicted position. No extensive matching or search process is necessary. Used in the way, the visual FM offers improved sensory processing and faster behavioral learning.

Furthermore, the application of the visual FM for target re-identification has a biological motivation. Several studies on human saccades show that small displacements of target objects during saccades go unnoticed (e.g., Bridgeman, 1975; Deubel et al., 1996). This finding implies that the mechanisms which implement the reafference principle (see Sect. 1.5.2) are not precise enough to detect such small target shifts. Deubel (2004) proposes that visual stability between saccades is maintained by matching visual landmarks before and after the saccade. According to the robot model, we suggest that landmark re-identification is based on a visual FM which predicts approximately the position of the landmarks after the saccade. Such an FM is not precise enough for the detection of small displacements, but it suffices to point at a search region. If the target is not found within this region, the mismatch is detected (in the study of Bridgeman (1975), target shifts of 4 degrees are detected by the subjects in at least 40% of the trials).

6.5 Arm Controller

The purpose of the arm controller is to generate the motor command for the final grasping movement. As input, it receives the orientation of the target object, a red wooden block on the table surface (see Fig. 6.6), and the position of the cameras $\mathrm{s}_{\mathrm{KIN}}^{(t+1)}$ after a successful fixation movement towards the target object. The camera position implicitly encodes the position of the red block. As output,

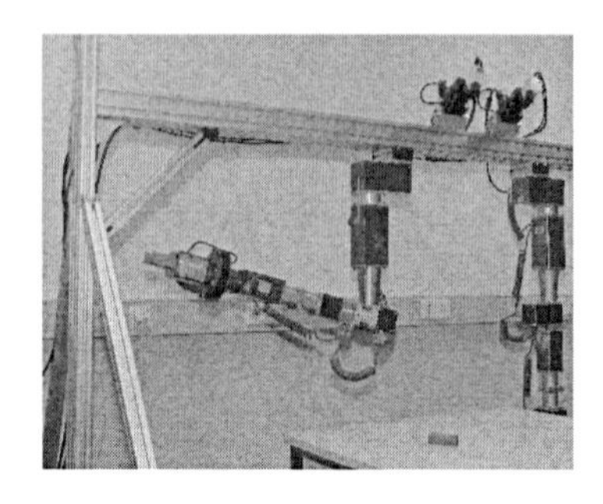
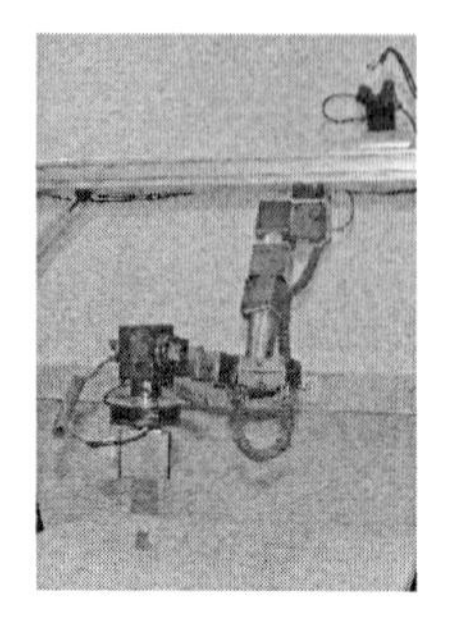
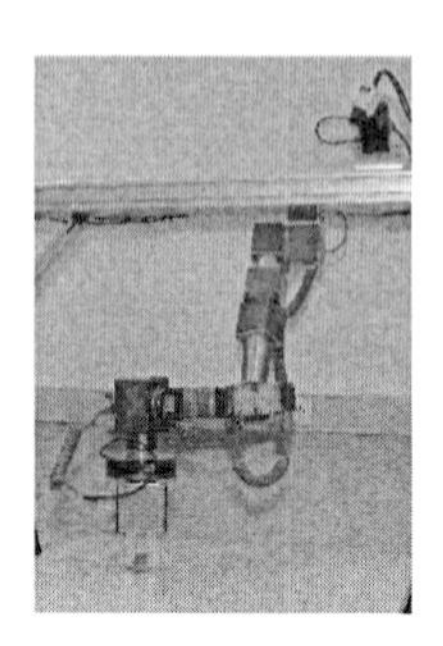

Figure 6.6 — Resting, pre-grasping and grasping posture (from left to right).

the arm controller produces two sets of joint angles, for the pre-grasping and the grasping posture. The pre-grasping posture serves as via point for the robot arm when it moves from its resting position to the final grasping position. This is necessary to avoid collisions with the environment and with the block before it is grasped. Figure 6.6 shows the resting, the pre-grasping, and the grasping posture for a single grasping trial. In a perfect grasping movement, the approach direction of the gripper is perpendicular to the table surface. Because of the geometry of the robot arm, this movement is only possible over a restricted area of the table. Here, a rectangular region of 380×250 mm for the placement of target objects is used.

6.5.1 Data preprocessing

The arm controller is implemented as abstract recurrent neural network (details follow in Sect. 6.5.3). To achieve maximum learning success with this approach, certain preprocessing of the controller input and output is necessary. We use similar methods as in the study by Hoffmann et al. (2005).

The visual input of the arm controller is the orientation of the red block. To determine this orientation, the left retinal image *after* the successful fixation saccade towards the target object is used as default (although this is later varied in the experiments). A color filter is used to generate an image where the block appears as single white segment on a completely black background. In the next step, four compass filters enhance the edges in four different directions ($0°$, $45°$, $90°$, and $135°$) (see Fig. 6.7). After thresholding, the remaining pixels in each image are counted to give a value for the distribution of edges in a given direction. The resulting four values are normalized so that there sum yields 1.0. These normalized values form a "compass filter histogram" which uniquely encodes the orientation of the block independent of its size.

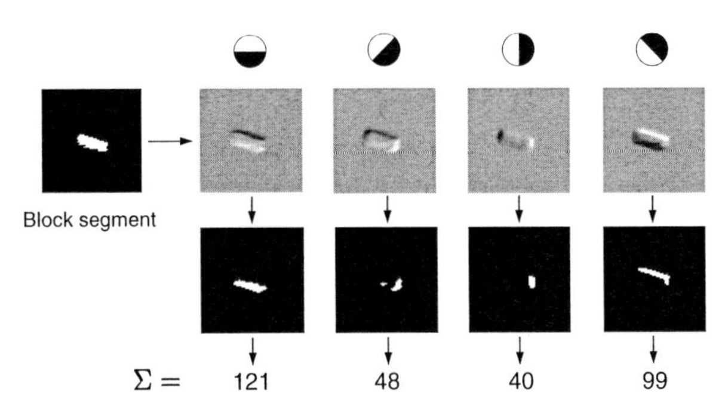

Figure 6.7 — Image processing to encode the block's orientation. On the very left is the block segment. In each column, the preprocessing steps for one compass filter (top) are shown, the edge-image, the threshold image, and the sum of white pixels in the threshold image (adapted from Hoffmann et al., 2005, © Springer).

All postural variables (camera position, arm joint angles) are encoded by tuning curves: A variable x is represented by the values of four Gaussian functions $f_i(x) = \exp(-(x - c_i)^2/(2\sigma^2))$ whose centers c_i are uniformly distributed within the maximal range of the variable. σ equals the distance between two neighboring centers (Fig. 6.8). Overall, there are 20 input values for the arm controller (4 compass filter values and 4×4 values for the camera position), and 48 output values (2 arm postures with 6×4 values each).

6.5.2 Collection of training data

A single training example for the arm controller network is collected in the following way (similar to the procedure suggested by Hoffmann et al., 2005): First, a random block position and orientation on the table surface are generated. By the analytical solution of the inverse kinematics of the robot arm, a corresponding pre-grasping and grasping posture are determined. If the inverse kinematics yields several applicable solutions, one of them is chosen at random. The robot arm is moved to this position with the red block held by the gripper, and releases the red block on arrival. Afterwards, the arm returns to its resting position. The saccade controller from Sect. 6.3 is used to fixate the red block; to enhance precision, a second corrective saccade is carried out if the radial target distance r is larger than 0.015 after the first saccade. Afterwards, the image of the left camera is recorded and mapped to the retinal image. From the information which is gathered in this sequence, a full learning example with input (camera position and block orientation) and output (pre-grasping and grasping

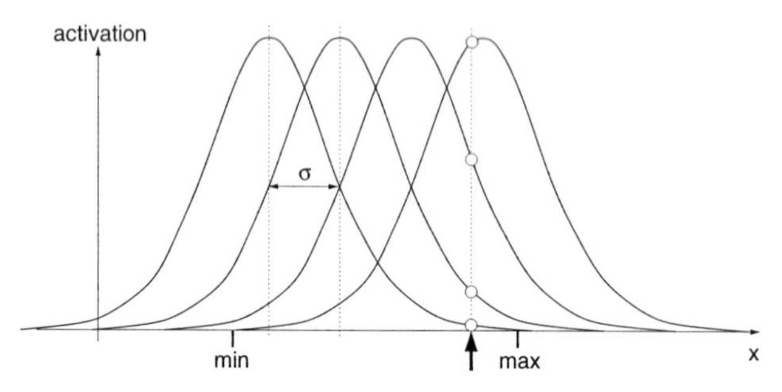

Figure 6.8 — Four Gaussians functions with different centers encode the value (thick arrow) of a variable x with four values (circles) (adapted from Hoffmann et al., 2005, © Springer).

posture) is constructed. This way of collecting learning examples is a technical solution and not intended for biological modeling.

Altogether, 3213 learning examples were collected. All input and output dimensions were normalized to mean 0.0 and variance 1.0. The postural dimensions were normalized *before* the encoding to tuning curve values.

6.5.3 Neural network algorithm

Since one of the possible solutions of the inverse kinematics is chosen at random, the training data represents a one-to-many mapping. For this reason, function approximator networks like MLPs are not suitable for the implementation of the controller. This is a general problem found in many motor control tasks. Möller and Hoffmann (2004) suggested so-called "abstract recurrent neural networks" as solution. These networks consist of a set of hyperellipsoids in the sensorimotor space which comprises both the input and output dimensions. The hyperellipsoids decribe the training data manifold with considerably fewer parameters than the original training data contains. To determine the center and shape of the hyperellipsoids, different algorithms are suggested in the literature (for an overview, see Hoffmann, 2004). In this study, the NGPCA method by Möller and Hoffmann (2004) was applied. A detailed description of this neural network algorithm is provided in Sect. 3.4.

For the arm controller, an NGPCA network consisting of 100 hyperellipsoids with 4 dimensions was used. The learning parameters were $T = 100000$, $T_{ortho} = 10000$, $\epsilon(0) = 0.5$, $\epsilon(T) = 0.05$, $\rho(0) = 1.0$, $\rho(T) = 0.01$, $\sigma^2(0) = 0.0$, and $\lambda(0) = 10.0$ (for an explanation of the parameters see Table 3.3). From the collected 3213 learning examples, 2900 randomly selected examples were used for training (and 313 for the test set). After training, the arm controller showed the following performance figures on the test set: The average horizontal distance between the gripper tip and the block on the table surface amounted to 8.7 mm. The average difference between the orientation of the gripper and the block orientation amounted to 3.4 degrees (for a more detailed specification of these performance indicators see Sect. 6.7).

6.6 Experiments

In the final experiments, it was tested if the overall robot model shows the hypothesized performance in different task conditions. These conditions vary with respect to the experimental sequence within a single trial. In general, an experimental trials starts by generating a block position and orientation at random. The robot arm is used to place the red block on the table surface at exactly this position and in this orientation. Afterwards, the robot arm returns to its resting posture.

Task condition `WW` represents grasping to properly fixated targets.[2] In this task condition, the saccade controller is used for a very precise fixation movement towards the red block. A maximum of five correction saccades is allowed to reduce the radial target distance to less than $r = 0.015$. The left camera image after the last saccade is used to compute the retinal image and the compass filter values as input for the arm controller. Moreover, the final camera position is used as input $s_{\text{KIN}}^{(t+1)}$ for the arm controller.

In task condition `OW`, grasping to extrafoveal target objects is carried out.[3] After the red block has been placed on the table surface, the cameras are moved to a random position where (1) the block is visible in both camera images as input to the saccade controller, and (2) the full shape of the block is visible in the left retinal image as input to the visual FM. Afterwards, the saccade controller is used to generate one saccade towards the red block, but this saccade is never carried out, only $s_{\text{KIN}}^{(t+1)}$ is computed. The visual FM predicts the hypothetical retinal image after the saccade, and from this image the compass filter values are determined as input for the arm controller.

Task condition `WW1` serves as comparison: It is equal to `WW`, but only *one* saccade is carried out, accepting a less than optimal target fixation.[4] This allows a more fair comparison with `OW`, where only a single hypothetical saccade is determined, but no correction saccade. This reduces the quality of the cam-

[2] `WW`: With saccade execution, With proper retinal image

[3] `OW`: withOut saccade execution, With proper retinal image

[4] `WW1`: With saccade execution, With proper retinal image, only 1 saccade

era position input $\mathbf{s}_{\mathrm{KIN}}^{(t+1)}$ of the arm controller, and the retinal images after the saccade may differ slightly from the images which were used during training (where a corrective saccade was carried out if necessary).

Task condition `OO` is used as a control experiment to demonstrate that the extraction from orientation information from the retinal images is not trivial and depends actually on the position of the block in the retinal image.[5] Here, the sequence is similar to condition `OW`, but the visual FM is not used. Instead, the retinal image before the hypothetical fixation saccade is used to compute the compass filter values as input for the arm controller.

In all conditions, the grasping movement which is finally generated as output from the arm controller was carried out at the end of every sequence, and the grasping success was evaluated. Overall, 100 trials were performed in every task condition.

6.7 Results

To evaluate the grasping success, the most important indicator is the percentage of successful grasping trials; a trial was rated as success if the gripper of the robot arm was able to grasp the red block firmly and to lift it. This measure tolerates small position and orientation errors since the distance between the gripper jaws amounted to 60 mm when it approached the red block. The red block itself had a horizontal cross section of 74×23 mm. Moreover, the following indicators are used to evaluate the grasping precision:

- Block position error: The Euclidean distance between the center of mass of the red block and the center of the open gripper, projected onto the table surface.

- Vertical position error: The difference between the ideal height of the gripper tip above the table surface (held constant for all learning examples) and the actual height.

- Block orientation error: The difference between the block's orientation on the table surface and the orientation of the perpendicular to the line that connects both gripper jaws, projected onto the table surface.

- Vertical orientation error: In all learning trials, the approach direction of the gripper is exactly perpendicular to the table surface. The vertical

[5] `OO`: withOut saccade execution, withOut proper retinal image

Task condition			
WW	WW1	OW	OO
97 %	89 %	85 %	40 %

Table 6.1 — Success rate (over 100 grasping trials in each experimental condition).

orientation error is the difference between this ideal approach orientation and the actual approach orientation.

6.7.1 Grasping success

First of all, the percentage of successful grasping trials clearly shows that the model of extrafoveal grasping actually works as expected (see Table 6.1): In condition OW, the success rate amounts to 85%. As expected, the success rate in grasping towards precisely fixated target objects (condition WW) is higher, amounting to 97%. Condition WW1 (target objects only fixated with one saccade) has a success rate of 89%, which shows that the performance difference between conditions WW and OW is largely attributable to the less precise camera position information if only one saccade is scheduled. The baseline condition OO has only a success rate of 40%, clearly indicating that the prediction of the retinal image is not just a trivial add-on, but instead crucial for successful grasping towards extrafoveal targets. The pairwise differences are statistically significant on the $p < 0.01$ level, expect of the differences WW vs. WW1 (only $p < 0.05$) and WW1 vs OW (not significant) (four cell Chi-square test).

6.7.2 Grasping precision

The indicators for grasping precision show results which are consistent with the grasping success rate. Table 6.2 and Fig. 6.9 present the average position and orientation errors for the grasping posture of the robot arm. In addition to the error results for all trials, Fig. 6.9 also presents separate results for successful and failed trials. Regarding the mean value of all trials, condition WW shows always the best precision, closely followed by WW1, and with a certain distance by OW. OO is always the worst performer, especially with regard to the block orientation error. The last result illustrates very clearly the impact of the missing visual FM.

The vertical position and orientation error are much smaller than the block position and orientation error in all conditions. This is no surprise since the distance between gripper tip and table surface and the default approach direction

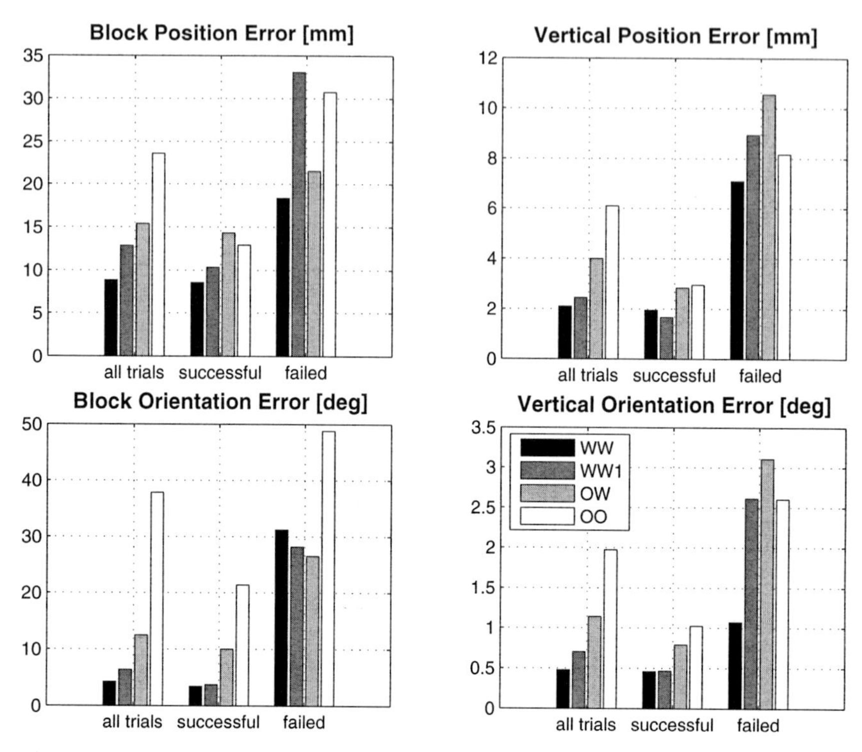

Figure 6.9 — Average position and orientation errors for the different task conditions. The exact values for all trials are listed in Table 6.2.

of the gripper are constant for all learning examples and thus rather easy to learn by the adaptive arm controller. The distinction between successful and failed trials (Fig. 6.9) reveals that the mean position and orientation errors within one task condition are always higher for the failed trials (as it has to be expected).

The statistical tests were restricted to a pairwise comparison of the mean values for all trials. For each error measure, pairwise t-tests (two-sided) were computed for independent samples between the four task conditions. The degrees of freedom were corrected to compensate for the unequal estimated population variances (Bortz, 1993). Most of the 24 tests yielded significant results at least on the $p < 0.05$ level with the following exceptions: block position error: `WW1` vs. `OW`; vertical position error: `WW` vs. `WW1`, `WW1` vs. `OW`, `OW` vs. `OO`; block orientation error: `WW` vs. `WW1`; vertical orientation error: `WW` vs. `WW1`, `WW1` vs. `OW`, `WW1` vs. `OO`, `OW` vs. `OO`.

Error	Task condition			
	`WW`	`WW1`	`OW`	`OO`
Block position [mm]	8.9 (7.2)	12.9 (12.2)	15.4 (9.7)	23.6 (27.1)
Vertical position [mm]	2.1 (2.9)	2.4 (4.8)	4.0 (6.4)	6.1 (12.1)
Block orientation [deg]	4.2 (9.4)	6.4 (13.2)	12.5 (13.1)	37.9 (23.0)
Vertical orientation [deg]	0.5 (0.4)	0.7 (1.2)	1.1 (1.5)	2.0 (5.1)

Table 6.2 — Average position and orientation errors for the different task conditions. Standard deviations are given in brackets.

6.7.3 Saccadic precision

In condition `WW`, the average radial target distance after the final saccade amounts to $r = 0.012$, while in condition `WW1` with only one saccade it amounts to $r = 0.018$. This shows that the lower saccadic precision in condition `WW1` is actually the most plausible source of the larger mean block position error and smaller success rate found in `WW1` compared to `WW`. Furthermore, in condition `WW1` the correlation between saccade length (for the left camera) and radial target distance amounts to $r_{Corr} = 0.16$. Inspired by this finding, it was investigated if there is a direct relationship between saccade length and grasping precision. For conditions `WW1`, `OW`, and `OO`, the correlations between saccade length and the different position and orientation errors were computed. The largest absolute correlation coefficient is found between saccade length and block orientation error in the `OO` condition ($r_{Corr} = 0.11$). However, even this correlation value is not significantly different from zero ($t = 1.1; df = 98$), thus one cannot draw any firm conclusions from these correlation coefficients.

6.7.4 Visualization

Figures 6.10 to 6.13 show some exemplary retinal images which are used for the computation of the compass filter values for the different task conditions. In the first row of each figure (white background), the images from four successful trials are shown, in the second row (gray background) from three or four failed trials. In addition to the retinal image, the segment which is identified as red block is shown together with the compass filter histogram which is computed from this segment. Moreover, Fig. 6.12 for task condition `OW` shows on top of this information the retinal image which is used as input for the visual FM. The predicted retinal image is depicted underneath together with the identified block segment and compass filter histogram.

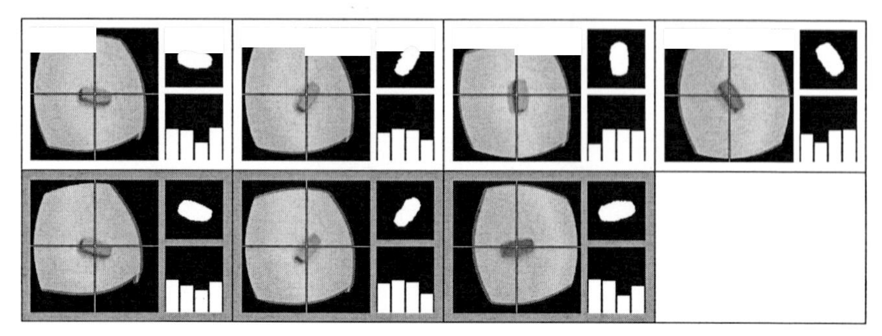

Figure 6.10 — This figure shows the retinal images which are used to extract the block orientation for the WW task condition. A cross marks the center of each retinal image. In addition, the block segment and the corresponding compass filter histogram are shown for each retinal image. The first row with white background depicts successful trials, the second row with gray background failed trials. There are only three failed trials in the WW condition.

Figure 6.10 shows the examples for the WW condition. The block is well centered in the retinal image, indicating good saccadic accuracy. For the first row with successful trials, four different block orientations have been chosen. The compass filter histogram reflects the block orientation by the position of the minimum within the histogram.

Figure 6.11 is dedicated to the WW1 condition. The retinal images reveal that even in the successful trials the red block is not as well centered as in the WW condition, resulting in slightly banana-shaped segments (top left trial).

Figure 6.12 displays exemplary trials of the OW condition. The difference between the depicted retinal images which are input and output of the visual FM illustrate its performance. Shape and orientation of the block in the retinal images differs considerably between input and output. But it becomes also clear that the prediction is sometimes not accurate enough and generates atypical block shapes like in the second trial in the first row (nevertheless successful) or in the second trial in the second row (failed). The predicted block shape in the latter is nearly quadratic resulting in a compass filter histogram with equal values. Histograms like this don't occur in the learning examples for the arm controller network, thus this input is outside the learned data manifold and causes erratic extrapolation and failure.

Finally, Fig. 6.13 shows exemplary trials for the OO condition. Here, no saccade and no prediction takes place, and the retinal images which are recorded at the initial camera position are used to generate the compass filter histograms from the block segment. The shapes of the block segment differ strongly from

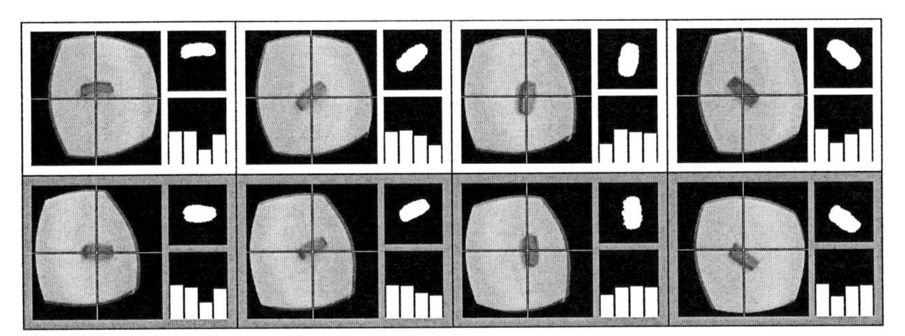

Figure 6.11 — Retinal images of the WW1 task condition (for further explanation see caption of Fig. 6.10).

the ideal shapes shown in Fig. 6.10 in the context of the WW condition. Accordingly, the compass filter histograms are sometimes ill-shaped (especially showing too large differences between minima and maxima). Moreover, the correction of the orientation of the block segment, which is accomplished by the visual FM in condition OW, is missing. These findings correlate well with the low grasping success rate in the OO condition.

6.8 Discussion

6.8.1 Evaluation of the results

The most important goal of this robotics study was to show that the model of grasping towards extrafoveal targets actually works as expected. All important components of the model — a saccade controller, a visual FM, and an arm controller — were implemented for the use with a a real-world robot setup for this grasping task. The results show that the suggested architecture is actually capable to fulfill this task. This supports the claim that spatial attention shifts are accompanied by the preparation of eye movements (as the premotor theory of attention states; Rizzolatti et al., 1994), and corroborates our specific hypothesis that a visual FM predicts how the target object would appear in the fovea, and that this prediction is used to extract precise information about orientation (or in a more general sense, about shape).

Furthermore, we expected that grasping towards precisely fixated target objects results in a better grasping performance than grasping towards extrafoveal target objects (as suggested by the literature on eye-arm coordination; Abrams et al., 1990; Vercher et al., 1994). This expectation was confirmed in a compari-

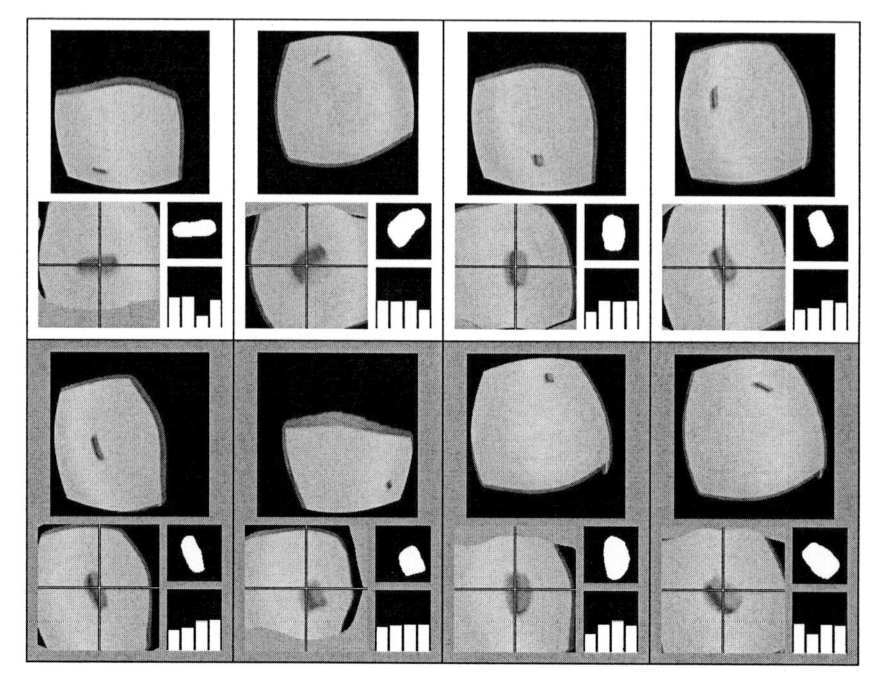

Figure 6.12 — Retinal images of the OW task condition (for further explanation see caption of Fig. 6.10; in addition, the retinal image which is used as input for the visual FM is shown for each trial on top of the other images).

son of the respective task conditions with regard to the overall grasping success and with regard to the grasping precision. In an additional task condition, the influence of saccadic accuracy on grasping success was explored. The fixation movement was restricted to one saccade regardless of the resulting accuracy (like in the extrafoveal condition). This revealed that the superior performance of grasping towards precisely fixated targets compared to extrafoveal targets can be attributed to a large extent to the inferior saccadic accuracy. Only part of the performance difference has to be explained by insufficient visual prediction.

The baseline condition without saccade execution and without visual prediction was used to show that the retinal mapping causes non-trivial changes of object shape and orientation depending on the position in the retinal image. As expected, directly extracting orientation information from the non-predicted retinal images and feeding it to the arm controller resulted in low grasping success. Furthermore, especially the block orientation error in this task condition was very high compared to the other conditions.

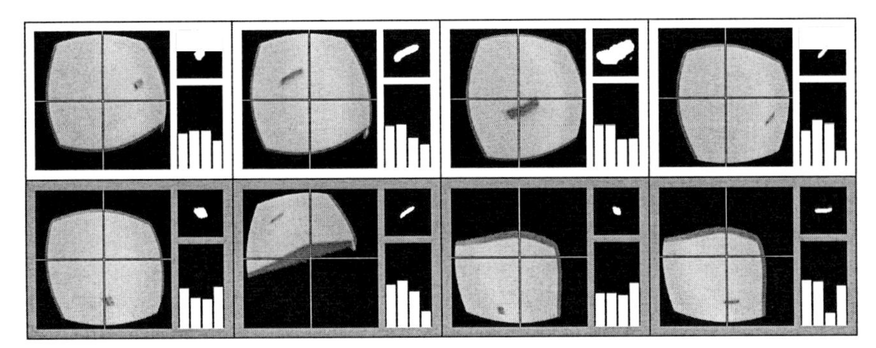

Figure 6.13 — Retinal images of the OO task condition (for further explanation see caption of Fig. 6.10).

6.8.2 Model compared to neurophysiological findings

Overall, the model operates on a rather abstract level. The input and output of the different components are intended to model the information flow on a level of abstraction which is feasible to be implemented on a robot setup. Nevertheless, in the following some correspondences between the model and neurophysiological findings are pointed out.

An important input to the arm controller is the camera position which corresponds to the gaze direction of the eyes in biological systems. In the literature on eye-arm coordination, this information is often referred to as "extraretinal eye position information" (EEPI) (e.g., in Bockisch and Miller, 1999). EEPI plays in important role in the localization of targets which appear as visual stimuli on the retina (e.g., Battaglia-Mayer et al., 2003; Bock, 1986). EEPI allows the transformation from eye-centered to head-centered coordinates; head position information is needed to further localize the target in body-centered coordinates. As simplification, the latter is omitted in the model, assuming a fixed head position. Both kinesthetic eye position information and the efference copy of the eye positioning commands are plausible sources for EEPI (Bridgeman, 1995; Weir, 2006). Here it is supposed that both sources are compatible with each other and can be added up to compute the hypothetical eye position information after a non-executed saccade. Otherwise, an additional internal model would be necessary for this transformation. Despite this simplification, it is important for the plausibility of our model that the "hypothetical EEPI" is actually available to the nervous system. Actually, experimental studies show that EEPI starts to change before saccade onset (Bockisch and Miller, 1999; Dassonville et al., 1992; Matin et al., 1970); thus, the nervous system has the means to update

EEPI before the eye movement takes place and therefore before the kinesthetic eye position information can change.

Many studies on eye-arm coordination emphasize the necessity of coordinate transforms for the localization of target objects in body- or arm-centered space after they have been perceived as visual input on the retina. For example, Snyder (2000) presents the finding that in some cortical regions the locally represented retinal position is modulated by the population code of the gaze direction (this modulation has been termed "gain fields"). Coordinate transforms like this allow to compute the position of a target object in head- or body-centered coordinates, but they do not explain how the overall object shape is transformed. Here, the visual FM of our model offers a plausible mechanism.

The output of the arm controller in our model are final arm postures, not trajectories. This is consistent with the result of Graziano et al. (2002) in a study on monkeys, in which the stimulation of certain motor cortex neurons lead to hand locations independent of the initial arm posture. Thus, this level of encoding is biologically plausible. A more thorough discussion of kinematic control is provided in Sect. 2.1.2.

6.8.3 Alternative solutions

The retinal mapping in the model is used to change the pattern of sensor activation depending on the position of a visual stimulus in a retinal image (like in biological systems as pointed out in Sect. 6.1). These activation differences are still relevant on the next processing level where compass filters are used to detect edge orientations (like the simple cells in the visual cortex; Hubel and Wiesel, 1962). In consequence, the arm controller which has been adapted to orientation information gained from the foveal region of the retina cannot work successfully with extrafoveal target objects. As solution, visual prediction of the foveal region by an FM was suggested. This has the advantage that the system can solve the task by a single arm controller which is also used for grasping to precisely fixated targets.

Within our framework, there exist two additional approaches which offer alternative solutions. Both cause considerably more overhead on part of the arm controller. First, instead of visual prediction, one might use a multitude of arm controllers, each adapted to a certain region of the retina. This might work in theory, but would require a lot of storage effort for the parameters of the large number of arm controllers. Moreover, each arm controller would need its own learning examples in which the target object is exactly depicted on the retinal region for which the controller is responsible. This would result in considerable additional effort in the collection of learning examples and in the adaptation process. As second alternative solution, one might use a single arm controller which also takes the retinal position of the target as input. This seems to be a straightforward solution, but it suffers from the disadvantage that the manifold of training data becomes considerably more complex. The relation between compass filter values and joint angles would be mediated by retinal position in a non-linear fashion which basically adds two non-redundant dimensions to the data manifold. Thus, the interpolation task of any neural network algorithm (whether biological or artificial) would become more difficult and would require more complex network structures. Moreover, the required amount of training data for such a network would be much larger: Enough learning examples for all retinal regions would be needed to allow for adequate interpolation.

6.8.4 Remarks on the components of the model

The components of the overall model — the saccade controller, the visual FM, and the arm controller — are not pre-wired. Instead, they are acquired by different learning strategies. These strategies have been presented and discussed in previous publications (Hoffmann et al., 2005; Schenck and Möller, 2007, 2006) and in preceding chapters of this thesis. For this reason, the discussion is restricted to a few remarks at this point.

In the present implementation, the saccade controller works with non-distorted camera images, while the forward model and the arm controller rely on retinal images. This inconsistency does not affect the validity of the overall model and might be resolved as suggested in Sect. 6.4.2. For saccade adaptation, learning by averaging is used although many authors favor feedback error learning for saccade control in humans and primates (Dean et al., 1994; Gancarz and Grossberg, 1999). Nevertheless, learning by averaging offers a new way of adaptive motor control which is genuinely simple and low-level in its algorithmic structure and therefore a viable candidate for biological modeling. Future research has to show for which motor tasks it is suited as biologically plausible mechanism.

The visual FM is learned by matching regions in the retinal images before and after the saccade. Over many learning trials, correspondences emerge during the matching process. From these correspondences, a mapping between pixel positions in the retinal images before and after the saccade is constructed, and non-predictable image regions are detected by the lack of any clear correspondence. This learning process is restricted to low-level visual processing and therefore a plausible candidate for biological modeling. Studies on predictive remapping (see Sect. 1.3.4.2) support the claim that visual prediction takes place in the brain. In these studies, neurons which shift their visual receptive fields in anticipation of an upcoming saccade were discovered in various brain areas (Duhamel et al., 1992; Umeno and Goldberg, 1997; Walker et al., 1995).

For the arm controller, the NGPCA method is used as abstract recurrent neural network. This learning algorithm is capable to cope with one-to-many mappings which arise in many motor control tasks (especially, when postural control is concerned as in our model). Alternative methods for this purpose are the "mean of multiple computations (MMC)" network (Cruse and Steinkühler, 1993; Steinkühler and Cruse, 1998) and the "parametrized self-organizing map (PSOM)" (Ritter, 1993; Walter et al., 2000). Furthermore, Hoffmann (2004) compared the performance of Kernel PCA and a mixture of local PCA on a kinematic arm model.

6.8.5 Final conclusion

The overall model offers a novel functional framework for grasping to extrafoveal targets based on the premotor theory of attention which has gained a lot of experimental support in the past (e.g., Eimer et al., 2006). It identifies visual prediction as an important putative component of eye-hand coordination in this task domain. Moreover, its applicability to a real-world setup is successfully demonstrated, which corroborates its plausibility for biological modeling.

Chapter 7

Visuokinesthetic Prediction for a Block-Pushing Task[1]

7.1 Introduction

In this chapter, the idea of "visual perception through anticipation" is explored on a robot arm setup on the basis of a block-pushing task under visual control. The theoretical background of the anticipation approach (Möller, 1999) is described in detail in Sect. 1.5.3. The basic hypothesis states that visual perception of shape and space is based on an internal simulation process which relies on the anticipation of the sensory consequences of motor actions, and which is closely related to motor planning. Starting from the current sensory situation, an inverse model (IM) suggests several motor actions. A corresponding forward model (FM) predicts the sensory consequences of all these actions. On the basis of the predicted sensory situations, further motor actions are suggested, afterwards their consequences are predicted as well, and so on, until a maximum step size is reached or at least some simulated action sequences have led to sensory results with a clear positive or negative meaning to the agent.

As stated in Sect. 1.5.3, the main problem of the anticipation approach is how to restrict the number of simulated movement sequences to a feasible number despite the combinatorial explosion that occurs when a large number of motor commands is tested in parallel for each predicted sensory state. In the present study, we solve this problem by the application of an optimization method which guides the search for the "right" movement sequence. This approach is similar to the one in the study by Hoffmann and Möller (2004) on motor planning and perception with a mobile robot (this study is described as well in more detail in Sect. 1.5.3).

The present study starts from a motor control perspective. The first task

[1] This chapter is an extended version of the publication by Schenck et al. (2008). The results presented by Schenck et al. (2008) are more recent and demonstrate a better performance.

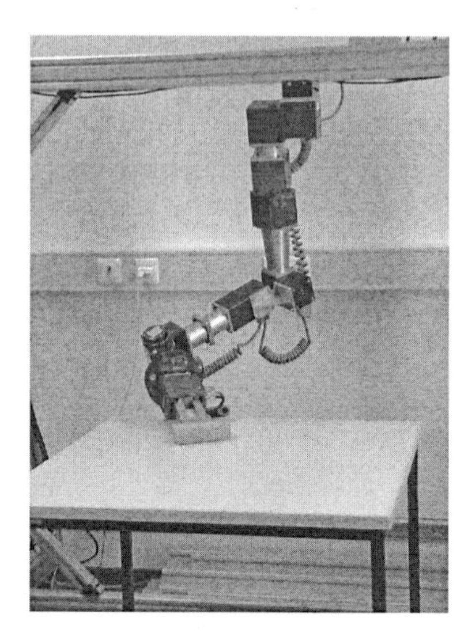

Figure 7.1 — Left: The robot arm in a typical posture for the pushing task; the red block appears in this monochrome image in light gray color in front of the gripper. Right: Tool that is held by the gripper during pushing (adapted from Sinder, 2006).

is to generate a sequence of motor commands so that the block-pushing task can be successfully fulfilled. Once the motor planning capability is established, we will re-interprete in the discussion section this process as a perceptual one for the location of blocks in the working area of the robot, as the anticipation approach hypothesizes.

The work on this study was carried out in cooperation with Dennis Sinder, who wrote his diploma thesis under the author's supervision. For thorough information about technical details, we would like to refer the reader to his thesis (Sinder, 2006) and restrict the presentation here to the main ideas.

7.2 Method

7.2.1 Setup and task

The robot arm setup which is used in this study is described in detail in App. A. For the block-pushing task, movements of the robot arm are restricted to a 2D plane directly above the surface of the white table (see Fig. 7.1, left). With the help of a special tool (Fig. 7.1, right), which is held by the gripper, the robot arm pushes a red block made from foam material (size: $135 \times 45 \times 45\,\mathrm{mm}$) around

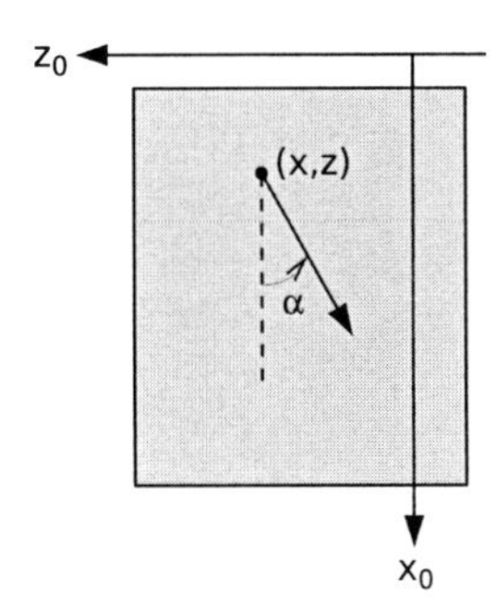

Figure 7.2 — Sketch of the working area (gray) of the robot arm for pushing movements on the table surface. A robot arm posture is defined by the gripper position (x, z) and the pushing orientation α.

the table surface. The posture of the robot arm is defined by the workspace coordinates x and z of the gripper tip (the axes of the world coordinate system are depicted in Fig. A.1 in App. A), and by an angle α that indicates the pushing orientation of the gripper; an angle $\alpha = 0°$ indicates a pushing direction parallel to the x axis of the world coordinate system, away from the base joint of the robot arm (see Fig. 7.2). The two remaining degrees of freedom with regard to the gripper orientation are fixed as well as the y position of the gripper tip, resulting in robot arm postures as shown in Fig. 7.1 (left) with constant distance between table surface and gripper. Collision-free operation is only possible for a restricted area of the table surface defined by $x \in [330\,\text{mm}; 730\,\text{mm}]$ and $z \in [-69.5\,\text{mm}; 250.5\,\text{mm}]$. In this working area, the operating range for α amounts to $[-40°; +40°]$.

Visual data is collected with the right camera of the robot camera head shown in Fig. B.1 in App. B. The camera is held in a fixed position in which the camera image records the entire white table surface (see Fig. 7.3, left). From the original camera image (in RGB color) with a size of 320×240 pixels, a central region with a size of 200×200 pixels is extracted which covers the effective working area for the block-pushing task. The region is denoted as "camera image" in the following for simplicity.

The task of the robot arm is to push the red block from a start position to a goal position within the working area. Because of the geometry of the robot arm, the general pushing direction is directed away from its base joint. The orientation of the block generally varies between start and goal. In the current implementation, the red block is first placed at its goal position by the operator and the first camera image is recorded; afterwards the operator moves the robot

Figure 7.3 — Input and output of the image processing for the encoding of the position of the red block. Left: Original full-sized camera image (in this monochrome version, the red block appears in dark gray in front of the gripper). Right: Activation values of the 3×3 "neurons" with Gaussian receptive fields encoding the block position (white: maximum activation; black: minimum; the depicted activation values are simulated for illustrative purposes).

arm to an arbitrary start posture (considering the general pushing direction) and places the red block in front of the pushing tool. A second camera image is collected, showing the red block at the start position, and from the first and the second camera image a sequence of motor commands is determined, by which the robot arm manages to push the red block from the start to the goal.

To generate the sequence of motor commands by an interal simulation process, a visuokinesthetic FM is required. The FM predicts visual data (position and orientation of the red block in the camera image) and kinesthetic data (position and orientation of the gripper as indicators of the arm posture) resulting from a given movement. In the following, we describe the training and the application of this FM.

7.2.2 Image processing

Visual prediction as performed by the visuokinesthetic FM is a difficult task because of the high dimensionality of visual data. Instead of applying a general solution to the prediction task as in Chapt. 5, we decided to drastically reduce the dimensionality of the visual data. This is possible since we only have to encode the position and orientation of the red block on the white table surface.

First, the camera image is converted into a monochrome image, in which all pixels of the red block get maximum intensity and all other pixels zero intensity. From this image, a population code of the block's position is obtained by superimposing a grid of 3×3 "neurons" with Gaussian receptive fields, the centers of which cover the image uniformly. The activation of a neuron is the weighted sum over all pixels, with weight factors taken from the corresponding Gaussian function. The resulting 9 activation values encode the position of the block (see Fig. 7.3, right).

The orientation of the block is encoded by a compass filter histogram as in Sect. 6.5.1. Four compass filters enhance the edges of the block segment in the monochrome image in four different directions ($0°$, $45°$, $90°$, and $135°$) (see Fig. 6.7). After thresholding, the remaining pixels in each image are counted to give a value for the distribution of edges in a given direction. The resulting four values encode the orientation of the block. Both the encoding of the position and of the orientation are inspired by the work by Hoffmann et al. (2005).

7.2.3 Network structure and training

The visuokinesthetic FM for the internal simulation process has the following inputs:[2] First, the current gripper position and orientation as kinesthetic input $\mathbf{s}_{\mathrm{KIN}}^{(t)}$ (t denotes the time step) with $\mathbf{s}_{\mathrm{KIN}}^{(t)} = (x_t, z_t, \alpha_t)$; second, a nine-dimensional vector $\mathbf{s}_{\mathrm{POS}}^{(t)}$ comprising the activation of the 3×3 grid of neurons encoding the position of the block in the camera image; third, a four-dimensional vector $\mathbf{s}_{\mathrm{OR}}^{(t)}$ that contains the compass filter values encoding the orientation of the block; and fourth, a motor command $\mathbf{m}_t$ with $\mathbf{m}_t = (\Delta x_t, \Delta z_t, \Delta \alpha_t)$. Although the motor command $\mathbf{m}_t$ and the kinesthetic state $\mathbf{s}_{\mathrm{KIN}}^{(t)}$ share the same coordinate system and the computation of $\mathbf{s}_{\mathrm{KIN}}^{(t+1)}$ is straightforward with $\mathbf{s}_{\mathrm{KIN}}^{(t+1)} = \mathbf{s}_{\mathrm{KIN}}^{(t)} + \mathbf{m}_t$, it is important to note that this addition is never carried out in the model, in which the motor space and the kinesthetic space are conceptually different entities. The compatibility between $\mathbf{m}_t$ and $\mathbf{s}_{\mathrm{KIN}}^{(t)}$ is only exploited at the level of software implementation for the functions which evaluate learning success and which move the robot arm. Moreover, it facilitates the training of precise visuokinesthetic FMs with basic artificial neural network techniques (which is also a technical design decision). The output of the visuokinesthetic FM consists of the kinesthetic and the visual state of the next time step: $\widehat{\mathbf{s}}_{\mathrm{KIN}}^{(t+1)}$, $\widehat{\mathbf{s}}_{\mathrm{POS}}^{(t+1)}$, and $\widehat{\mathbf{s}}_{\mathrm{OR}}^{(t+1)}$. Overall, the input-output relationship is

$$\mathbf{s}_{\mathrm{KIN}}^{(t)}, \mathbf{s}_{\mathrm{POS}}^{(t)}, \mathbf{s}_{\mathrm{OR}}^{(t)}, \mathbf{m}_t \longrightarrow \widehat{\mathbf{s}}_{\mathrm{KIN}}^{(t+1)}, \widehat{\mathbf{s}}_{\mathrm{POS}}^{(t+1)}, \widehat{\mathbf{s}}_{\mathrm{OR}}^{(t+1)} \quad .$$

Learning this relationship is a function-approximation task; for this reason, the FM is implemented by a multi-layer perceptron (MLP; see Sect. 3.1). 37500

[2] In contrast to Chapts. 2 and 4, the symbol $\mathbf{m}$ is used in this chapter to denote motor commands, and the symbol $\mathbf{s}$ to denote sensory states. This change is applied to avoid confusion with the workspace coordinate x.

learning examples for the MLP were generated by systematically moving the gripper of the robot arm along different trajectories through the working area, while it was pushing the red block. The movements were either translations in the current gripper direction α of a size of 10, 20, or 30 mm or rotations by a small angle $\Delta\alpha = 5°$. At the beginning and the end of each movement step, a camera image was recorded, so that a full learning example consisting of kinesthetic and visual data could be constructed. The systematic approach and the large number of learning examples ensured a uniform distribution of learning examples in the whole working area (including the gripper orientation α).[3] This process can be interpreted as a "motor babbling" stage, in which the system learns the relevant sensorimotor relationships through its own experience.

In the thesis by Sinder (2006), different MLP topologies, types of input encoding, pattern set sizes, and learning algorithms were compared. The best performance with regard to the precision of the predicted data was obtained by three separate MLPs for each output $\widehat{\mathbf{s}}_{\text{KIN}}^{(t+1)}$, $\widehat{\mathbf{s}}_{\text{POS}}^{(t+1)}$, and $\widehat{\mathbf{s}}_{\text{OR}}^{(t+1)}$, with the already described input encoding and pattern set size, and with plain online backpropagation (see Sect. 3.1.2) as learning algorithm. The final networks from which the visuokinesthetic FM is composed have a single hidden layer with 10 units with hyperbolic tangent as activation function. The activation function of the input and output units is the identity function. The input for the MLP for the kinesthetic prediction is restricted to $\mathbf{s}_{\text{KIN}}^{(t)}$ and $\mathbf{m}_t$, while the other MLPs get the full input. The learning rates amount to $\eta = 0.0022$ for the $\mathbf{s}_{\text{KIN}}$-MLP, to $\eta = 0.002$ for the $\mathbf{s}_{\text{POS}}$-MLP, and to $\eta = 0.003$ for the $\mathbf{s}_{\text{OR}}$-MLP. After 300 epochs of network training, Sinder (2006) reports for this type of visuokinesthetic FM the following prediction accuracy on a test set: The average absolute percentage difference between the correct output and the network-generated output amounts to less than 3% for the position and orientation output units and to nearly 0% for the kinesthetic output units (the percentage values are computed in relation to the desired output). This precision is sufficient for an iterative prediction several time steps ahead in the future, as we will see in the following.

7.2.4 Motor planning as optimization problem

The internal simulation process for motor planning and perception requires an iterative application of the visuokinesthetic FM. For $t = 1$ with known sensory input, an adequate motor command $\mathbf{m}_1$ has to be generated (without executing

[3] Since the data range in the different input and output dimensions varies considerably, all dimensions were normalized to a mean value of 0.0 and a variance of 1.0 before MLP training.

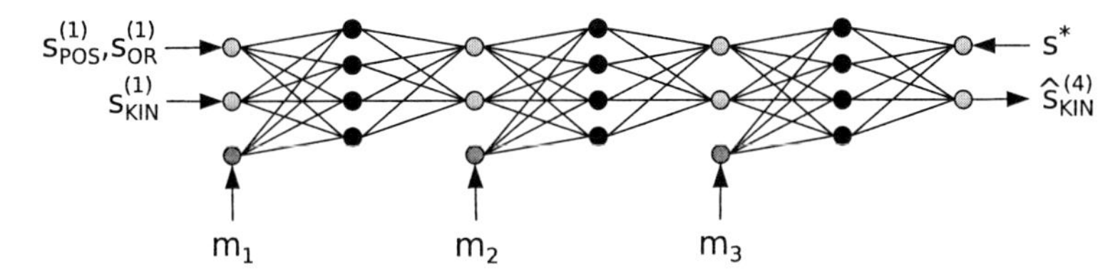

Figure 7.4 — The iterative application of the visuokinesthetic FM, depicted exemplary as chain of three FMs. The initial sensory state is used as input to the chain, the final output $(\widehat{s}_{POS}^{(4)}, \widehat{s}_{OR}^{(4)})$ of the last FM is constrained to be as close to the desired sensory goal state s^* as possible (indicated by the left-pointing arrow from s^*). The motor commands $\mathbf{m}_1$ to $\mathbf{m}_3$ are free parameters in the optimization process. Network units are only shown schematically to indicate the implementation of the FM through MLPs (adapted from Hoffmann, 2004; Sinder, 2006).

it). The FM predicts the sensory state $\widehat{s}_2 = (\widehat{s}_{KIN}^{(2)}, \widehat{s}_{POS}^{(2)}, \widehat{s}_{OR}^{(2)})$ of the next time step $t = 2$, a second motor command $\mathbf{m}_2$ is generated (without execution), the FM predicts the sensory state for $t = 3$ on the basis of the input $(\widehat{s}_2, \mathbf{m}_2)$, and repeatedly so, until the sequence of motor commands $\{\mathbf{m}_t\}$ $(t = 1..N)$ results in a predicted sensory state that either indicates failure or success, or until the number of prediction steps N is equal to a predefined maximum. Such an iterative application of an FM is illustrated in Fig. 7.4 for three prediction steps.

In the block-pushing task, the initial sensory state $s_1 = (s_{KIN}^{(1)}, s_{POS}^{(1)}, \widehat{s}_{OR}^{(1)})$ is determined from the initial posture of the robot arm and the camera image showing the red block at the start position. The sensory goal state $s^* = (s_{POS}^*, s_{OR}^*)$ is determined from the camera image that shows the red block at its goal position. It is important to note that s_{KIN} is *not* part of the sensory goal state. The system has no direct way to determine the kinesthetic state at the goal position. A movement sequence $\{\mathbf{m}_t\}$ is successful if the difference between $(\widehat{s}_{POS}^{(N)}, \widehat{s}_{OR}^{(N)})$ and s^* is very small. If the sequence $\{\mathbf{m}_t\}$ is actually executed afterwards, the final real sensory state $(s_{POS}^{(N)}, s_{OR}^{(N)})$ may differ considerably from s^*, depending on the precision of the prediction by the visuokinesthetic FM. Thus, a precise visuokinesthetic FM is an important precondition for a realistic internal simulation process.

The anticipation approach to visual perception hypothesizes that many movement sequences $\{\mathbf{m}_t\}$ are simulated in parallel. In previous studies, the motor commands were either generated by an IM with additional random variation of its motor output (Möller and Schenck, 2008), they were determined on the basis of a movement heuristic or by recursive search (Hoffmann, 2007), or they were computed by an optimization process (Hoffmann and Möller, 2004). For the present study, we decided to use an optimization method as well because it leaves the generation of motor commands in a kind of "black box". One may even argue that the optimization process acts like an IM which produces an entire movement sequence from a given initial sensory state and a given sensory goal state.

The optimization problem is stated as follows. The initial sensory state is given by $\mathbf{s}_1 = (\mathbf{s}_{\text{KIN}}^{(1)}, \mathbf{s}_{\text{POS}}^{(1)}, \mathbf{s}_{\text{OR}}^{(1)})$, the sensory goal state by $\mathbf{s}^* = (\mathbf{s}_{\text{POS}}^*, \mathbf{s}_{\text{OR}}^*)$. The number of iteration steps is set to a fixed number N. The free parameters in the optimization process are the motor parameters in the sequence $\{\mathbf{m}_t\}$ ($t = 1..N$). The optimization criterion is the minimization of the difference between $(\widehat{\mathbf{s}}_{\text{POS}}^{(N)}, \widehat{\mathbf{s}}_{\text{OR}}^{(N)})$ and $\mathbf{s}^*$.

In each internal simulation step, the visuokinesthetic FM predicts first a translational movement in direction of the current estimated gripper orientation with length r_t, afterwards it predicts on the basis of the new estimated sensory state the rotation by an angle $\Delta\alpha_t$. Thus, in each iteration step i a double prediction is carried out to avoid that the MLPs have to operate in an untrained part of the input data space, since they were only trained with purely translational and purely rotational movements. For the same reason, Δx_t and Δz_t are not allowed to vary freely, but are instead computed as $\Delta x_t = r_t \cos(\widehat{\alpha}_t)$ and $\Delta z_t = -r_t \sin(\widehat{\alpha}_t)$ since the training data only contains purely translational movements in direction of the current gripper orientation. In summary, the free motor parameters are $(r_t, \Delta\alpha_t)$ for each movement step. Overall, this defines an optimization problem with $2N$ free parameters.

Differential evolution (DE) is used as optimization method (see Sect. 3.5) with the following parameter settings: $N_{\text{DE}} = 50$, $\lambda = 0.7$, $\gamma = 0.7$, $p_{\text{CR}} = 0.95$, $G_{\max} = 15$, and $E_{\min} = 1.0^{-10}$ (see Table 3.4 for the parameter definition). The energy E of a movement sequence indicates the "fitness" of the specific solution in the process of differential evolution (the smaller the better; E is designated as f_{opt} in Sect. 3.5). E is computed by the following equation on the basis of unnormalized sensory values:

$$E = \|\widehat{\mathbf{s}}_{\text{POS}}^{(N)} - \mathbf{s}_{\text{POS}}^*\| \cdot \left(1 + 0.0005 \cdot \|\widehat{\mathbf{s}}_{\text{OR}}^{(N)} - \mathbf{s}_{\text{OR}}^*\|\right)$$

The formula defines a tradeoff between position and orientation accuracy with priority to the former. Moreover, penalty terms are added to E if any motor parameter r_t or $\Delta\alpha_t$ is outside the range that the MLPs of the visuokinesthetic FM have encountered during training. Penalties are added as well if any estimated kinesthetic state $\widehat{\mathbf{s}}_{\text{KIN}}^{(t)}$ during the simulation of the movement sequence is outside the working area. In this way, the energy function prevents the optimization process from drifting into areas in which the MLPs of the FM do not work properly because they have to extrapolate.

Since the distance between the start and the goal is not known beforehand, the optimization process has to be carried out with different numbers of iteration steps N. In our experiments, we varied N between 7 and 15. Considering the population size $N_{\text{DE}} = 50$ and the maximum number of generations $G_{\max} = 15$, the computation of the best movement sequence required the internal simulation of 6750 different movement sequences.

The movement sequence which resulted from the optimization trial with the lowest final energy E was picked as overall best movement sequence. However, for a fair comparison between optimization trials with a different iteration depth N, we multiplied E before the comparison with an "increase factor" equal to 1.2^N. In this way, the energy of the movement sequences increases exponentially by a factor of 1.2 for every movement step. This is motivated by the fact that the precision of the final prediction gets worse the more internal simulation steps have to be carried out. Thus, before the multiplication with the increase factor a solution with $N = 15$ may have resulted in a smaller energy than a solution with $N = 7$, but ultimately the shorter sequence yields the better performance because the accumulated prediction error is smaller, which in turn implies a more precise energy value E. As solution to this problem, energy values were compared not until after multiplication with the increase factor.

7.3 Results

Simulation runs The results that are reported here were generated in a simulation study, in which 100 movement tasks with different start and goal positions were solved. The start and goal positions and their accompanying visual sensory data were retrieved from the pattern set for the training of the visuokinesthetic FM with 37500 learning examples. For each of the 100 tasks, two learning examples were drawn at random. Since originally each of these learning examples represents a small movement step, the sensory data for the start position was extracted from the part of the *first* learning example that encodes the sensory state *before* the movement, the sensory data for the goal position from the part of the *second* learning example that encodes the sensory state *after* the movement. Certain constraints were applied to these randomly generated movement tasks to ensure that they are geometrically possible (e.g., the start position has to be closer to the base joint of the robot arm than the goal position), that the overall orientation difference is not too large, and that start and goal are not placed at the very border of the working area.

For each movement task, the optimization process generates a movement sequence by the algorithm described in the preceding section. By executing this sequence, the red block would be ideally pushed to the goal position which is encoded by $\mathrm{s}^* = (\mathrm{s}^*_{\mathrm{POS}}, \mathrm{s}^*_{\mathrm{OR}})$. The corresponding desired final arm posture is denoted as $\mathrm{s}^*_{\mathrm{KIN}} = (x^*, z^*, \alpha^*)$, the actual final arm posture after the simulated movement as $\mathrm{s}^{(N)}_{\mathrm{KIN}} = (x_N, z_N, \alpha_N)$. Since $\mathrm{s}^*_{\mathrm{KIN}}$ is part of the learning examples in the pattern set, it can be used for evaluation purposes. However, if an operator places the red block at a random goal position in the working area for a real-world test, the accompanying arm posture and thus $\mathrm{s}^*_{\mathrm{KIN}}$ would be unknown.

The mean position error, defined as the Euclidean distance between (x_N, z_N) and (x^*, z^*), amounts to 27.9 mm for the 100 movement tasks ($\sigma = 18.2$ mm), the mean orientation error, defined as absolute value of the difference between α_N and α^*, amounts to 6.0 degrees ($\sigma = 6.3$ degrees).[4] The mean movement distance, defined as the distance between the start and the goal position of a movement task in the (x, z) space, amounts to 175 mm. The percentage ratio between the mean position error and the mean movement distance amounts to 16.0%. On average, a movement sequence has a length of 8.6 steps. The movement distance is correlated to the length of the corresponding movement sequence ($r = 0.33$, $p < 0.001$) and to the resulting position error ($r = 0.73$, $p < 0.001$). The orientation error shows no significant correlations with other important variables.

Figure 7.5 shows the movement sequences that were generated in 16 of the 100 movement tasks. Each panel depicts the complete working area, the x-axis pointing in the vertical, the z-axis in the horizontal direction. The goal position (x^*, z^*) is indicated by a circle with a diameter of 20 mm in each panel. The longer bar of the cross at the center of the circle points into the goal orientation α^*. Successive movement steps within a sequence are separated by bars that are orthogonal to the movement direction. The upper nine examples show rather precise solutions over various movement distances, while the four examples in the last row illustrate failed solutions. The mean position error of the 16 examples is slightly larger than the overall position error, while the mean orien-

[4] An additional simulation run with 100 different movement tasks was carried out to explore the influence of the population size N_{DE} on the movement precision. The population size was set to $N_{\mathrm{DE}} = 30$, reducing the number of simulated movement sequences per movement task to 4050 (compared to 6750). This parameter configuration results in a position error of 29.9 mm and an orientation error of 9.0 degrees. The difference to the run with $N_{\mathrm{DE}} = 50$ is statistically significant only with regard to the orientation error ($p < 0.01$). This shows that the number of movement sequences that are simulated in parallel can be considerably reduced without a major loss in movement precision, most likely because many of the simulated sequences are very similar.

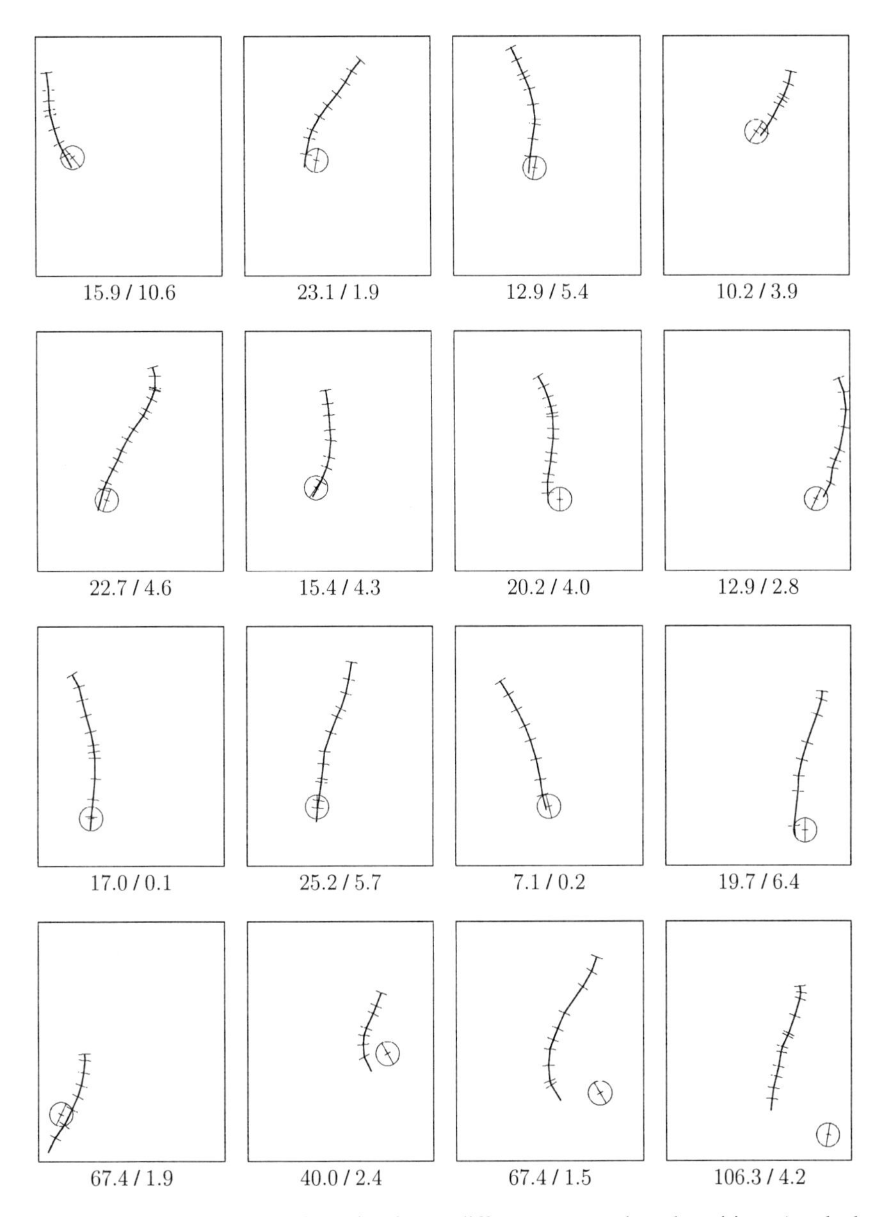

Figure 7.5 — Simulated trajectories for 16 different start and goal positions (marked with a cross within a circle; the longer bar of the cross indicates the goal orientation); unsatisfactory trials are shown in the last row. The figures underneath each trajectory indicate the final position error (left; in mm) and the final orientation error (right; in degrees).

tation error is slightly lower for the shown examples. Thus, Fig. 7.5 provides a representative illustration of the performance of the suggested motor planning scheme. The failed solutions are remarkable insofar as the orientation error is kept small although the position error is large. This may be caused by a bad performance of the MLP that predicts $\widehat{s}_{POS}$ in some regions of the input space.

Practical tests Tests with the robot arm, in which the red block is placed by an operator at random start and goal positions within the working area, revealed a similar performance as the simulation runs.

7.4 Discussion

The performance of the internal simulation process for motor planning is rather good, given the prediction accuracy of the MLPs which constitute the visuokinesthetic FM. While the kinesthetic prediction is nearly flawless, the visual position and orientation data is predicted with an average accuracy of 3%. Although this sounds like a tolerable error, even small prediction errors accumulate easily, rendering the final output useless. This is reflected by the strong correlation between movement distance and final position error. Nevertheless, the internal simulation process can successfully generate even movement sequences with a length of 15 steps with tolerable final position and orientation errors (for example, in the first column/second row of Fig. 7.5), although success is not guaranteed, as the four examples in the last row of Fig. 7.5 show. As already stated, a likely reason for failed trials is a sub-average prediction accuracy of the visuokinesthetic FM in some regions of its input space. Thus, further research has to concentrate on techniques for the training of even more precise FMs.

The trajectories which are generated in the movement planning process are not defined by their via-points first, as suggested by many models for arm trajectory generation (e.g., Rosenbaum et al., 2001; Wada and Kawato, 2004). Instead, the motor commands for every movement step are specified first, and the locations of the via-points are determined afterwards by updating the estimated kinesthetic state iteratively through the visuokinesthetic FM. Thus, our model offers an alternative approach to trajectory formation, in which the positions of the via-points emerge from the best movement sequence. In our model, we work with motor commands which indicate a kinematic change of the arm posture. Future research has to address the question if this method of trajectory generation is actually feasible for more complex motor commands and for the dynamic domain. Moreover, one has to ask if there are "cheaper ways" of trajectory generation, since the application of an optimization method is rather expensive with regard to the number of simulated movement sequences for a single task. However, this method has the great advantage that there is no need for an explicit movement strategy or an IM. Instead, the optimization criterion implicitly defines the movement strategy.

One may critize from the cognitive and biological modeling perspective that the representation of the kinesthetic state is too abstract. However, the conversion between $\mathrm{s}_{\mathrm{KIN}}$ and the joint angles of the robot arm is purely kinematic, and the joint angles would be a plausible representation of the kinesthetic body state, even though their neural encoding in the brain may be rather complex. Moreover, we omitted a clear distinction between sensory and state variables, and the kinesthetic state has been subsumed into the sensory data. We feel that this simplification enhances the clarity of the description, and the distinction between sensory and state variables is conceptually not important in the scope of this study. The visual "sensory" representations are also rather abstract. However, the compass filters work by extracting edges of a specific orientation and therefore act in analogy to the simple cells in the primary visual cortex (Hubel and Wiesel, 1962).

A small twist in the formulation of the movement task reveals that the internal simulation process is not only the basis for movement planning, but also for the visual perception of space. Instead of placing the red block first at the goal position and afterwards at the start position, from which the pushing is initiated, the red block is *only* placed at the goal position. The perceptual task is to infer from the camera image the block's position in the working area in kinesthetic coordinates, and thus to perceive the location of the block in a way that is intimately linked to the body of the agent (here: the robot arm).

Since we work in a very restricted domain, we have to assume that the robot arm is in a valid posture for the pushing task, thus in a kinesthetic state $\mathrm{s}_{\mathrm{KIN}}^{(1)}$, which can be expressed by the variables x_1, z_1, and α_1. Now, we use the pattern set for the visuokinesthetic FM with its 37500 learning examples as lookup table. $\mathrm{s}_{\mathrm{KIN}}^{(1)}$ is the retrieval cue, $\mathrm{s}_{\mathrm{POS}}^{(1)}$ and $\mathrm{s}_{\mathrm{OR}}^{(1)}$ are recalled;[5] in this way, the complete initial input for the chain of FMs is obtained. The final desired output $(\mathrm{s}_{\mathrm{POS}}^{*}, \mathrm{s}_{\mathrm{OR}}^{*})$ is already known through the presentation of the red block at its current position. Using the mental simulation process, $\widehat{\mathrm{s}}_{\mathrm{KIN}}^{(N)}$ is generated, the internal estimate of the arm posture directly in front of the red block. The generation of $\widehat{\mathrm{s}}_{\mathrm{KIN}}^{(N)}$ can be interpreted as perception of the location of the red

[5] This lookup table is not implemented yet, but there is no reason why the recall from the pattern set should fail. Alternatively, one might use an NGPCA network (see Sect. 3.4) that is trained with these learning examples for the recall.

block in kinesthetic body coordinates.[6] The results of the simulation runs in the results section can be reinterpreted in this way as well.

One may argue that it would be easier within the presented framework to infer $\widehat{\mathrm{s}}_{\mathrm{KIN}}^{(N)}$ directly, using the lookup table approach on the basis of the pattern set for the visuokinesthetic FM, but this time applying $(\mathrm{s}^*_{\mathrm{POS}}, \mathrm{s}^*_{\mathrm{OR}})$ as retrieval cues. Although this direct visuokinesthetic association may work for the specific setup of this study, this does not question the applicability of the internal simulation approach. For example, in the study by Möller and Schenck (2008) dead ends and corridors were successfully classified by internal simulation, whereas the direct classification just by visual data would have been much more difficult, if not impossible. Thus, independent of the existence of a straight association between visual data and other perceptual categories (object classification, kinesthetic data), the internal anticipation approach allows for the perception of shape (dead end vs. corridor) or space (block location). This fits nicely to the core idea that action enables perception (Gibson, 1979; Hoffmann and Möller, 2004; Möller, 1999; Möller and Schenck, 2008; Noë, 2005).

Nevertheless, it would strengthen the argument in favor of the perception through anticipation approach if one was able to avoid the usage of a lookup table or similar recall mechanism at all. Future experiments will aim at generating estimates of $\mathrm{s}_{\mathrm{POS}}^{(1)}$ and $\mathrm{s}_{\mathrm{OR}}^{(1)}$ in the process of optimization as well (most desirably with biologically more plausible optimization methods). If this succeeds, the system would not only perceive the location of the block in kinesthetic coordinates, but would also get a visual image of how the object would look like in a position close to the gripper (given the current start posture of the arm). In more general terms, the imagined visual stimulation by an object close to the agent's body could trigger other associations, for example imagined tactile sensations, in this way providing a multisensory and body-related perception and understanding of an object which is only observed visually in the first place.

[6] The MLP that predicts $\widehat{\mathrm{s}}_{\mathrm{KIN}}^{(t+1)}$ works very precisely so that the difference between the real kinesthetic state $\mathrm{s}_{\mathrm{KIN}}^{(N)}$ and the estimated kinesthetic state $\widehat{\mathrm{s}}_{\mathrm{KIN}}^{(N)}$ at the end of a simulated movement sequence is completely negligible. Thus, the perception of space does not suffer from any additional inaccuracy which could theoretically arise from the iterative prediction of $\widehat{\mathrm{s}}_{\mathrm{KIN}}^{(t+1)}$.

Chapter 8
Summary and Outlook

The focus of this thesis were adaptive internal models for integrated sensorimotor processing. We explored motor learning strategies for inverse models and learning algorithms for visual forward models. In doing so, we pursued the goal to provide novel learning schemes which might be used as basis for biological modeling, but which could also serve as technical solution in the area of adaptive robotics. Furthermore, we developed a model for grasping to extrafoveal targets, composed from inverse and forward models, which is motivated by the premotor theory of attention (Rizzolatti et al., 1994). Finally, we demonstrated in the context of a block-pushing task that visual perception of space can be realized by an internal simulation of movement sequences as proposed by the "perception through anticipation" approach (Möller, 1999). As research method, robot experiments and computer simulations were used in all studies. In the following, the main results of each study are summarized, and an outlook of future research is given.

8.1 Kinematic Motor Learning

In the study on kinematic motor learning in Chapt. 4, we compared "learning by averaging" (Schenck and Möller, 2004, 2006), "feedback error learning" (FEL; Kawato, 1990), "distal supervised learning" (DSL; Jordan and Rumelhart, 1992), and "direct inverse modeling" (DIM; e.g., Kuperstein, 1988). Learning by averaging was developed by the author; it was tested in a staged (SLbA) and a continuous (CLbA) version. Furthermore, direct inverse modeling on the basis of the NGPCA algorithm for abstract recurrent neural networks (Hoffmann and Möller, 2003; Möller and Hoffmann, 2004) was included in the comparison study (DIM_NGPCA).

The performance of these motor learning strategies was compared on different tasks for simulated robot setups. The first setup was a robot camera head with stereo vision; this system had to learn to fixate target objects on a table surface or within a 3D working area. The second setup was a simulated planar arm with its number of links varying between two and four; this system had to learn the correct arm postures for reaching movements. In the study on the planar arm, we introduced different task conditions: without constraint, with a "maximum symmetry" constraint, and with a "minimum energy" constraint. Furthermore, for both the camera head and the planar arm, there was a task condition with additional sensory noise. In this way, the goal was to find out which learning strategies perform best for linear plants (camera head), for non-linear plants (planar arm), with additional learning constraints (planar arm), with additional noise, and how these strategies can cope with one-to-many mappings (planar arm).

The performance measure in these comparisons was the required number of "exploration trials" (actual movements of the cameras or of the arm for the generation of a learning example) for controller training by the respective learning strategy. Training was halted as soon as the motor output of the controller (alias inverse kinematics model) exceeded a very good pre-defined precision level (called controller quality in Chapt. 4).

Main results

- For linear plants, local linear approximation techniques like FEL and DSL showed a very good performance, as did DIM and DIM_NGPCA.

- For non-linear plants, DIM_NGPCA was the clear winner. Depending on the number of links of the planar arm, either SLbA or FEL/DSL came in second. CLbA failed for the planar arm (most likely because of the negative impact of catastrophic interference).

- SLbA coped very well with the additional learning constraints in the planar arm task. It showed the most consistent and partly the best performance in this respect. Furthermore, SLbA is well suited to incorporate additional learning constraints through its quality function.

- Sensory noise had no considerable impact on the performance ranking order of the learning strategies (with the exception of SLbA in one task condition for the planar arm).

- DIM cannot cope at all with one-to-many mappings unless it is used in combination with abstract recurrent neural networks (DIM_NGPCA). In this combination, DIM_NGPCA has the theoretical advantage to store all applicable controller outputs for a given controller input simultaneously.

In practice, this advantage was as beneficial as expected for the learning of one-to-many mappings in the planar arm task. FEL/DSL were also able to cope with one-to-many mappings by converging to a single controller output for a given input. SLbA only performed well with one-to-many mappings as long as the number of dimensions of the sensorimotor space was kept low or additional learning constraints were applied.

- Overall, DIM_NGPCA showed the best performance. It can be generally recommended for kinematic motor learning in technical applications (for restrictions, see the discussion in Sect. 4.3 and the following outlook). FEL and DSL are also well suited for many applications, especially for linear plants. If the plant is known analytically (and not too complex), FEL might be preferred because the needed feedback controller can be analytically determined, whereas DSL requires the prior training of a forward model. SLbA is recommendable if a straight implementation of additional learning constraints is desired.

- With regard to biological plausibility, the most promising candidates are FEL (for simple plants that do not require any analytical knowledge) and CLbA (see Table 2.1). The other learning strategies can be critized for various reasons. The biological plausibility of DIM is often questioned in the literature (e.g., Kawato, 1990) because of the supposed need for "neural rewiring". In Sect. 4.3, we briefly suggested a neural architecture based on neural maps which would resolve this problem for DIM and DIM_NGPCA.

Outlook

- In this comparison study, the performance of each learning strategy was determined via the required number of exploration trials until a predefined controller quality was achieved. In a more thorough analysis, one might vary this quality level systematically to incorporate the learning speed profile in the comparison. Furthermore, one might vary the sensory noise levels or might even include motor noise. However, all of these additional variations will require a lot of additional computational effort if they are combined with an as thorough exploration of the parameter space for each learning strategy as in the present study.

- The main weakness of DIM_NGPCA is the lacking goal-directedness in the collection of learning examples for controller training. If the operating range of the controller defines only a small subset of the overall sensorimotor space, this can result in a performance drop of DIM_NGPCA. To overcome this weakness, it would be worthwhile to develop goal-directed search strategies in motor space for the collection of learning examples. This idea is related to the work by Bongard et al. (2006) on directed exploration (see Sect. 1.4).

- One might further investigate on the applicability of learning by averaging for biological modeling. SLbA for the planar arm worked best when additional constraints were applied. The kinematic control of the human arm can be modeled on the basis of cost functions (Cruse et al., 1990). Using these cost functions as constraints for SLbA might result in a plausible model for the human learning of kinematic arm control.

8.2 Visual Prediction

In Chapt. 5 on visual prediction, we presented two versions of a learning algorithm for visual forward models in the context of saccade-like camera movements. This learning algorithm was able to overcome the two main difficulties of visual prediction: first, the high dimensionality of the input and output space (both image data) and, second, the need to detect which part of the visual output is non-predictable (because it is out of view in the visual input of the forward model). To demonstrate the robustness of the presented learning algorithm, we did not work on plain camera images, but on distorted "retinal images" with a decreasing resolution towards the corners.

In the proposed visual forward model, the direct prediction of output pixel intensities was replaced by a mapping between output pixel positions and input pixel positions. This mapping depends on the motor command; it encodes the change of pixel positions induced by the motor action (a relative pan-tilt movement of the camera). This mapping was implemented by a so-called "mapping model" (MM). Furthermore, we introduced a "validator model" (VM) which generates a binary output for each output pixel position, indicating if the respective output pixel is predictable at all, given the scheduled motor command.

The main challenge was to develop a learning algorithm for the MM and the VM. The presented algorithm worked by carrying out systematic camera movements from many different initial camera positions. After each movement, possibly matching pixels in the input image were determined for many different output pixel positions. Over many learning trials, the possibly matching input pixel positions for each specific camera movement and output pixel position were accumulated, yielding a density distribution which was stored in a so-called "cumulator unit". These cumulator units were generated at specific positions in the input space of the MM and VM, defined by a regularly spaced grid. After learning, the output of the MM corresponded to the density maximum in each cumulator unit, because it indicated the best matching input pixel position. If no clear maximum arose during learning, the respective output pixel was non-predictable; in this way, the binary output of the VM was determined. To interpolate the output of the MM and the VM between the grid points, radial basis function networks were applied. Finally, we demonstrated that the trained MM and VM networks can be successfully used for the visual prediction task.

The main function of the cumulator units was to store the density distribution over the input pixel position space. Afterwards, the modal value of this distribution was determined as MM output. To avoid the storage of the cumulator units, we developed an additional iterative learning algorithm for modal values which relied on the idea of fitting an ellipsoid to the density distribution so that the center of the ellipsoid settles down at the modal value. Moreover, we suggested a method of how to determine if a clear modal value exists at all (for the output of the VM). First results indicated the success of these online learning methods although there was still room for improvement.

The proposed learning algorithms offered two core ideas for biological modeling in the area of visual forward models: first, learning the input-output relationship by matching low-level visual features, and second, identifying predictable regions by detecting that a good match emerges during the learning process.

Outlook

- The proposed learning algorithm is restricted to visuomotor relationships in which depth information is not relevant. For head or whole-body movements it is neccessary to extend the algorithm so that it includes depth information.

- The online learning of modal values suffers from occasional outliers. To overcome this problem in the context of the MM, one might couple the learning process of neighbouring units in the grid like in a self-organizing map (Kohonen, 1995).

- To get rid of the fixed grid in the input space of the MM and the VM during learning, one might combine a neural network algorithm like "supervised growing neural gas" (SGNG; Fritzke, 1998), which features flexible and adaptive unit placement, with the online learning of modal values.

Each unit of the SGNG network might hold the necessary parameters for modal value learning in its region of the input space. The resulting learning scheme would be more adaptive and better suited for high-dimensional motor commands as input of the visual forward model.

- One might extend the core ideas of the presented algorithms (matching low-level visual features, detecting regions without good match) to a biologically inspired model of the learning of shifting visual receptive fields as in predictive remapping (e.g., Umeno and Goldberg, 1997).

8.3 Grasping to Extrafoveal Targets

In Chapt. 6, we presented a model for grasping to extrafoveal targets which was implemented on a robot setup, consisting of a camera head and a robot arm. This model was based on the premotor theory of attention (Rizzolatti et al., 1994) and added one specific hypothesis: Attention shifts caused by saccade programming imply a prediction of the retinal foveal images after the saccade. For this purpose, a visual forward model was used. Without visual prediction, grasping towards extrafoveal target objects would be difficult because of the heavy distortions found in retinal images. Whenever an object is depicted at an extrafoveal position, its picture is significantly different from its foveal representation, rendering the reliable extraction of object features like orientation very difficult. Thus, the predicted foveal images were required to determine movement parameters for the manual interaction with objects at the target location of the attention shift.

The model consisted of three parts. First, a saccade controller acquired through "continous learning by averaging"; second, a visual forward model predicting retinal images (with decreasing image resolution towards the corners in analogy to the sensor distribution on the human retina); and third, an arm controller for grasping movements which received the output of the saccade controller and the orientation of the target object as inputs.

We compared the grasping precision of four different versions of the robot model: (1) for grasping towards target objects which were precisely fixated by a series of saccades, using the actual retinal images instead of the predicted ones; (2) for grasping towards target objects which were fixated just by one saccade, also using the actual retinal images instead of the predicted ones; (3) for grasping towards non-fixated target objects using visual prediction; (4) for grasping towards non-fixated target objects without visual prediction.

The first task condition produced the best results, followed by the second and the third condition. The fourth condition caused considerable orientation errors. These results demonstrated that grasping towards extrafoveal targets is less precise than grasping towards properly fixated targets (which is consistent with the literature on eye-hand coordination), and furthermore that grasping towards extrafoveal targets requires the application of a visual forward model. Otherwise, grasping towards extrafoveal targets is bound to fail.

In conclusion, the proposed model offered a functional framework for grasping to extrafoveal targets, identifying visual prediction as an important component of eye-hand coordination in this task domain. Moreover, the model's applicability to a real-world setup was successfully demonstrated.

Outlook

- In the present implementation of the overall model, the saccade controller works with unmapped camera images, while the forward model and the arm controller rely on retinal images. Although this inconsistency does not impair the validity of the overall model, it would be desirable to use retinal images for the saccade controller as well. In addition, one might apply the visual forward model for target re-identification during the adaptation and operation of the saccade controller as suggested in Sect. 6.4.2.

- One might modify the model to account for the finding that human subjects overestimate the distance to peripheral targets in pointing tasks (Bock, 1993). Most likely, this would require additional mechanisms which are independent of the basic claim that visual prediction is necessary to avoid errors in shape recognition (e.g., orientation) and subsequent grasping.

8.4 Perception by Anticipation

The goal of the robot study in Chapt. 7 was to show that the visual perception of space might be based on an internal simulation process as supposed by the "perception through anticipation" approach (Möller, 1999). The test bed for this study was a block-pushing task for a robot arm. Since the internal simulation for perception is closely related to movement planning, we started from a motor control perspective, facing the task to generate the motor commands to push a small wooden block over a table surface from a start to a goal position with the help of the robot arm. To accomplish this task, a visuokinesthetic forward model for the next state prediction and a method for the computation of the right motor commands were needed.

The visuokinesthetic forward model had to predict the new visual and kinesthetic state after a small movement step of the robot arm. Its input were the previous visual and kinesthetic state. The visual state indicated the position and orientation of the block on the table surface (encoded by Gaussian units and a compass filter histogram), the kinesthetic state the position and orientation of the gripper of the robot arm. The forward model was implemented by a collection of multi-layer perceptrons. The training data for these networks were generated by executing a large number of small pushing movements with the robot arm within the workspace. During this "motor babbling" stage, the visual and kinesthetic data were collected for the construction of the learning examples.

The visuokinesthetic forward model was used for the internal prediction of the sensory consequences of a sequence of small pushing movements. For this purpose, the prediction had to be carried out in an iterative way for all elements of the movement sequence. To generate a movement sequence for the successful accomplishment of a specific pushing task (from a specific start to a specific goal), we used the optimization method "differential evolution". The optimization goal was to minimize the difference between the desired visual state at the goal position and the predicted visual state at the end of the movement sequence. The number of movement steps was fixed within a single optimization trial, as were the visual and kinesthetic state at the start. The free parameters in the optimization process were the motor commands of each movement step. Since the optimal number of movement steps is not known beforehand, it was varied systematically over several optimization trials, finally executing the sequence with the best fit between the desired and the final predicted visual state.

In our experiments, this optimization process proved to be quite successful for the generation of movement sequences for block-pushing. The average final position and orientation error were within an acceptable range. Some outliers were attributed to an insufficient precision of the visuokinesthetic forward model in certain parts of the workspace.

Finally, the internal simulation was re-interpreted as perception of space. Instead of showing the block at the goal position and afterwards at the start position, it was proposed to show the block only at the goal position. In this setting, the perceptual task would be to infer from the visual goal state the block's position in the workspace in kinesthetic coordinates, and thus to perceive the location of the block in "body coordinates". This perceptual task might be solved on the basis of the presented motor planning method, only the inital visual state would have to be inferred from the current kinesthetic state of the robot arm to provide the full initial input to the iterative prediction. To generate the visual state from the kinesthetic state, a lookup table approach was suggested. The data for the lookup table might be extracted from the learning examples for the visuokinesthetic forward model.

In conclusion, this study demonstrated how the visual perception of space can be accomplished by an internal simulation process which is closely related to motor planning. In this way, the study contributed to the embodied approach to perception and cognition.

Outlook

- Prediction errors of forward models easily accumulate, rendering the final output of an iterative sequence useless. For this reason, one should continue in finding ways for the training of forward models which result in more precise predictions. Such research has to include neural network and machine learning algorithms and methods of training data acquisition such that good interpolation is possible over the entire workspace.

- It would be desirable to accomplish perception through internal simulation without the need for table lookup at the beginning. For this purpose, one might modify the optimization process such that the inital visual state is not fixed, but also generated during optimization.

- One might extend the visual state of the system such that it also includes the position and orientation of the gripper of the robot arm. In addition, the visuokinesthetic should be trained also with data showing "empty" pushing movements without block. Afterwards, the perceptual task might be solved by a visuo-visual association between the block at the goal position and the imagined impression of the gripper in close vicinity to the block. The generated movement sequence would rely on small movement steps without considering the block, but only the visual state of the gripper. In this way, the table lookup between kinesthetic and visual data might be avoided as well.

- The generation of movement sequences by internal simulation is related to trajectory formation. One might pursue the question if it is possible to enforce certain types of trajectories by the goal criterion of the optimization process. This might offer a new approach to model human arm trajectories in different tasks.

8.5 Final Remarks

The title of this thesis is "Adaptive Internal Models for Motor Control and Visual Prediction". In the following, we will briefly summarize the contributions across all chapters to the two main topics — motor learning and visual prediction — and to the guiding theoretical framework of embodiment.

Motor learning

The main contribution to motor learning was presented in Chapt. 4, in which we thoroughly compared the performance of various learning strategies — partly developed on our own, partly taken from the literature — for kinematic motor learning on different simulated setups. In addition, continuous learning by averaging was applied to real-world saccade learning within the model for grasping to extrafoveal targets (Chapt. 6). The results of the comparison study are both of interest for technical applications and for biological modeling.

Motor learning in the domain of dynamics was briefly addressed with a small experiment on feedback error learning (FEL) for trajectory generation in Sect. 2.1.4.4 (comparing FEL to a learning strategy related to direct inverse modeling).

The study on block-pushing (Chapt. 7) offered a very different approach to the generation of motor commands based on optimization. No inverse model was needed there, only an adaptive forward model underlying the internal simulation process.

Visual prediction

To the best of our knowledge, we developed the first learning algorithm for visual forward models which scales up to input and output images of nearly arbitrary size (within reasonable limits). Two versions of this algorithm were presented in Chapt. 5. This algorithm offers a technical solution for visual prediction in the area of saccade-like camera movements. Furthermore, its basic idea — matching of low-level visual features — is a promising starting point for biological modeling in the area of eye movements.

In Chapt. 6, we used the resulting visual forward model as important component of the overall model of grasping towards extrafoveal targets and proposed its application for target re-identification after saccadic eye movements. Both applications were biologically motivated.

Finally, we trained a visuokinesthetic forward model for the block-pushing task in Chapt. 7. In contrast to the visual forward model of Chapt. 5, it was restricted to visual prediction in a specific domain and worked only with low-dimensional visual data. However, the visuokinesthetic forward model was especially taylored for the application in the internal simulation process, a more general application was not intended.

Embodiment

The research in this thesis was pursued with the research paradigm of embodiment in mind. According to its premises, we used real robot setups as our main reseach tool. Furthermore, we concentrated on integrated sensorimotor processing. Although we did not build any complete agent, our models on grasping to extrafoveal targets and on the visual perception of space implemented the complete sensorimotor loop for a specific behavioral or perceptual ability.

The comparison study on motor learning and the algorithms for visual prediction are not necessarily linked to the embodied approach. From a technical point of view, the offered solutions are of value independent of any theoretical approach. However, from the cognitive science perspective, algorithms for motor learning and adaptive sensory prediction are the building blocks of autonomous embodied systems. In this way, these studies also contribute to embodied cognition and action.

Finally, we demonstrated in the study on block-pushing that action might enable perception, in this case the visual perception of space. This research was based on simulation theories of perception and cognition, which in turn are closely related to embodiment.

Appendix A
Experimental Setup

The experimental setup which is used in the studies throughout this thesis consists of a robot arm and a robot camera head (see Fig. A.1).[1] The robot arm has six rotatory degrees of freedom (DOF) and a two-finger gripper. It is composed of PowerCube modules from the company Schunk. The model numbers of the modules are: PR090 (2$\times$ for the "shoulder"), PR070 (2$\times$ for the "elbow"), PW070 (2-DOF wrist module), and "Gripper 70". The modules are connected via CAN bus with a controller card in a desktop computer. The PowerCube robot arm is well suited for kinematic control since each module can precisely adjust and hold its commanded position.

The origin of the world coordinate system is placed in the center of the right block of the uppermost joint module of the left robot arm (see Fig. A.1, left). In this thesis, only the left robot arm is used. Its Denavit-Hartenberg parameters are reported in Table A.1. Except in singularities, the inverse kinematics of the robot arm allows for eight different solutions for every gripper position and orientation within the working range. In practice, considering collisions of the robot arm with its environment or with itself, usually only two or four different solutions are applicable. The white table in Fig. A.1 in front of the robot arm is used to hold the target objects for fixation movements with the robot camera head and for grasping and pushing with the robot arm. The robot camera head is described in detail in App. B.

[1] Figure 1.1 in the introduction and Fig. A.1 show slightly different versions of this setup at different points in time. A precise definition of the robot arm as it was used throughout this thesis is given by the Denavit-Hartenberg parameters in Table A.1.

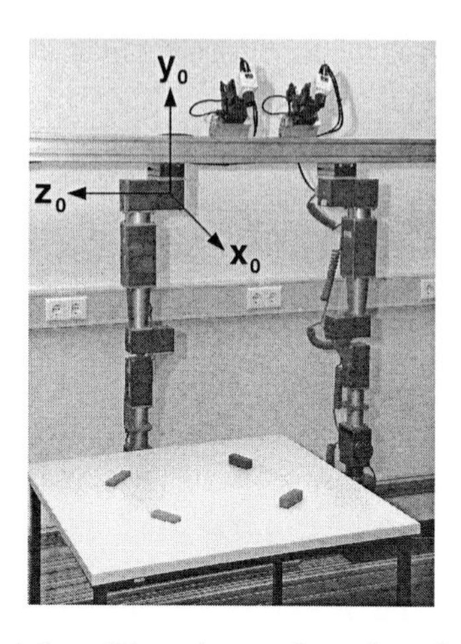

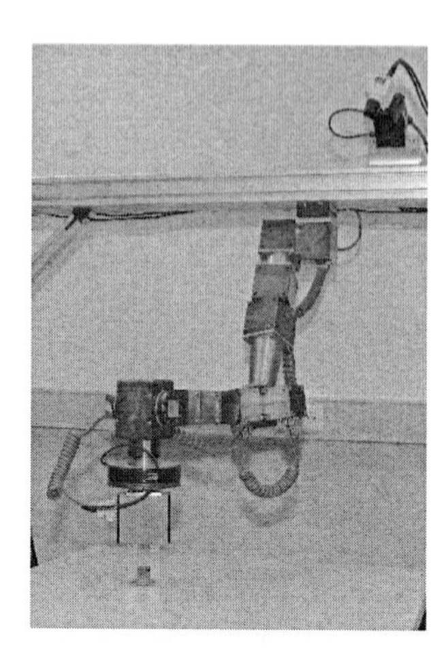

Figure A.1 — These image show the robot setup used throughout the experimental studies. The world coordinate system K_0 is shown with black axes in the left image. The origin of K_0 is exactly in the center of the right block of the uppermost joint module of the left robot arm. The x_0-axis points towards the viewer (left image adapted from Schenck and Möller, 2007, © Springer).

Link i	θ_i^{Null} [deg]	d_i [mm]	α_i [deg]	a_i [mm]
1	0	90.5	90	0.0
2	0	440.5	90	0.0
3	180	0.0	90	0.0
4	90	407.0	90	0.0
5	180	0.0	90	0.0
6	0	323.5	90	0.0

Table A.1 — Denavit-Hartenberg parameters of the robot arm (for the definition and the notation, see e.g. Spong and Vidyasagar, 1989). θ_i^{Null} is the value for θ_i for the null position of the robot arm when it points straight downwards like in the left image in Fig. A.1. Since the robot arm has only rotatory degrees of freedom, only the parameters θ_i are variable.

Appendix B

The Robot Camera Head

The robot camera head of the experimental setup consists of two cameras, each mounted on its own pan-tilt unit (camera body: Imaging Source DFK 50H13/P; lens: Computar T 0412 FICS-3; pan-tilt unit: Directed Perception PTU 46-17.5). Figure A.1 shows the position of the cameras within the overall setup, Fig. B.1 provides a close-up view. Throughout the thesis, the left/right assignment is based on the viewpoint of the cameras. Thus, the camera that appears on the left side in Fig. B.1 is designated as the *right* camera, and vice versa.

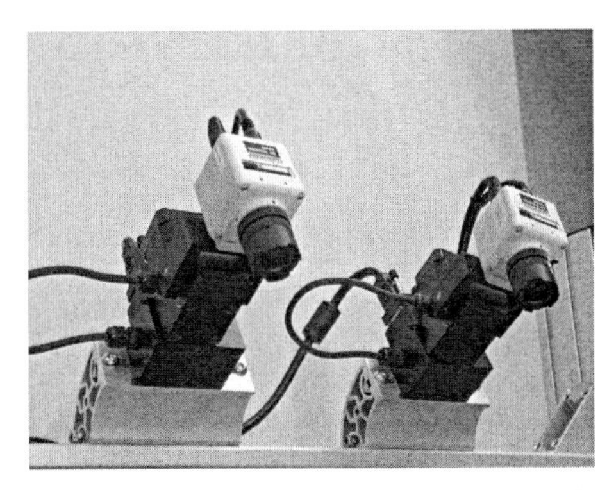

Figure B.1 — This stereo-vision robot camera head was used throughout the experimental studies. Each camera is mounted on its own pan-tilt unit.

B.1 Basic Geometry

Figure B.2 illustrates the geometry of each pan-tilt unit (in the following: PTU) with attached camera. Starting from the world coordinate system K_0 with axes x_0, y_0, and z_0 (as shown in Fig. A.1 as well), the following transformations

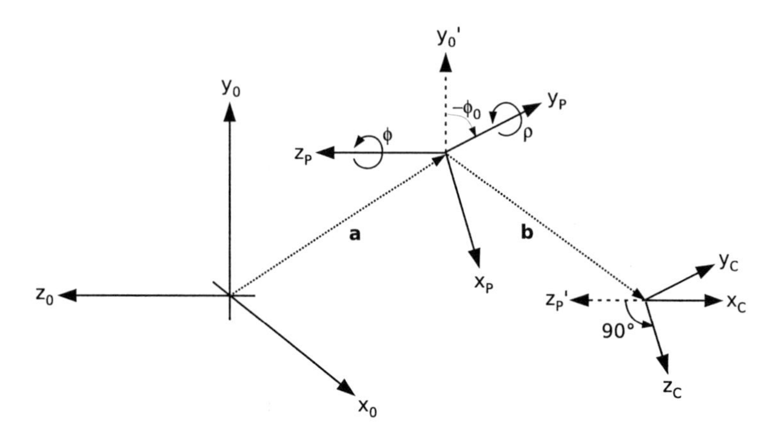

Figure B.2 — Geometry of a single pan-tilt unit with attached camera.

are necessary to arrive in the camera coordinate system K_C (transformation matrices in homogeneous coordinates are given in brackets):

1. Translation by a fixed vector $\mathbf{a}$ ($\mathbf{T_a}$)

2. Rotation around the shifted z-axis by a fixed angle $-\theta_0$ ($\mathbf{R}_{z,-\theta_0}$) $\longrightarrow$ resulting in the PTU base coordinate system K_P with axes x_P, y_P, and z_P

3. Rotation around the y_P-axis by the variable pan angle ρ ($\mathbf{R}_{y,\rho}$)

4. Rotation around the current z_P-axis (moved by the preceding rotation) by the variable tilt angle θ ($\mathbf{R}_{z,\theta}$)

5. Translation by a fixed vector $\mathbf{b}$ ($\mathbf{T_b}$) $\longrightarrow$ the origin of the resulting coordinate system is already at the center of the image plane of the camera, but the axes do not point in the right directions yet

6. Rotation around the current y-axis by $90°$ ($\mathbf{R}_{y,90°}$) $\longrightarrow$ resulting in the camera coordinate system K_C with axes x_C, y_C, and z_C

Altogether, the transformation matrix $\mathbf{H}$ from the camera to the world coordinate system is computed as:

$$\mathbf{H} = \mathbf{T_a}\mathbf{R}_{z,-\theta_0}\mathbf{R}_{y,\rho}\mathbf{R}_{z,\theta}\mathbf{T_b}\mathbf{R}_{y,90°} \tag{B.1}$$

The inverse can be written as:

$$\mathbf{H}^{-1} = \mathbf{R}_{y,90°}^{-1}\mathbf{T_b}^{-1}\mathbf{R}_{z,\theta}^{-1}\mathbf{R}_{y,\rho}^{-1}\mathbf{R}_{z,-\theta_0}^{-1}\mathbf{T_a}^{-1}$$

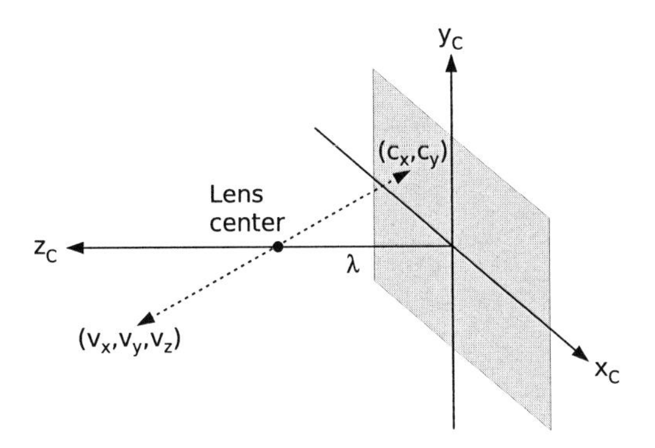

Figure B.3 — Illustration of the perspective transformation. A point $\mathbf{v} = (v_x, v_y, v_z)$ in the camera coordinate system is projected onto position (c_x, c_y) in the image plane (gray area). λ is the focal length of the lens. The optical axis points along the z_C-direction.

Knowing the matrix $\mathbf{H}^{-1}$, a point $\mathbf{w}$ in the world coordinate system can be transformed into the camera coordinate system by $\mathbf{v} = \mathbf{H}^{-1}\mathbf{w}$. To finally compute the point (c_x, c_y) on the image plane of the camera on which $\mathbf{v}$ is projected, a perspective transformation has to be carried out (see Fig. B.3) (Gonzalez and Woods, 1992). The perspective transformation matrix is defined as

$$\mathbf{P} = \begin{pmatrix} 1 & 0 & 0 & 0 \\ 0 & 1 & 0 & 0 \\ 0 & 0 & 1 & 0 \\ 0 & 0 & \frac{-1}{\lambda} & 1 \end{pmatrix}.$$

λ is the focal length of the lens. The product $\mathbf{Pv}$ yields a vector denoted $\tilde{\mathbf{c}}$:

$$\tilde{\mathbf{c}} = \mathbf{Pv} = \begin{pmatrix} 1 & 0 & 0 & 0 \\ 0 & 1 & 0 & 0 \\ 0 & 0 & 1 & 0 \\ 0 & 0 & \frac{-1}{\lambda} & 1 \end{pmatrix} \begin{pmatrix} v_x \\ v_y \\ v_z \\ 1 \end{pmatrix} = \begin{pmatrix} v_x \\ v_y \\ v_z \\ \frac{-v_z}{\lambda} + 1 \end{pmatrix}$$

The final point (c_x, c_y) on the image plane is obtained by:

$$c_x = \frac{\tilde{c}_1}{\tilde{c}_4} = \frac{\lambda v_x}{\lambda - v_z} \quad \text{(B.2)}$$

$$c_y = \frac{\tilde{c}_2}{\tilde{c}_4} = \frac{\lambda v_y}{\lambda - v_z} \quad \text{(B.3)}$$

B.2 Specifications

In the following, some important specifications are listed:

- Cameras and lenses:
  - Size of the imaging chip (CCD): 1/3 inch (4.8 mm × 3.6 mm)
  - Focal length λ of the lens: 4.0 mm
  - Angle of view of the camera-lens combination: $61.9°$ horizontally / $48.5°$ vertically

- Working range of each single PTU:
  - Horizontally (pan): $\rho_{min} = -60.4°$, $\rho_{max} = 23.8°$
  - Vertically (tilt): $\theta_{min} = -42.9°$, $\theta_{max} = 21.4°$

  The used PTUs are capable of a much larger working range, but this range was restricted by software to keep camera movements in the region above the table shown in Fig. A.1.

- Fixed parameters of the geometry of the robot camera head:
  - $\theta_0 = 60°$
  - Vector $\mathbf{a}$ is different for the left and right PTU (left/right assignment from the viewpoint of the cameras):
    $\mathbf{a}_{left} = \begin{pmatrix} 154\,\mathrm{mm} & 273\,\mathrm{mm} & -422\,\mathrm{mm} \end{pmatrix}^T$
    $\mathbf{a}_{right} = \begin{pmatrix} 154\,\mathrm{mm} & 273\,\mathrm{mm} & -182\,\mathrm{mm} \end{pmatrix}^T$
  - $\mathbf{b} = \begin{pmatrix} 150\,\mathrm{mm} & 64\,\mathrm{mm} & 0\,\mathrm{mm} \end{pmatrix}^T$

  The values for θ_0, $\mathbf{a}_{left}$, $\mathbf{a}_{right}$, and $\mathbf{b}$ were computed on the basis of manufacturer's specifications and own measurements. Only b_1 (the position of the CCD along the optical axis) was adjusted afterwards to obtain a better match between the simulated results and real-world results. This value is obviously too large but might compensate for other small measurement errors. A camera calibration was omitted because there was no need for an extremely precise camera model in any of the experimental studies.

B.3 Independence from the Pan Angle

In the following it is shown that the effect of a change of the PTU position $(\Delta\rho, \Delta\theta)$ does not depend on the current value of the pan angle ρ. The final effect of a PTU movement is a shift of the points on the image plane. A single point (c_x, c_y) as projection from an unmoved point $\mathbf{w}$ in the world coordinate system moves to a new position (c'_x, c'_y). To compute the new position, first the corresponding point $\mathbf{v}$ in the camera coordinate system before the movement has to be determined. Equations (B.2) and (B.3) are solved for the components of $\mathbf{v}$ with v_z as unknown variable:

$$\begin{aligned} v_x &= \frac{c_x}{\lambda}(\lambda - v_z) \\ v_y &= \frac{c_y}{\lambda}(\lambda - v_z) \end{aligned}$$

The corresponding point $\mathbf{w}$ in the world coordinate system is computed as $\mathbf{w} = \mathbf{H}\mathbf{v}$ (for $\mathbf{H}$ see Eqn. (B.1)). After the movement, the rotation matrices for the variable pan and tilt angles have changed to $\mathbf{R}_{y,\rho+\Delta\rho}$ and $\mathbf{R}_{z,\theta+\Delta\theta}$. The new overall transformation matrix $\mathbf{H}'^{-1}$ from the world to the camera coordinate system can be written as:

$$\mathbf{H}'^{-1} = \mathbf{R}_{y,90^\circ}^{-1}\mathbf{T}_{\mathbf{b}}^{-1}\mathbf{R}_{z,\theta+\Delta\theta}^{-1}\mathbf{R}_{y,\rho+\Delta\rho}^{-1}\mathbf{R}_{z,-\theta_0}^{-1}\mathbf{T}_{\mathbf{a}}^{-1}$$

After the movement, the constant point $\mathbf{w}$ in the world coordinate system has the new position $\mathbf{v}'$ in the camera coordinate system:

$$\begin{aligned} \mathbf{v}' &= \mathbf{H}'^{-1}\mathbf{w} \\ &= \mathbf{H}'^{-1}\mathbf{H}\mathbf{v} \\ &= \mathbf{R}_{y,90^\circ}^{-1}\mathbf{T}_{\mathbf{b}}^{-1} \cdot \mathbf{R}_{z,\theta+\Delta\theta}^{-1} \cdot \mathbf{R}_{y,\rho+\Delta\rho}^{-1} \cdot \mathbf{R}_{z,-\theta_0}^{-1}\mathbf{T}_{\mathbf{a}}^{-1}\mathbf{T}_{\mathbf{a}}\mathbf{R}_{z,-\theta_0}\mathbf{R}_{y,\rho}\mathbf{R}_{z,\theta}\mathbf{T}_{\mathbf{b}}\mathbf{R}_{y,90^\circ}\mathbf{v} \\ &= \mathbf{R}_{y,90^\circ}^{-1}\mathbf{T}_{\mathbf{b}}^{-1}\overbrace{\mathbf{R}_{z,\Delta\theta}^{-1}\mathbf{R}_{z,\theta}^{-1}}\overbrace{\mathbf{R}_{y,\Delta\rho}^{-1}\mathbf{R}_{y,\rho}^{-1}}\mathbf{R}_{y,\rho}\mathbf{R}_{z,\theta}\mathbf{T}_{\mathbf{b}}\mathbf{R}_{y,90^\circ}\mathbf{v} \\ &= \mathbf{R}_{y,90^\circ}^{-1}\mathbf{T}_{\mathbf{b}}^{-1}\mathbf{R}_{z,\Delta\theta}^{-1}\mathbf{R}_{z,\theta}^{-1}\mathbf{R}_{y,\Delta\rho}^{-1}\mathbf{R}_{z,\theta}\mathbf{T}_{\mathbf{b}}\mathbf{R}_{y,90^\circ}\mathbf{v} \end{aligned} \quad \text{(B.4)}$$

The new position (c'_x, c'_y) on the image plane can be computed from $\mathbf{v}'$ as in Eqns. (B.2) and (B.3). Equation (B.4) shows that $\mathbf{v}'$ does not depend on ρ. Thus, the shift of points on the image plane during a pan-tilt movement does not depend on the current pan angle.

B.4 Influence of the Tilt Angle

As shown in the previous section, the pan angle ρ has no influence how a PTU movement translates into the shift of points on the image plane. However, the tilt angle θ has a certain influence according to Eqn. (B.4). To get an estimate of the impact of θ on the displacement $\mathbf{d} = \begin{pmatrix} c'_x \\ c'_y \end{pmatrix} - \begin{pmatrix} c_x \\ c_y \end{pmatrix}$, the following procedure was carried out:

- For a certain point (c_x, c_y), vary $\Delta\rho_i$, $\Delta\theta_j$, and θ_k systematically (by three nested loops with iterators i, j, and k)
- For each combination of $\Delta\rho_i$ and $\Delta\theta_j$, compute $\mathbf{d}_{ijk}$ for all θ_k (if $(c'_x, c'_y)_{ijk}$ is outside the area of the imaging CCD exclude $\mathbf{d}_{ijk}$ from the following steps)
- Determine the smallest rectangle R_{ij} (aligned to the coordinate system of the image plane) in which all vectors $\mathbf{d}_{ijk}$ can be inscribed; this rectangle describes the variation due to θ_k for the PTU movement $(\Delta\rho_i, \Delta\theta_j)$; the rectangle has edge lengths $r_{x,ij}$ and $r_{y,ij}$
- Determine $r_x^{max} = \arg\max_{i,j} r_{x,ij}$ and $r_y^{max} = \arg\max_{i,j} r_{y,ij}$ (r_x^{max} and r_y^{max} represent the largest variation in x- and y-direction caused by different values of θ_k for the displacement of the point (c_x, c_y))

B.4 shows the resulting displacement rectangles with edge lengths r_x^{max} and r_y^{max} for various positions (c_x, c_y) on the image plane (both c_x and c_y are varied in the range [-1.8 mm; 1.8 mm] — this corresponds to the vertical size of the imaging CCD of the used cameras). Each rectangle is centered at its corresponding point (c_x, c_y). θ_k was varied in the range between $\theta_{min} = -42.9°$ and $\theta_{max} = 21.4°$ (see Sect. B.2) in 13 steps.

For the top graphics in Fig. B.4, $\Delta\rho_i$ ran from $-12.6°$ to $+12.6°$ in 13 steps, $\Delta\theta_j$ from $-9.6°$ to $+9.6°$ in 13 steps. This pan-tilt range guarantees that (c'_x, c'_y) is always within the area of the imaging CCD for points (c_x, c_y) with $c_x \in [-0.6; 0.6]$ and $c_y \in [-0.6; 0.6]$ (close to the image center). The displacement rectangles (which show the worst case) are rather small and are not expected to have a large impact on the precision of the saccadic forward model presented in Sect. 5.1, when θ is omitted as input.

The bottom graphics in Fig. B.4 was generated with twice as large values for $\Delta\rho_i$ and $\Delta\theta_j$. Points (c_x, c_y) at the image center can be moved outside the area of the imaging CCD by pan-tilt movements of this magnitude. The displacement rectangles are considerably larger than in the left graphics, especially along the x-axis. Still, the results with the forward model in Sect. 5.1 show that these variations are not large enough to dominate the information loss in outer regions of the camera image caused by the retinal mapping.

The unknown variable v_z which is necessary to compute (c'_x, c'_y) (see Sect. B.3) is set to 1000.0 mm in these simulations. This is rough estimate of the average distance of objects in our robot setup from the robot camera head.

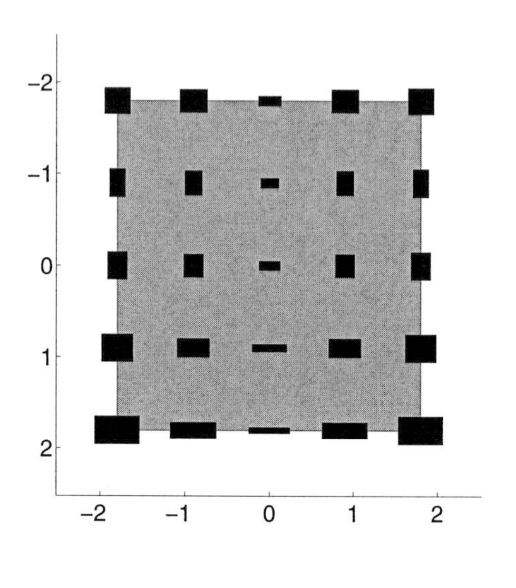

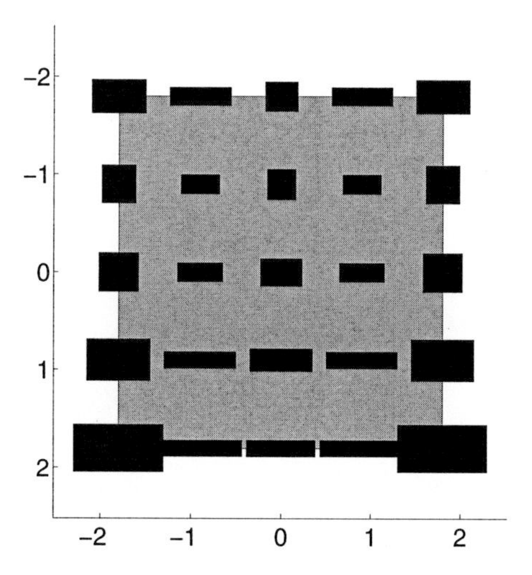

Figure B.4 — Displacement rectangles for different maximum pan-tilt movements on the image plane (axes scaled in mm) (top: $|\Delta\rho_i| \leq 12.6°$, $|\Delta\theta_j| \leq 9.6°$; bottom: $|\Delta\rho_i| \leq 25.2°$, $|\Delta\theta_j| \leq 19.2°$). The image plane is shown in the same orientation as it appears on a camera image (rotated by $180°$ and from behind). A detailed explanation is provided in the main text.

Appendix C

Experimental Settings of Chapter 4

In all tables, variable parameters are given in MATLAB vector notation:
start value : step size : final value.

C.1 Saccade Control

Learning strategy	Fixed parameters	Variable parameters
FEL	$T_{\max} = 2000$	$\eta = 0.08 : 0.02 : 0.36$
DSL	$T_{\max} = 5000$	$N_{\mathrm{FM}} = 10 : 10 : 100$
	$\eta_{\mathrm{FM}} = 0.025$	$\eta = 0.07 : 0.01 : 0.17$
DIM	$T_{\max} = 10000$	$N_{\mathrm{CON}} = 10 : 10 : 150$
	$\eta = 0.01$	
DIM_NGPCA	$T_{\max} = 100000$	$N_{\mathrm{CON}} = 20 : 10 : 150$
		$N = 1 : 1 : 4$
		$m = 4 : 1 : 16$
SLbA	$T_{\max} = 100$	$\sigma_0 = 1.0 : 0.2 : 2.8$
	$\eta = 0.01$	
CLbA	$T_{\max} = 3000$	$\sigma_0 = 0.1 : 0.2 : 2.5$
		$\eta = 0.04 : 0.04 : 0.32$

Table C.1 — Parameter values of the learning strategies for the **2D saccade learning task without retinal noise** with $Q^* = 0.985$.

Learning strategy	Fixed parameters	Variable parameters
FEL	$T_{\max} = 5000$	$\eta = 0.04 : 0.02 : 0.4$
DSL	$T_{\max} = 10000$ $\eta_{\mathrm{FM}} = 0.025$	$N_{\mathrm{FM}} = 10 : 10 : 150$ $\eta = 0.05 : 0.01 : 0.19$
DIM	$T_{\max} = 20000$ $\eta = 0.01$	$N_{\mathrm{CON}} = 10 : 10 : 300$
DIM_NGPCA	$T_{\max} = 100000$	$N_{\mathrm{CON}} = 20 : 10 : 150$ $N = 1 : 1 : 4$ $m = 4 : 1 : 16$
SLbA	$T_{\max} = 100$ $\eta = 0.01$	$\sigma_0 = 1.0 : 0.2 : 2.8$
CLbA	$T_{\max} = 3000$	$\sigma_0 = 0.3 : 0.2 : 2.1$ $\eta = 0.08 : 0.04 : 0.28$

Table C.2 — Parameter values of the learning strategies for the **2D saccade learning task with retinal noise** with $Q^* = 0.975$.

Learning strategy	Fixed parameters	Variable parameters
FEL	$T_{\max} = 5000$	$\eta = 0.04 : 0.02 : 0.4$
DSL	$T_{\max} = 10000$ $\eta_{\mathrm{FM}} = 0.025$	$N_{\mathrm{FM}} = 10 : 10 : 150$ $\eta = 0.05 : 0.01 : 0.2$
DIM	$T_{\max} = 10000$ $\eta = 0.01$	$N_{\mathrm{CON}} = 10 : 10 : 150$
DIM_NGPCA	$T_{\max} = 100000$	$N_{\mathrm{CON}} = 20 : 10 : 150$ $N = 1 : 1 : 4$ $m = 4 : 1 : 16$
SLbA	$T_{\max} = 100$ $\eta = 0.01$	$\sigma_0 = 0.2 : 0.2 : 3.0$
CLbA	$T_{\max} = 10000$	$\sigma_0 = 0.3 : 0.2 : 2.5$ $\eta = 0.04 : 0.04 : 0.32$

Table C.3 — Parameter values of the learning strategies for the **3D saccade learning task without retinal noise** with $Q^* = 0.98$.

Learning strategy	Fixed parameters	Variable parameters
FEL	$T_{\max} = 5000$	$\eta = 0.04 : 0.02 : 0.4$
DSL	$T_{\max} = 10000$	$N_{\text{FM}} = 10 : 10 : 150$
	$\eta_{\text{FM}} = 0.025$	$\eta = 0.05 : 0.01 : 0.2$
DIM	$T_{\max} = 20000$	$N_{\text{CON}} = 10 : 10 : 150$
	$\eta = 0.01$	
DIM_NGPCA	$T_{\max} = 100000$	$N_{\text{CON}} = 20 : 10 : 150$
		$N = 1 : 1 : 4$
		$m = 4 : 1 : 16$
SLbA	$T_{\max} = 100$	$\sigma_0 = 0.2 : 0.2 : 3.0$
	$\eta = 0.01$	
CLbA	$T_{\max} = 12000$	$\sigma_0 = 0.5 : 0.2 : 2.5$
		$\eta = 0.04 : 0.04 : 0.32$

Table C.4 — Parameter values of the learning strategies for the **3D saccade learning task with retinal noise** with $Q^* = 0.972$.

C.2 Planar Arm

SLbA requires a strategy how to increase the number of learning examples and training epochs in each stage. This strategy was varied depending on the number of links and the selected constraint. It is reported in Tables C.5 to C.14 in the format *LE: a-b-c / EP: a-b-c* with *a* being the start value, *b* the increase from stage to stage, and *c* the maximum value. *LE* indicates the number of learning examples, *EP* the number of epochs.

Learning strategy	Fixed parameters	Variable parameters
FEL ($\mathbf{J}^{+}$/$\mathbf{J}^{t}$)	$T_{\max} = 200000$	$\eta = 0.025 : 0.0125 : 0.125$
DSL	$T_{\max} = 200000$ $\eta_{\mathrm{FM}} = 0.005$	$N_{\mathrm{FM}} = 3000 : 3000 : 21000$ $\eta = 0.04 : 0.02 : 0.2$
DIM	$T_{\max} = 100000$ $\eta = 0.05$	$N_{\mathrm{CON}} = 25 : 25 : 300$
DIM_NGPCA	$T_{\max} = 100000$	$N_{\mathrm{CON}} = 25 : 25 : 300$ $N = 7 : 3 : 22$ $m = 1 : 1 : 4$
SLbA (a/b)	$T_{\max} = 50$ LE: $500 - 250 - 1000$ EP: $500 - 500 - 2000$ $\lambda_{\mathrm{SLbA}} = 0.005$ $\lambda_{\sigma} = 1.0$ $\eta = 0.05$	$\sigma_0 = 0.3 : 0.2 : 1.9$

Table C.5 — Parameter values of the learning strategies for the **2-link arm with quality function** Q_0 with $Q_0^* = 0.97$.

Learning strategy	Fixed parameters	Variable parameters
FEL (J^+/J^t)	$T_{\max} = 200000$	$\eta = 0.025 : 0.025 : 0.125$
DSL	$T_{\max} = 200000$	$N_{\text{FM}} = 1000 : 2000 : 11000$
	$\eta_{\text{FM}} = 0.005$	$\eta = 0.01 : 0.01 : 0.1$
DIM	$T_{\max} = 100000$	$N_{\text{CON}} = 100 : 50 : 650$
	$\eta = 0.05$	
DIM_NGPCA	$T_{\max} = 100000$	$N_{\text{CON}} = 50 : 50 : 600$
		$N = 7 : 3 : 22$
		$m = 1 : 1 : 4$
SLbA (a/b)	$T_{\max} = 50$	$\sigma_0 = 0.3 : 0.2 : 1.9$
	LE: $500 - 250 - 1000$	
	EP: $500 - 500 - 2000$	
	$\lambda_{\text{SLbA}} = 0.005$	
	$\lambda_\sigma = 1.0$	
	$\eta = 0.05$	

Table C.6 — Parameter values of the learning strategies for the **2-link arm with quality function Q_0 and additional sensor noise** with $Q^*_{0\text{N}} = 0.945$.

Learning strategy	Fixed parameters	Variable parameters
FEL/J^+	$T_{\max} = 200000$	$\eta = 0.015 : 0.005 : 0.075$
FEL/J^t	$T_{\max} = 200000$	$\eta = 0.005 : 0.005 : 0.05$
DSL	$T_{\max} = 200000$	$N_{\text{FM}} = 1000 : 2000 : 15000$
	$\eta_{\text{FM}} = 0.005$	$\eta = 0.01 : 0.01 : 0.05$
DIM_NGPCA	$T_{\max} = 100000$	$N_{\text{CON}} = 500 : 250 : 1500$
		$N = 40 : 20 : 100$
		$m = 2 : 1 : 5$
SLbA (a/b)	$T_{\max} = 50$	$\sigma_0 = 0.1 : 0.1 : 1.0$
	LE: $500 - 250 - 1500$	
	EP: $500 - 750 - 3500$	
	$\lambda_{\text{SLbA}} = 0.005$	
	$\lambda_\sigma = 1.0$	
	$\eta = 0.005$	

Table C.7 — Parameter values of the learning strategies for the **3-link arm with quality function Q_0** with $Q^*_0 = 0.96$.

Learning strategy	Fixed parameters	Variable parameters
FEL/J$^+$	$T_{\text{max}} = 200000$	$\eta = 0.015 : 0.005 : 0.075$
FEL/J^t	$T_{\text{max}} = 200000$	$\eta = 0.005 : 0.005 : 0.05$
DSL	$T_{\text{max}} = 200000$	$N_{\text{FM}} = 1000 : 2000 : 11000$
	$\eta_{\text{FM}} = 0.005$	$\eta = 0.01 : 0.01 : 0.05$
DIM_NGPCA	$T_{\text{max}} = 100000$	$N_{\text{CON}} = 1000 : 500 : 3000$
		$N = 20 : 20 : 100$
		$m = 2 : 1 : 5$
SLbA (a/b)	$T_{\text{max}} = 50$	$\sigma_0 = 0.1 : 0.1 : 1.0$
	LE: $500 - 250 - 1500$	
	EP: $500 - 750 - 3500$	
	$\lambda_{\text{SLbA}} = 0.005$	
	$\lambda_\sigma = 1.0$	
	$\eta = 0.005$	

Table C.8 — Parameter values of the learning strategies for the **3-link arm with quality function** Q_0 **and additional sensor noise** with $Q^*_{0\text{N}} = 0.93$.

Learning strategy	Fixed parameters	Variable parameters
DSL	$T_{\text{max}} = 200000$	$N_{\text{FM}} = 1000 : 2000 : 9000$
	$\eta_{\text{FM}} = 0.005$	$\eta = 0.002 : 0.004 : 0.038$
DIM_NGPCA	$T_{\text{max}} = 100000$	$N_{\text{CON}} = 1000 : 2000 : 9000$
		$N = 40 : 20 : 100$
		$m = 2 : 1 : 6$
SLbA (a/b)	$T_{\text{max}} = 50$	$\sigma_0 = 0.1 : 0.1 : 1.0$
	LE: $500 - 250 - 1500$	
	EP: $500 - 750 - 3500$	
	$\lambda_{\text{SLbA}} = 0.005$	
	$\lambda_\sigma = 1.0$	
	$\eta = 0.005$	

Table C.9 — Parameter values of the learning strategies for the **3-link arm with quality function** Q_1 with $Q^*_1 = 0.97$.

Learning strategy	Fixed parameters	Variable parameters
DSL	$T_{\max} = 200000$	$N_{\text{FM}} = 1000 : 2000 : 11000$
	$\eta_{\text{FM}} = 0.005$	$\eta = 0.01 : 0.01 : 0.05$
DIM_NGPCA	$T_{\max} = 200000$	$N_{\text{CON}} = 1000 : 10000 : 61000$
		$N = 20 : 100 : 720$
		$m = 1 : 1 : 6$
SLbA (a/b)	$T_{\max} = 50$	$\sigma_0 = 0.1 : 0.1 : 1.0$
	LE: $500 - 250 - 1500$	
	EP: $500 - 750 - 3500$	
	$\lambda_{\text{SLbA}} = 0.05$	
	$\lambda_\sigma = 0.9$	
	$\eta = 0.005$	

Table C.10 — Parameter values of the learning strategies for the **3-link arm with quality function** Q_2 with $Q_2^* = 0.86$.

Learning strategy	Fixed parameters	Variable parameters
FEL/J$^+$	$T_{\max} = 400000$	$\eta = 0.01 : 0.005 : 0.06$
FEL/J^t	$T_{\max} = 400000$	$\eta = 0.0025 : 0.0025 : 0.025$
DSL	$T_{\max} = 400000$	$N_{\text{FM}} = 1000 : 2000 : 11000$
	$\eta_{\text{FM}} = 0.005$	$\eta = 0.005 : 0.005 : 0.05$
DIM_NGPCA	$T_{\max} = 300000$	$N_{\text{CON}} = 2500 : 5000 : 17500$
		$N = 80 : 60 : 260$
		$m = 2 : 1 : 6$
SLbA (a/b)	$T_{\max} = 50$	$\sigma_0 = 0.1 : 0.15 : 1.15$
	LE: $500 - 250 - 3000$	
	EP: $500 - 500 - 5500$	
	$\lambda_{\text{SLbA}} = 0.005$	
	$\lambda_\sigma = 1.0$	
	$\eta = 0.005$	

Table C.11 — Parameter values of the learning strategies for the **4-link arm with quality function** Q_0 with $Q_0^* = 0.94$.

Learning strategy	Fixed parameters	Variable parameters
FEL ($\mathbf{J}^{+}$)	$T_{\max} = 400000$	$\eta = 0.01 : 0.005 : 0.06$
FEL ($\mathbf{J}^{t}$)	$T_{\max} = 400000$	$\eta = 0.0025 : 0.0025 : 0.025$
DSL	$T_{\max} = 400000$	$N_{\mathrm{FM}} = 1000 : 2000 : 11000$
	$\eta_{\mathrm{FM}} = 0.005$	$\eta = 0.005 : 0.005 : 0.05$
DIM_NGPCA	$T_{\max} = 300000$	$N_{\mathrm{CON}} = 2500 : 5000 : 22500$
		$N = 80 : 60 : 320$
		$m = 2 : 1 : 6$
SLbA (a/b)	$T_{\max} = 50$	$\sigma_0 = 0.05 : 0.05 : 0.6$
	LE: $500 - 250 - 3000$	
	EP: $500 - 500 - 5500$	
	$\lambda_{\mathrm{SLbA}} = 0.005$	
	$\lambda_{\sigma} = 1.0$	
	$\eta = 0.005$	

Table C.12 — Parameter values of the learning strategies for the **4-link arm with quality function Q_0 and additional sensor noise** with $Q^*_{0\mathrm{N}} = 0.9$.

Learning strategy	Fixed parameters	Variable parameters
DSL	$T_{\max} = 400000$	$N_{\mathrm{FM}} = 5000 : 10000 : 55000$
	$\eta_{\mathrm{FM}} = 0.005$	$\eta = 0.002 : 0.004 : 0.018$
DIM_NGPCA	$T_{\max} = 300000$	$N_{\mathrm{CON}} = 30000 : 30000 : 150000$
		$N = 40 : 80 : 360$
		$m = 2 : 1 : 6$
SLbA (a/b)	$T_{\max} = 50$	$\sigma_0 = 0.1 : 0.1 : 1.0$
	LE: $500 - 250 - 1500$	
	EP: $500 - 750 - 3500$	
	$\lambda_{\mathrm{SLbA}} = 0.005$	
	$\lambda_{\sigma} = 1.0$	
	$\eta = 0.005$	

Table C.13 — Parameter values of the learning strategies for the **4-link arm with quality function Q_1** with $Q^*_1 = 0.97$.

Learning strategy	Fixed parameters	Variable parameters
DSL	$T_{\mathrm{max}} = 400000$	$N_{\mathrm{FM}} = 1000 : 2000 : 11000$
	$\eta_{\mathrm{FM}} = 0.005$	$\eta = 0.005 : 0.005 : 0.05$
DIM_NGPCA	$T_{\mathrm{max}} = 200000$	$N_{\mathrm{CON}} = 1000 : 3000 : 34000$
		$N = 20 : 100 : 820$
		$m = 1 : 1 : 7$
SLbA (a/b)	$T_{\mathrm{max}} = 50$	$\sigma_0 = 0.1 : 0.1 : 1.0$
	LE: $500 - 250 - 1500$	
	EP: $500 - 750 - 3500$	
	$\lambda_{\mathrm{SLbA}} = 0.05$	
	$\lambda_\sigma = 0.9$	
	$\eta = 0.005$	

Table C.14 — Parameter values of the learning strategies for the **4-link arm with quality function** Q_2 with $Q_2^* = 0.86$.

Appendix D

Experimental Results of Chapter 4

D.1 Saccade Control

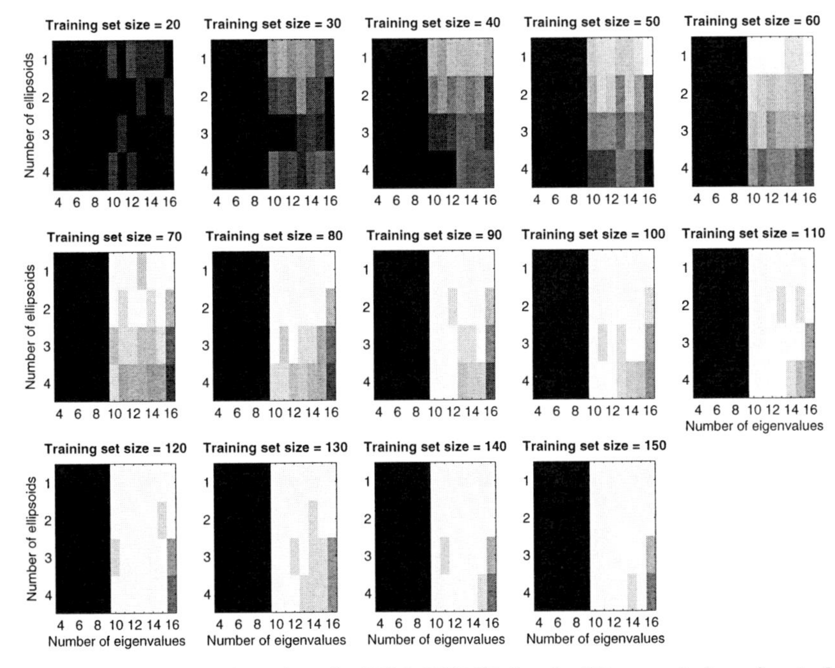

Figure D.1 — These plots show for DIM_NGPCA for the **2D saccade learning task without retinal noise** which combinations of the number of eigenvalues and the number of ellipsoids are successful throughout all 20 learning passes in exceeding the target quality Q^* for the respective sizes of the training set (white: successful in all passes; black: not successful in any pass).

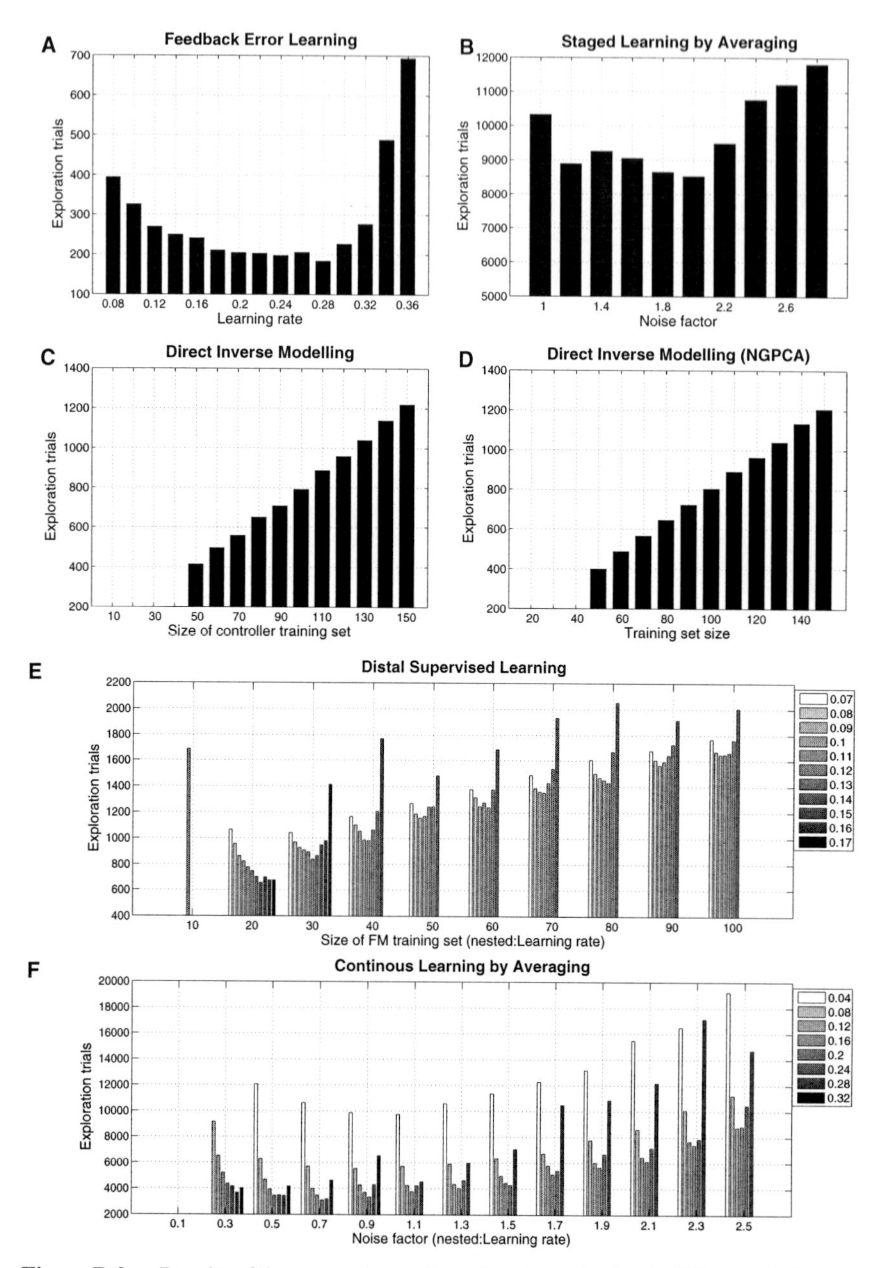

Figure D.2 — Results of the comparison of learning strategies for the **2D saccade learning task without retinal noise**. The length of the bars represents the number of required exploration trials. Bars are completely omitted whenever at least one of the 20 learning passes failed.

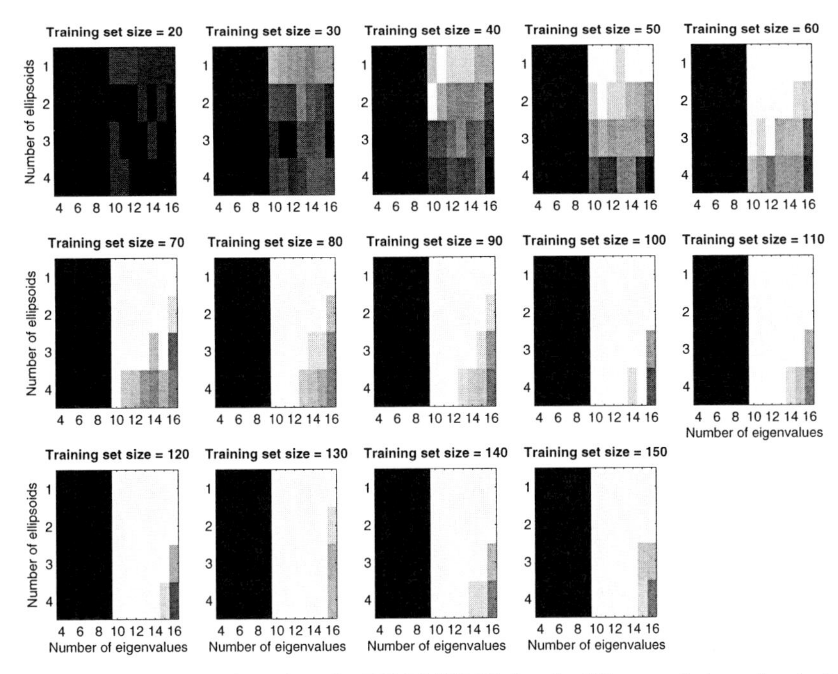

Figure D.3 — These plots show for DIM_NGPCA for the **2D saccade learning task with retinal noise** which combinations of the number of eigenvalues and the number of ellipsoids are successful throughout all 20 learning passes in exceeding the target quality Q^* for the respective sizes of the training set (white: successful in all passes; black: not successful in any pass).

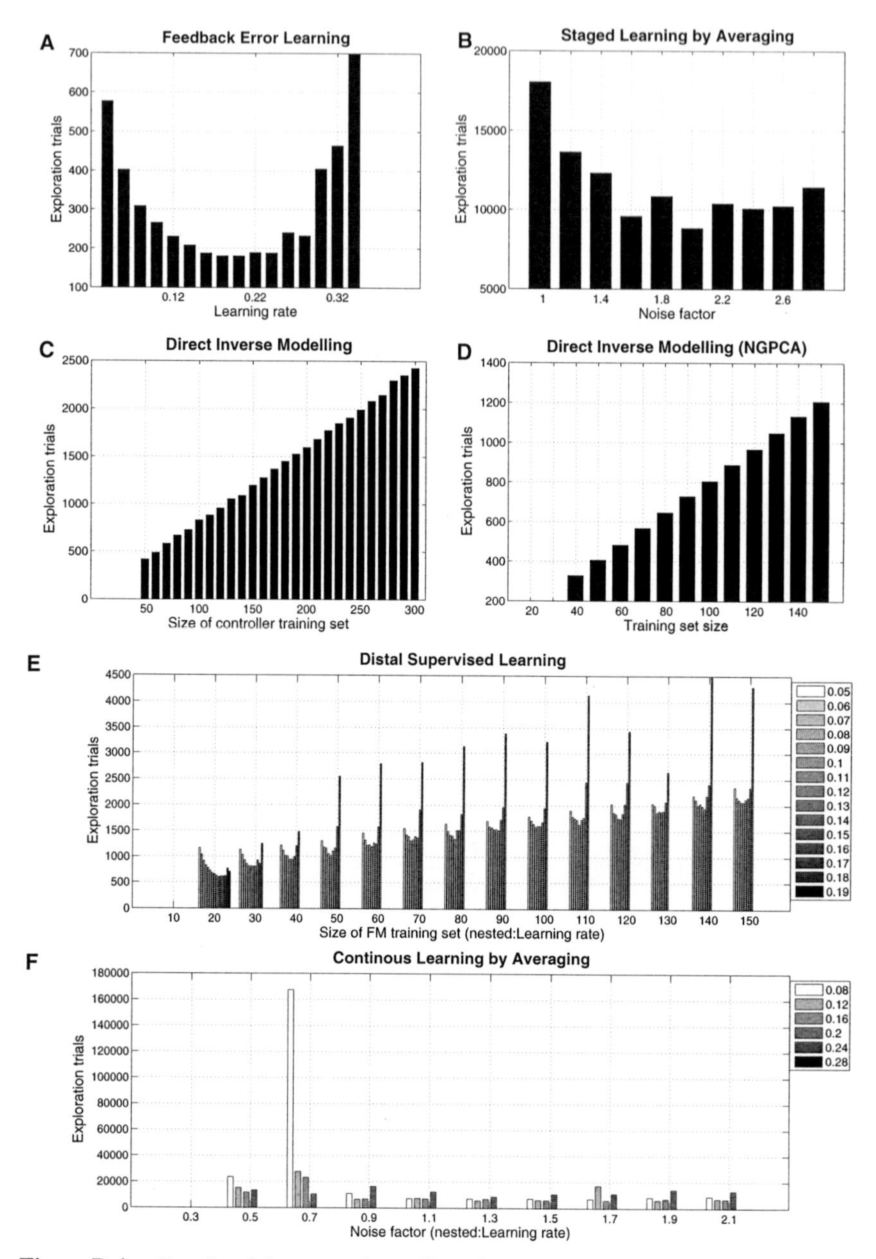

Figure D.4 — Results of the comparison of learning strategies for the **2D saccade learning task with retinal noise** (for further explanation see caption of Fig. D.2).

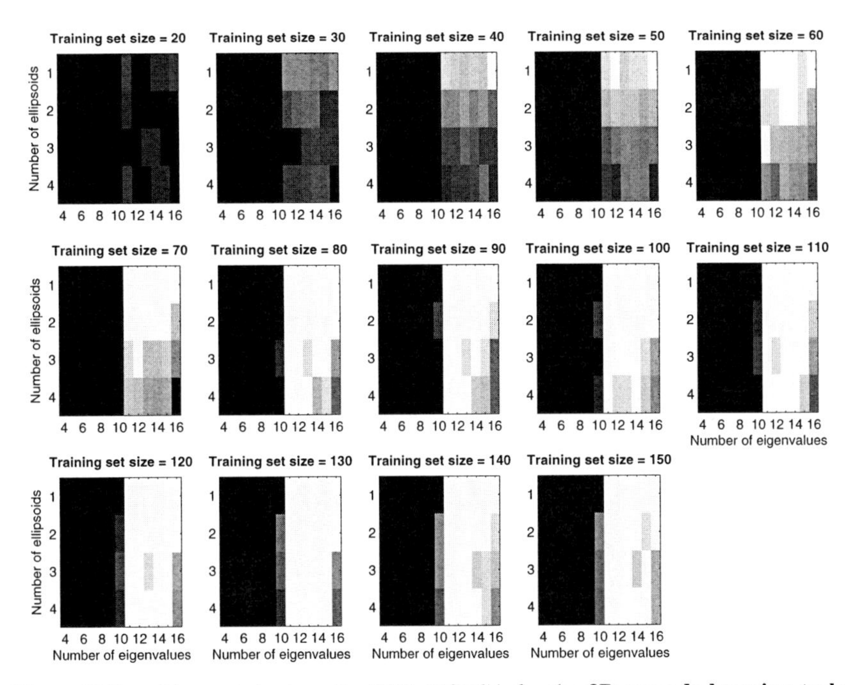

Figure D.5 — These plots show for DIM_NGPCA for the **3D saccade learning task without retinal noise** which combinations of the number of eigenvalues and the number of ellipsoids are successful throughout all 20 learning passes in exceeding the target quality Q^* for the respective sizes of the training set (white: successful in all passes; black: not successful in any pass).

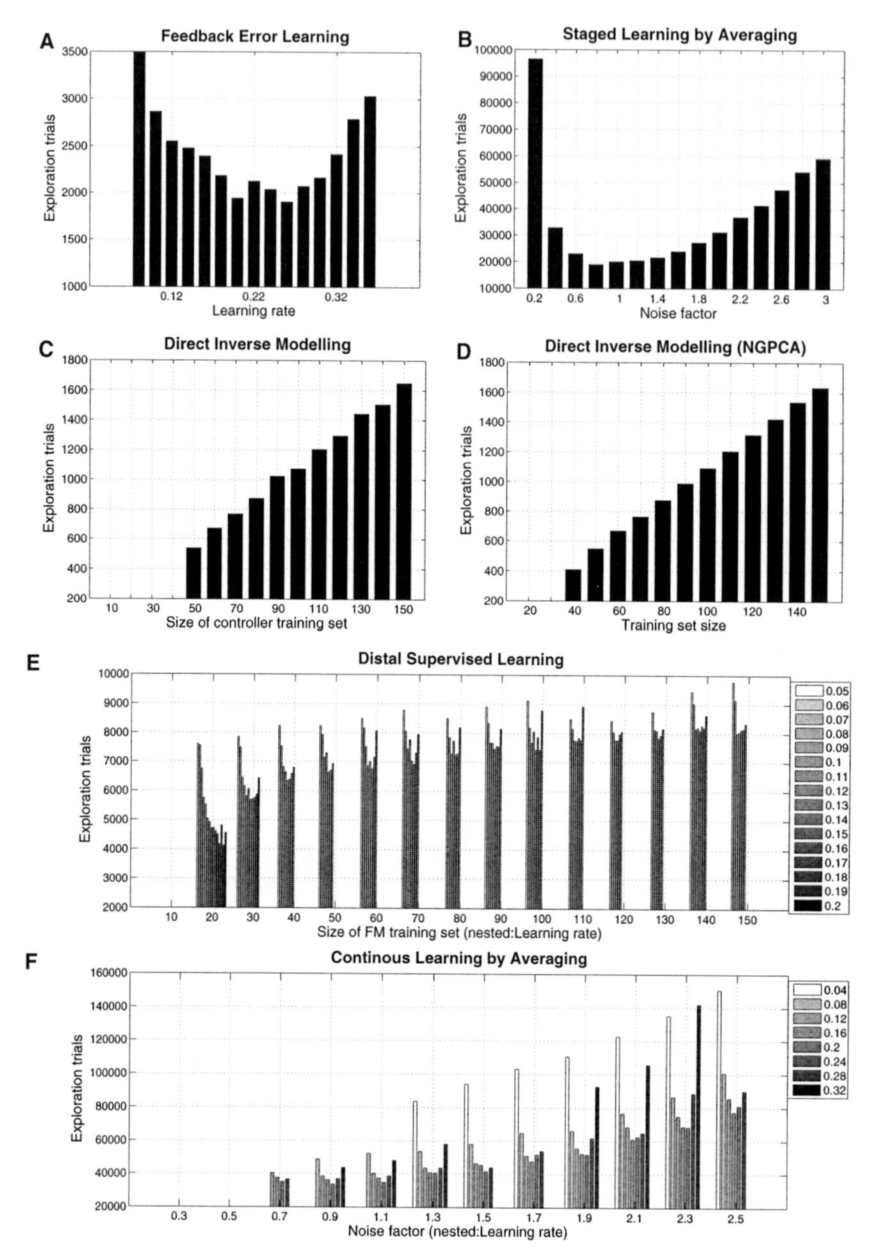

Figure D.6 — Results of the comparison of learning strategies for the **3D saccade learning task without retinal noise** (for further explanation see caption of Fig. D.2).

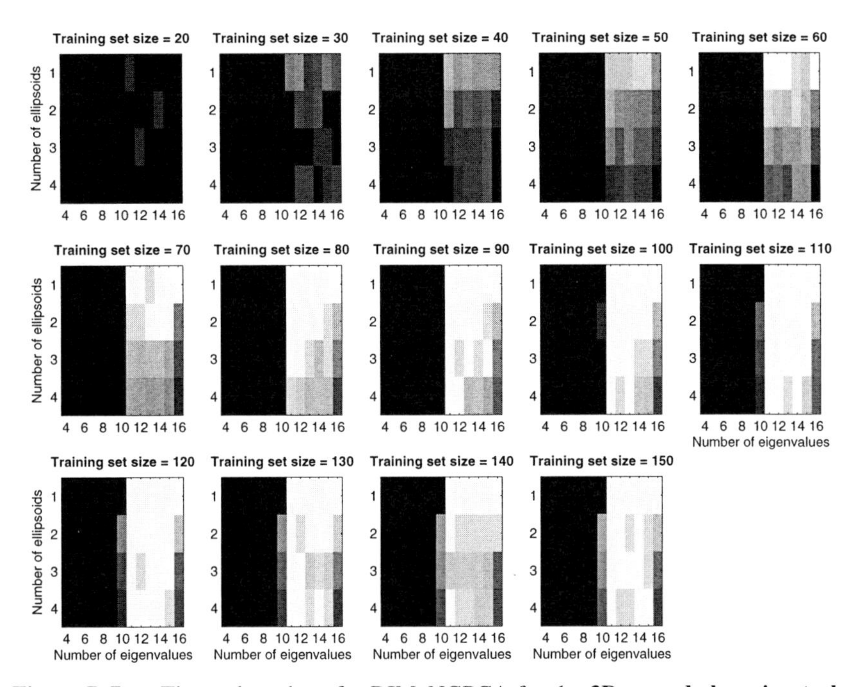

Figure D.7 — These plots show for DIM_NGPCA for the **3D saccade learning task with retinal noise** which combinations of the number of eigenvalues and the number of ellipsoids are successful throughout all 20 learning passes in exceeding the target quality Q^* for the respective sizes of the training set (white: successful in all passes; black: not successful in any pass).

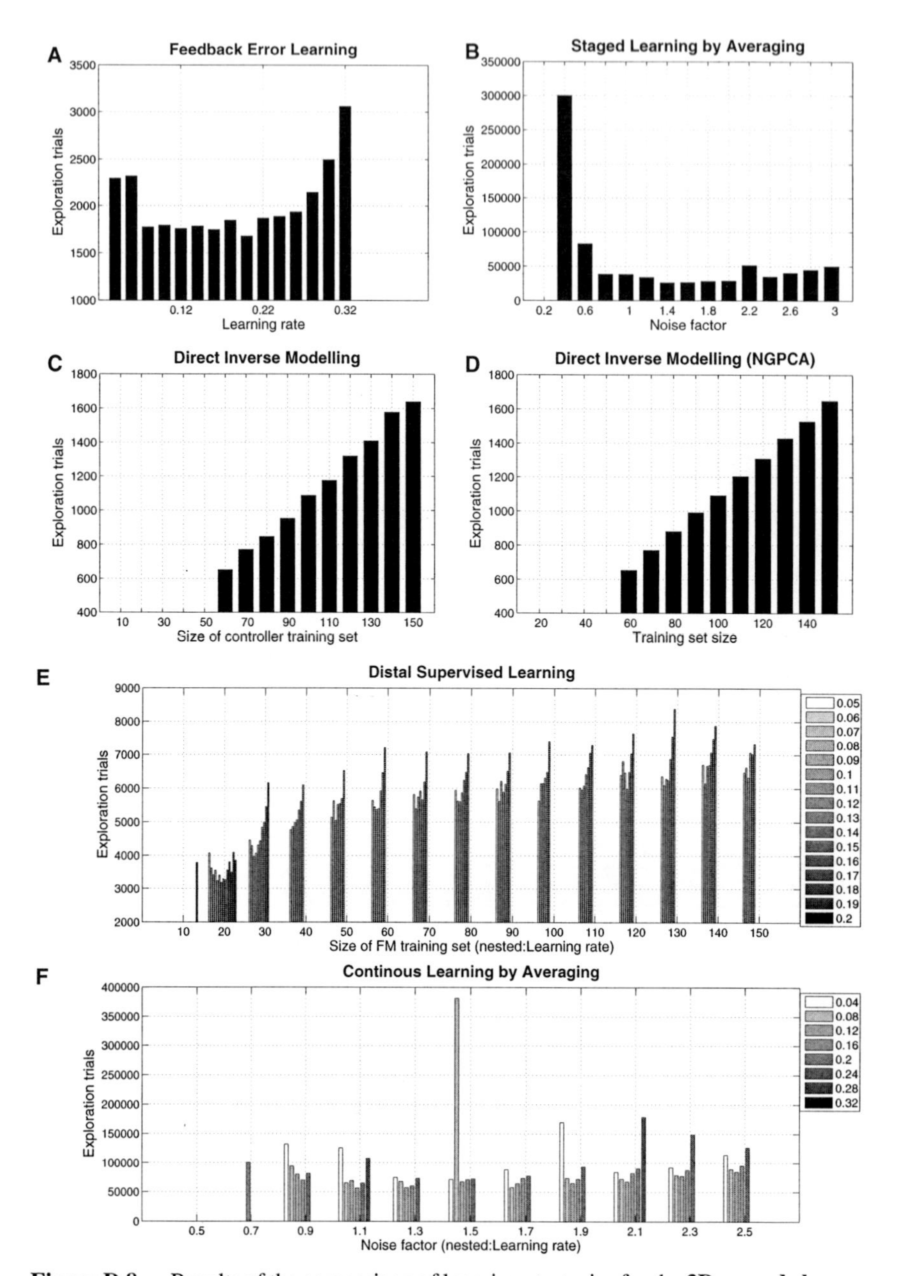

Figure D.8 — Results of the comparison of learning strategies for the **3D saccade learning task with retinal noise** (for further explanation see caption of Fig. D.2).

D.2 Planar Arm

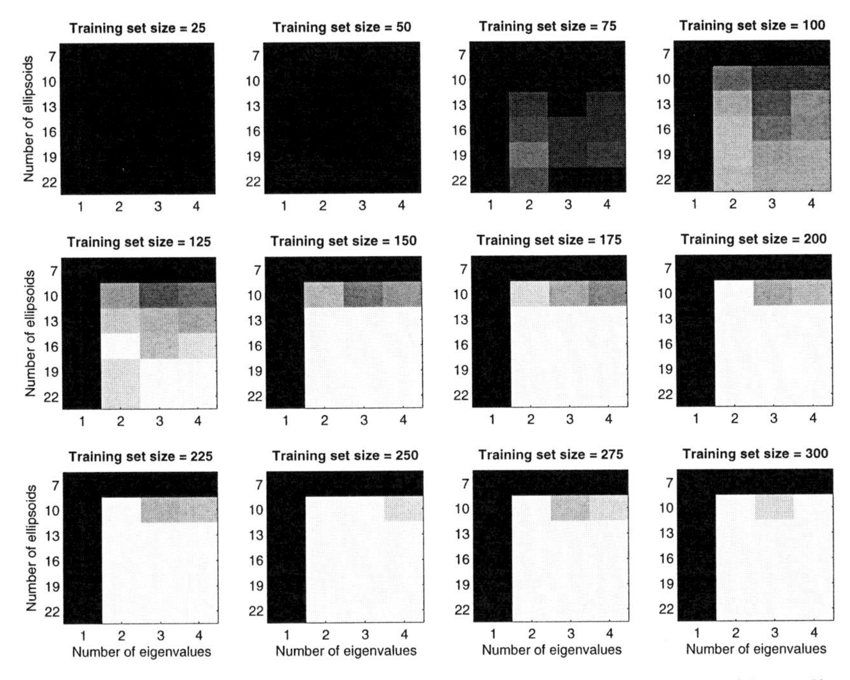

Figure D.9 — These plots show for DIM_NGPCA for the **2-link arm with quality function** Q_0 which combinations of the number of eigenvalues and the number of ellipsoids are successful throughout all 20 learning passes in exceeding the target quality Q_0^* for the respective sizes of the training set (white: successful in all passes; black: not successful in any pass).

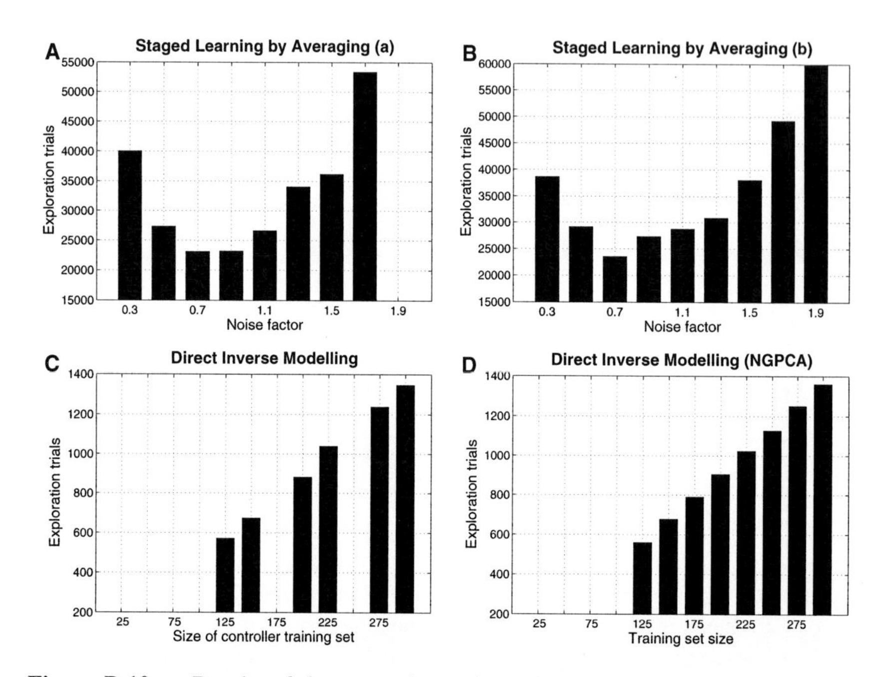

Figure D.10 — Results of the comparison of learning strategies for the **2-link arm with quality function** Q_0. The length of the bars represents the number of required exploration trials. Bars are completely omitted whenever at least one of the 20 learning passes failed.

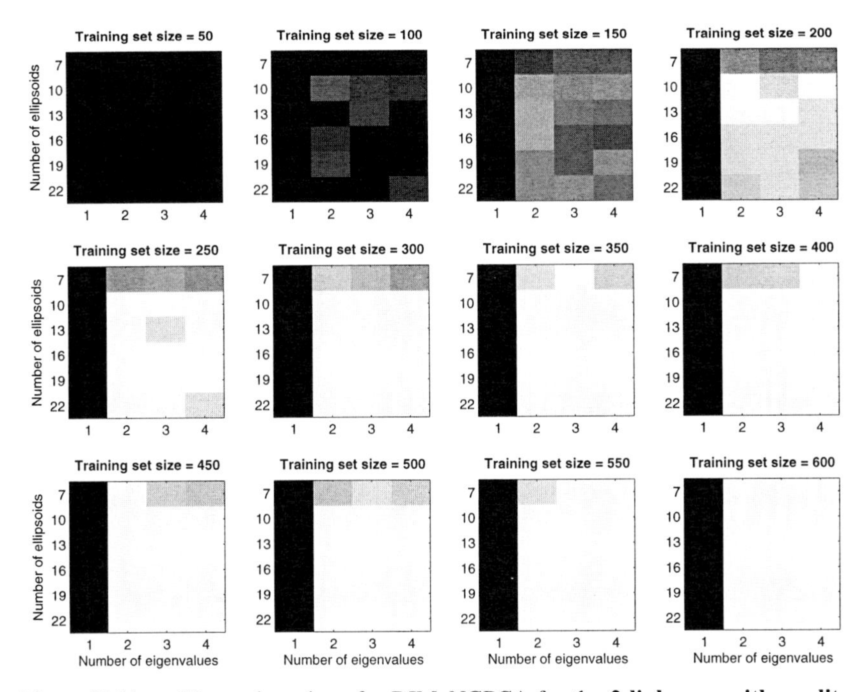

Figure D.11 — These plots show for DIM_NGPCA for the **2-link arm with quality function Q_0 and additional sensor noise** which combinations of the number of eigenvalues and the number of ellipsoids are successful throughout all 20 learning passes in exceeding the target quality $Q^*_{0\mathrm{N}}$ for the respective sizes of the training set (white: successful in all passes; black: not successful in any pass).

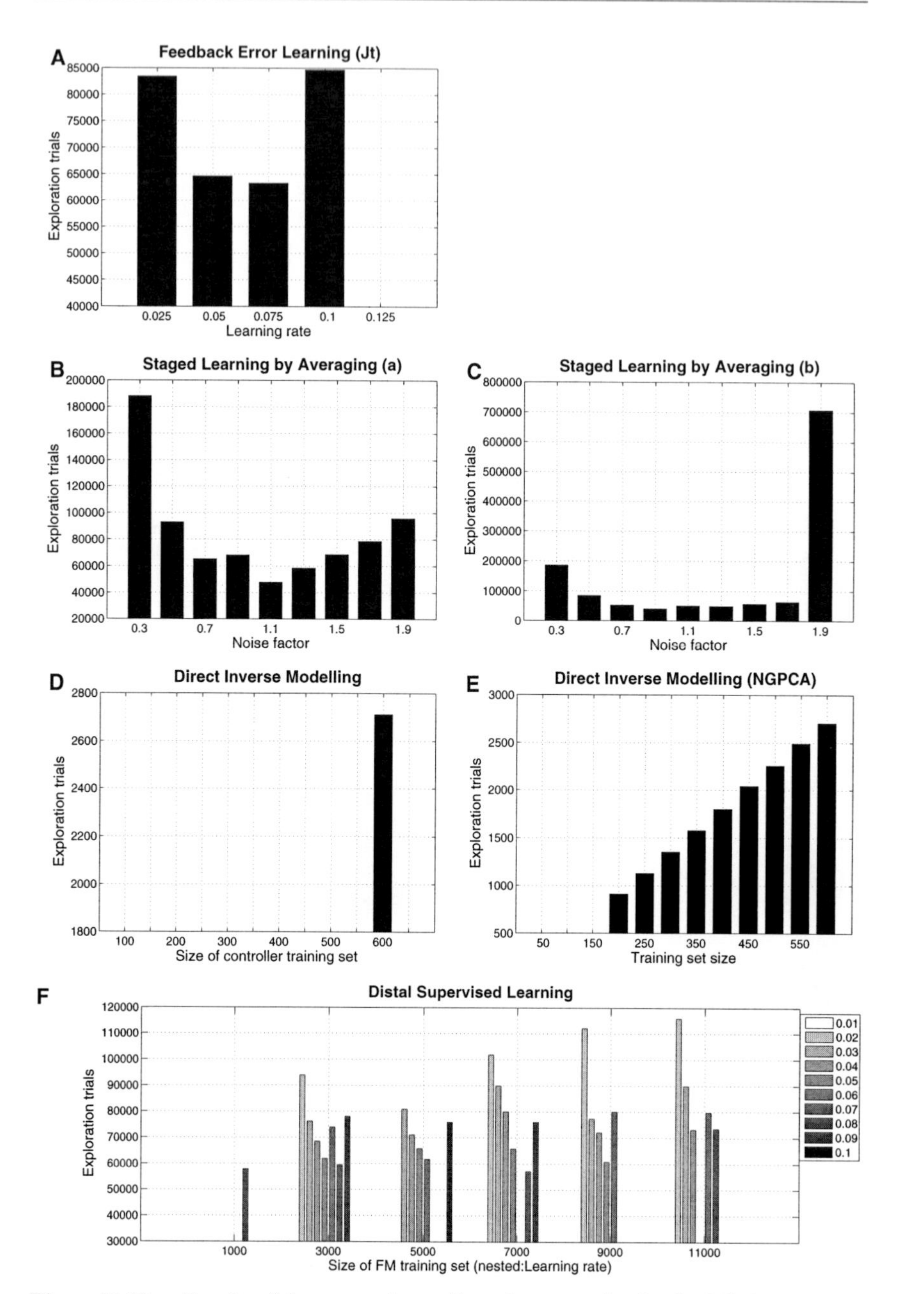

Figure D.12 — Results of the comparison of learning strategies for the **2-link arm with quality function Q_0 and additional sensor noise** (for further explanation see caption of Fig. D.10).

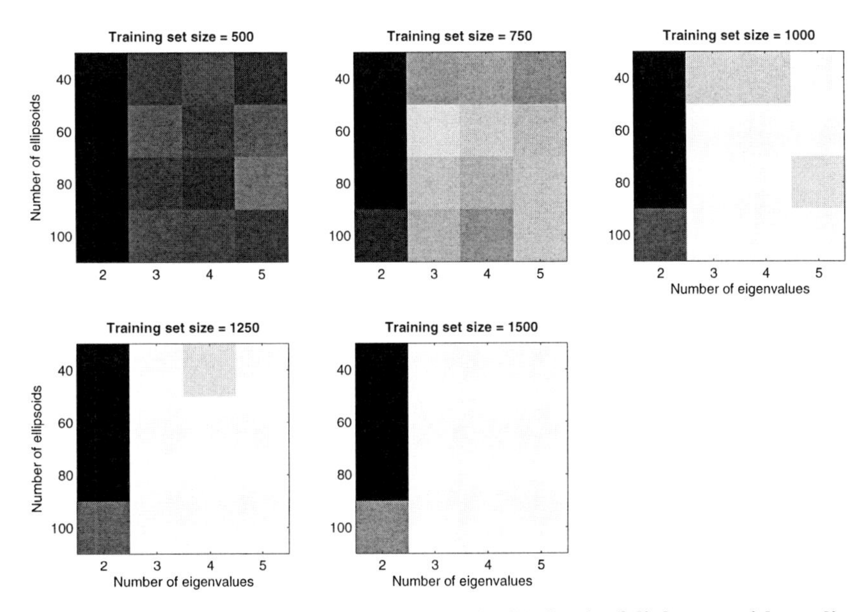

Figure D.13 — These plots show for DIM_NGPCA for the **3-link arm with quality function** Q_0 which combinations of the number of eigenvalues and the number of ellipsoids are successful throughout all 20 learning passes in exceeding the target quality Q_0^* for the respective sizes of the training set (white: successful in all passes; black: not successful in any pass).

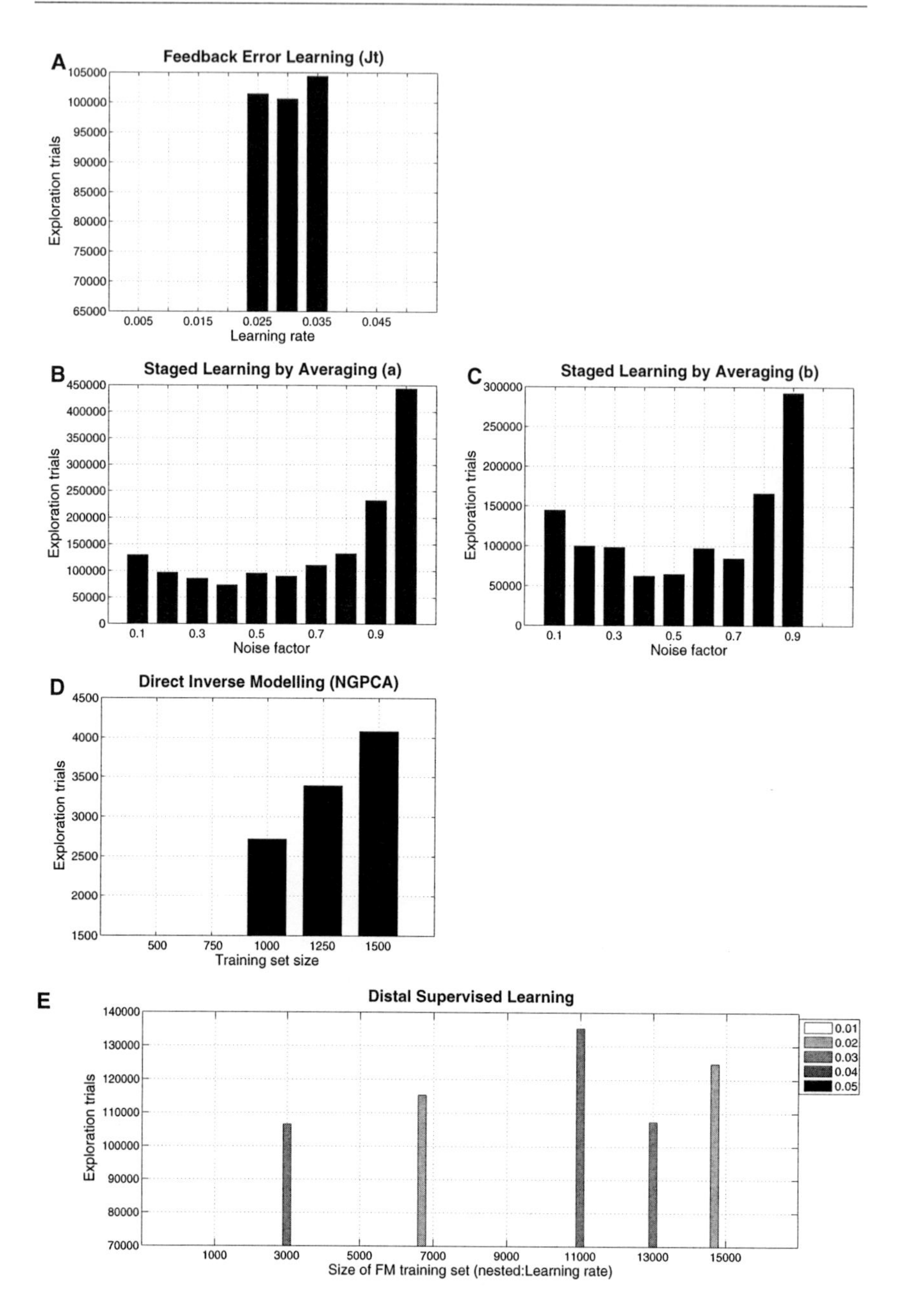

Figure D.14 — Results of the comparison of learning strategies for the **3-link arm with quality function** Q_0 (for further explanation see caption of Fig. D.10).

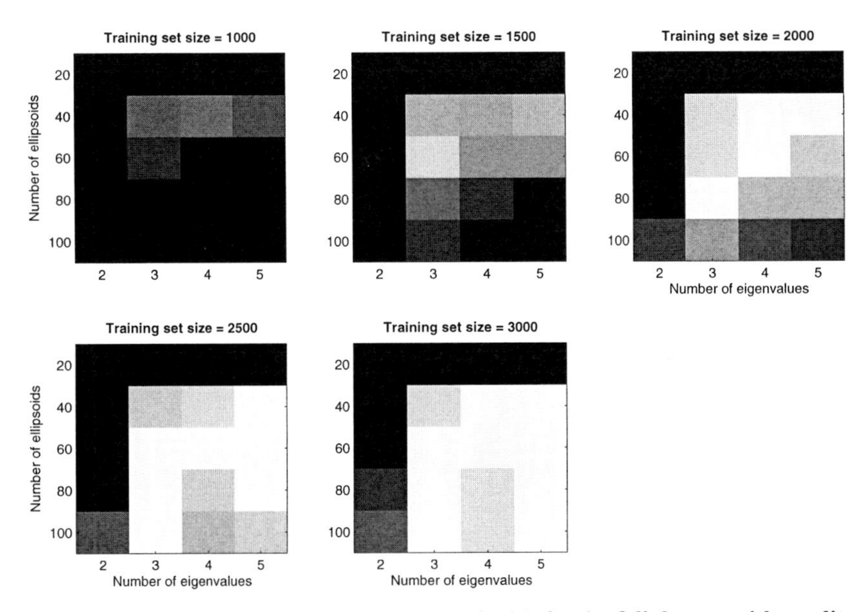

Figure D.15 — These plots show for DIM_NGPCA for the **3-link arm with quality function Q_0 and additional sensor noise** which combinations of the number of eigenvalues and the number of ellipsoids are successful throughout all 20 learning passes in exceeding the target quality $Q^*_{0\mathrm{N}}$ for the respective sizes of the training set (white: successful in all passes; black: not successful in any pass).

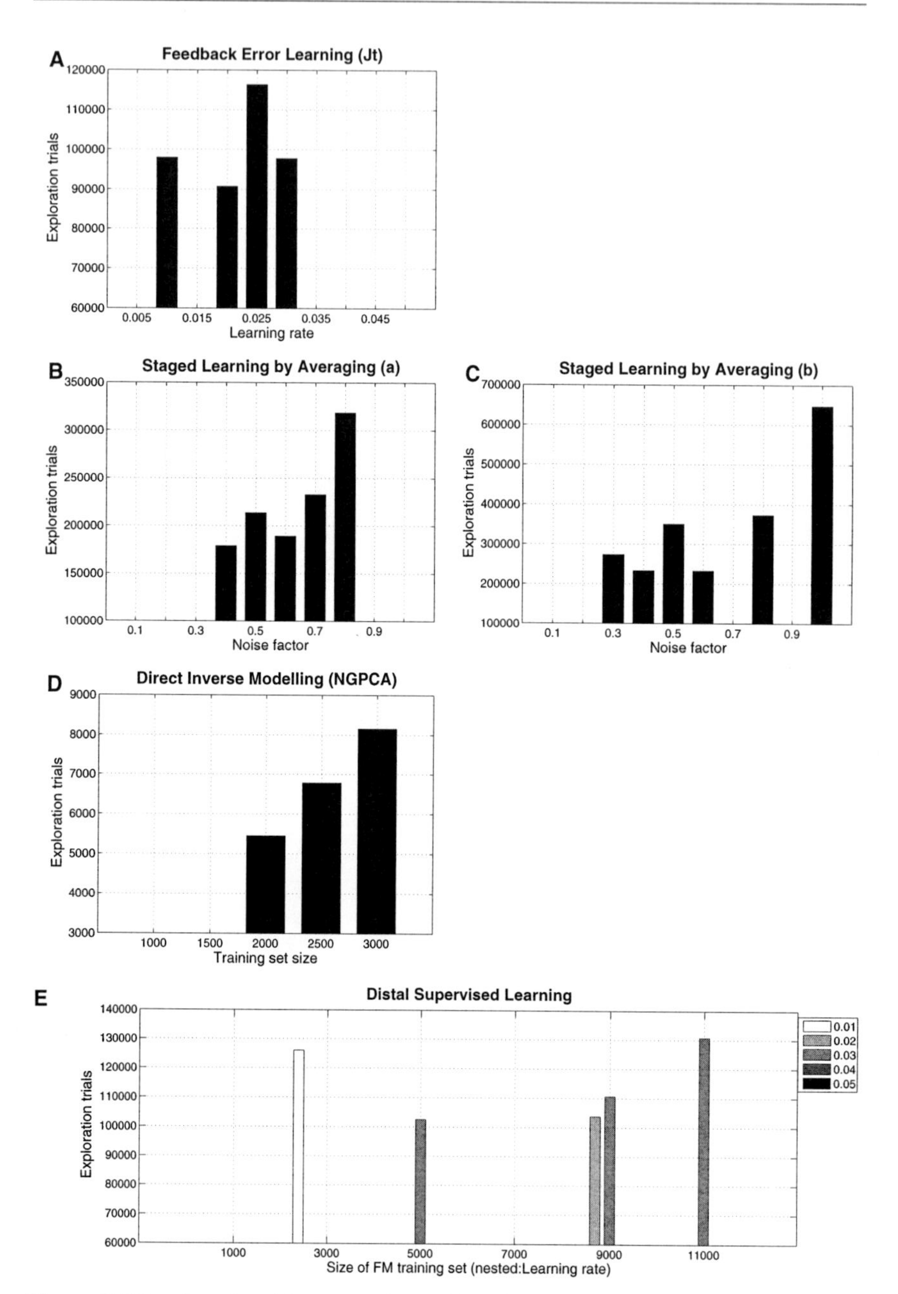

Figure D.16 — Results of the comparison of learning strategies for the **3-link arm with quality function Q_0 and additional sensor noise** (for further explanation see caption of Fig. D.10).

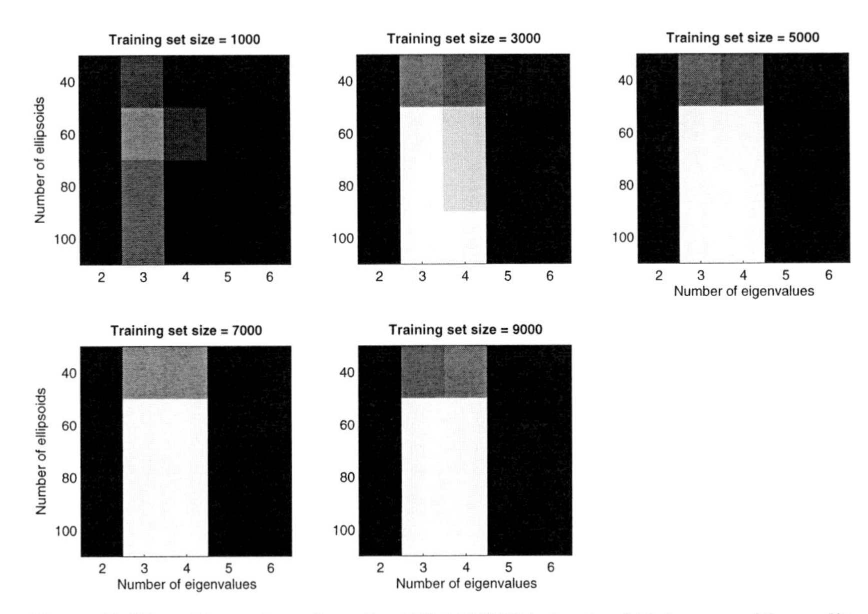

Figure D.17 — These plots show for DIM_NGPCA for the **3-link arm with quality function** Q_1 which combinations of the number of eigenvalues and the number of ellipsoids are successful throughout all 20 learning passes in exceeding the target quality Q_1^* for the respective sizes of the training set (white: successful in all passes; black: not successful in any pass).

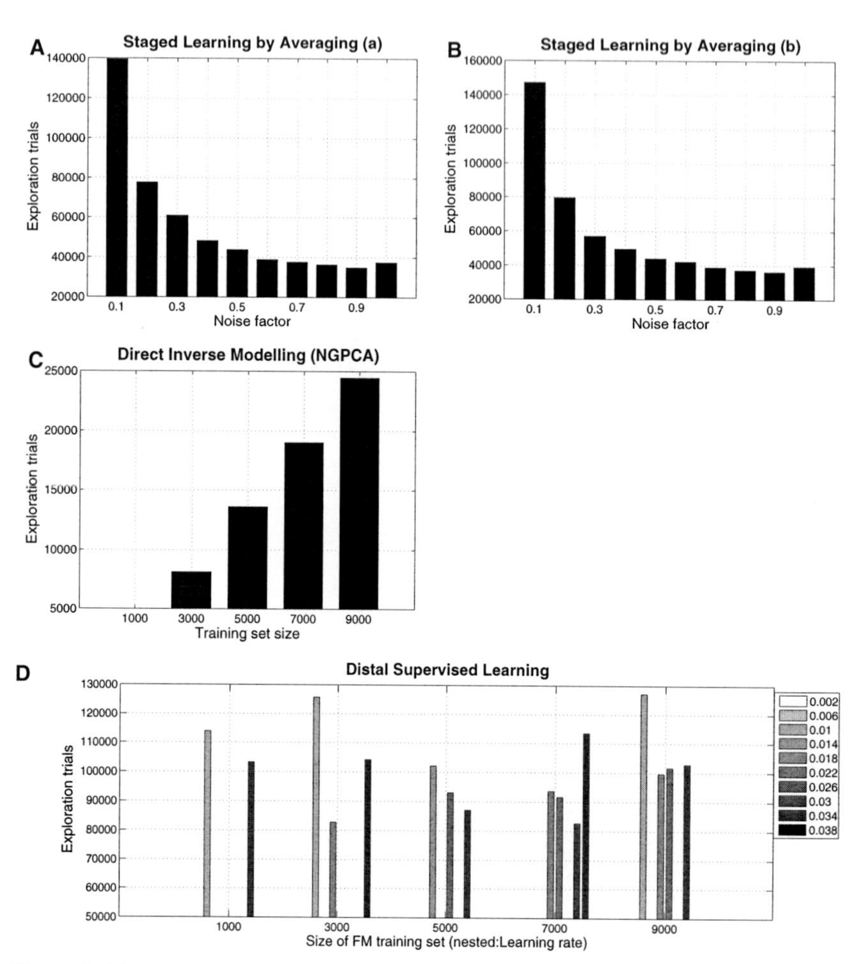

Figure D.18 — Results of the comparison of learning strategies for the **3-link arm with quality function** Q_1 (for further explanation see caption of Fig. D.10).

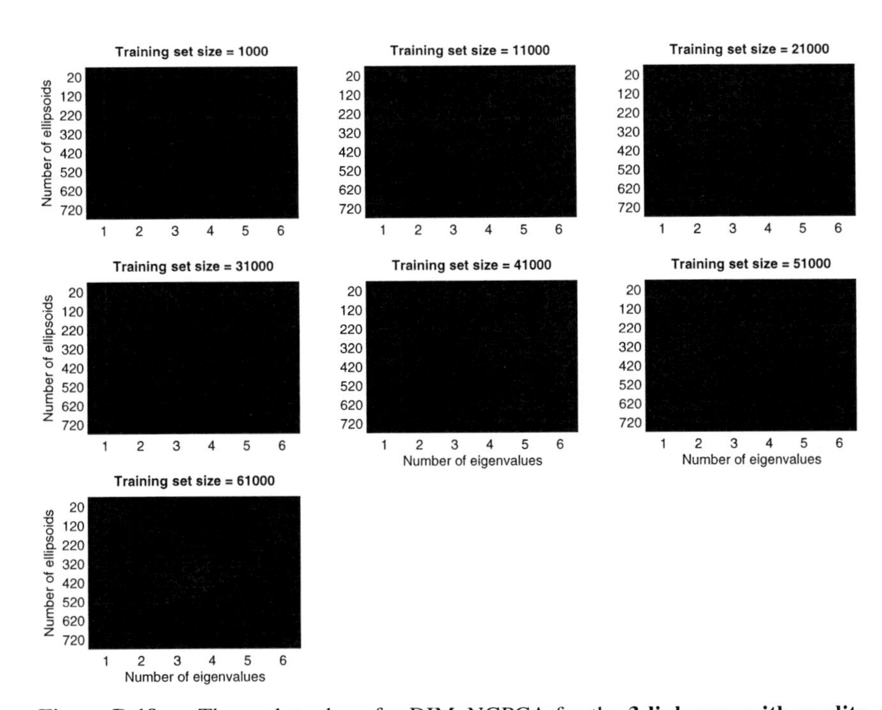

Figure D.19 — These plots show for DIM_NGPCA for the **3-link arm with quality function** Q_2 which combinations of the number of eigenvalues and the number of ellipsoids are successful throughout all 5 learning passes in exceeding the target quality Q_2^* for the respective sizes of the training set (white: successful in all passes; black: not successful in any pass).

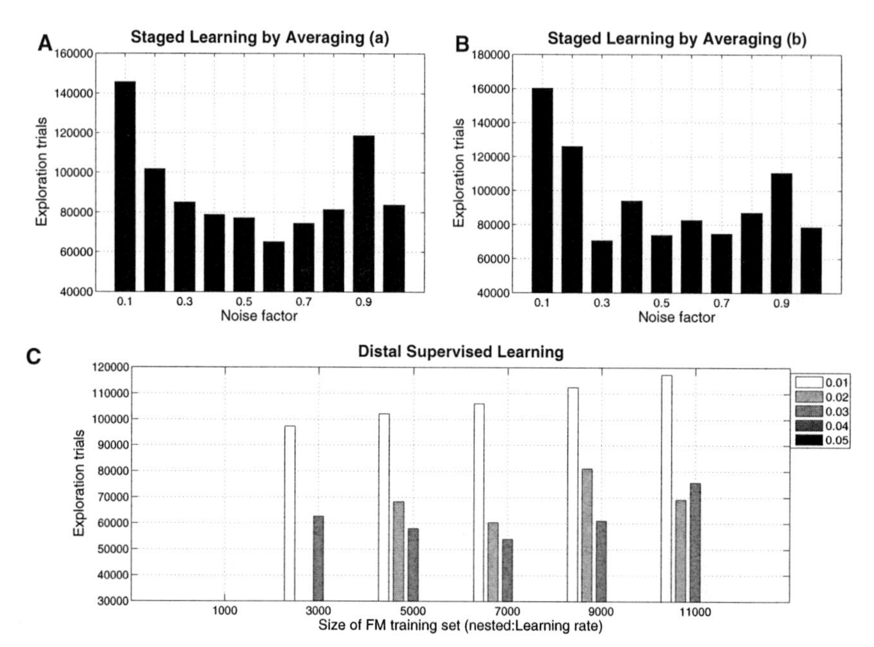

Figure D.20 — Results of the comparison of learning strategies for the **3-link arm with quality function** Q_2 (for further explanation see caption of Fig. D.10).

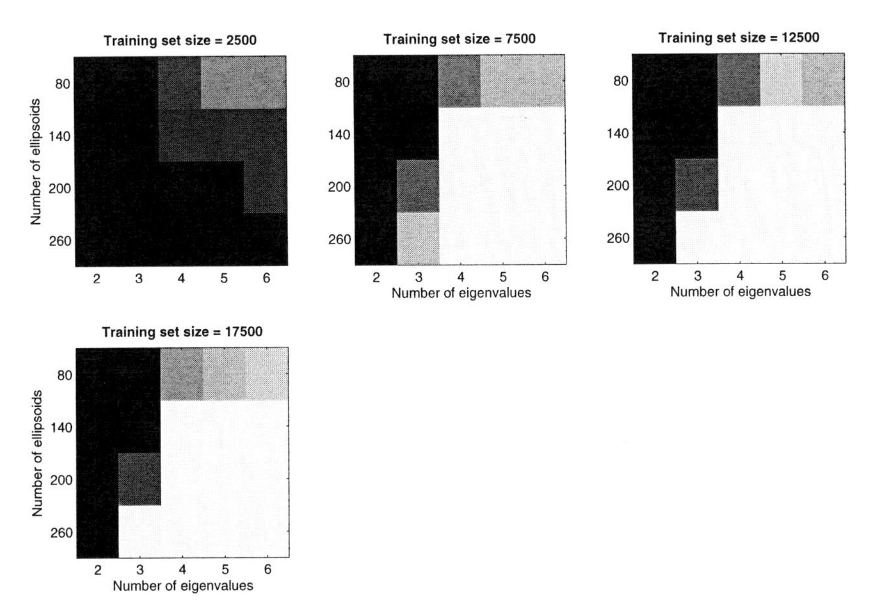

Figure D.21 — These plots show for DIM_NGPCA for the **4-link arm with quality function** Q_0 which combinations of the number of eigenvalues and the number of ellipsoids are successful throughout all 20 learning passes in exceeding the target quality Q_0^* for the respective sizes of the training set (white: successful in all passes; black: not successful in any pass).

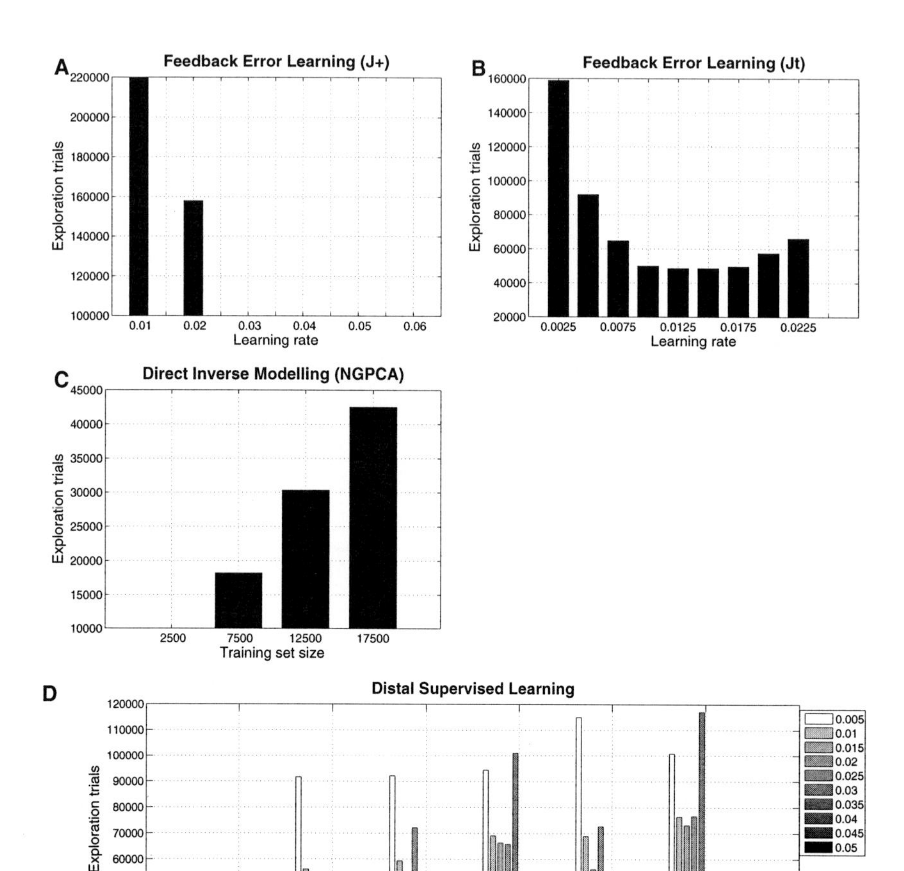

Figure D.22 — Results of the comparison of learning strategies for the **4-link arm with quality function** Q_0 (for further explanation see caption of Fig. D.10).

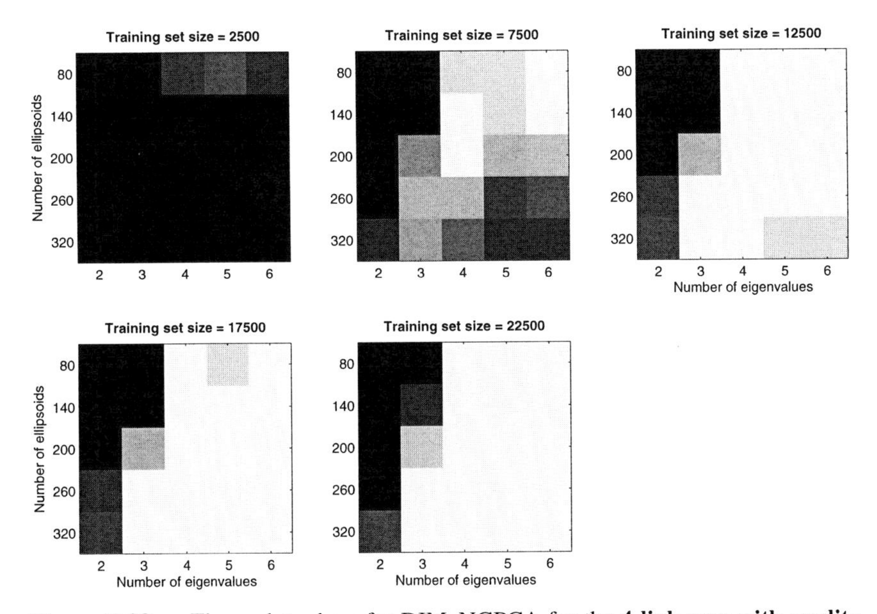

Figure D.23 — These plots show for DIM_NGPCA for the **4-link arm with quality function Q_0 with additional sensor noise** which combinations of the number of eigenvalues and the number of ellipsoids are successful throughout all 20 learning passes in exceeding the target quality Q^*_{0N} for the respective sizes of the training set (white: successful in all passes; black: not successful in any pass).

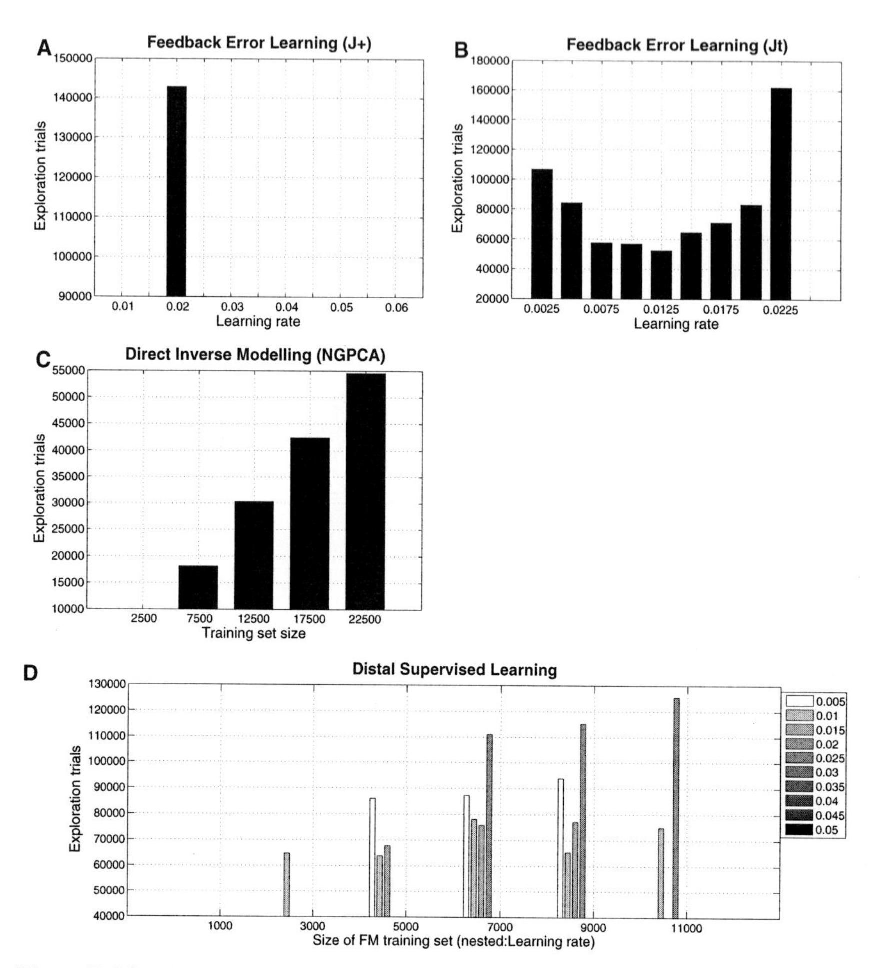

Figure D.24 — Results of the comparison of learning strategies for the **4-link arm with quality function Q_0 with additional sensor noise** (for further explanation see caption of Fig. D.10).

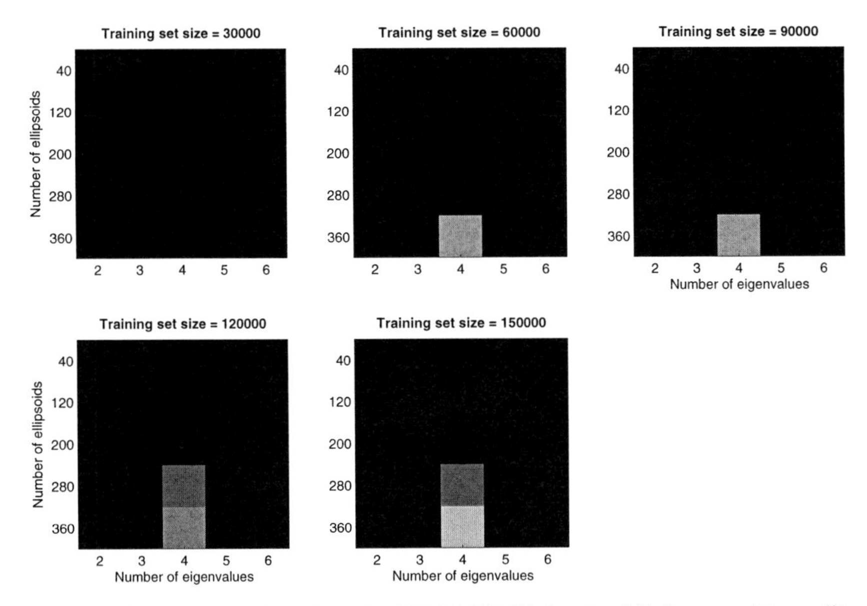

Figure D.25 — These plots show for DIM_NGPCA for the **4-link arm with quality function** Q_1 which combinations of the number of eigenvalues and the number of ellipsoids are successful throughout all 5 learning passes in exceeding the target quality Q_1^* for the respective sizes of the training set (white: successful in all passes; black: not successful in any pass).

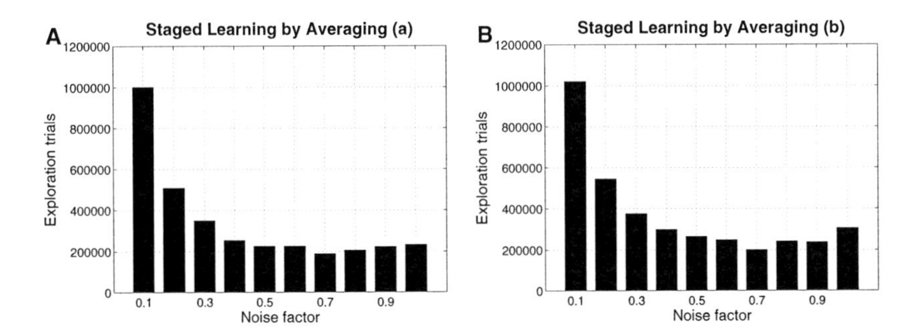

Figure D.26 — Results of the comparison of learning strategies for the **4-link arm with quality function** Q_1 (for further explanation see caption of Fig. D.10).

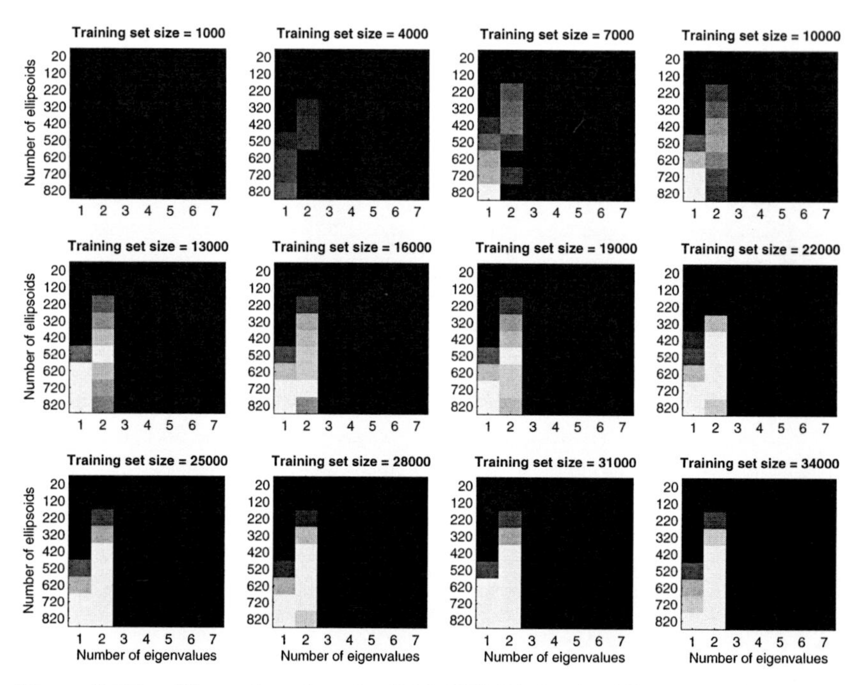

Figure D.27 — These plots show for DIM_NGPCA for the **4-link arm with quality function** Q_2 which combinations of the number of eigenvalues and the number of ellipsoids are successful throughout all 20 learning passes in exceeding the target quality Q_2^* for the respective sizes of the training set (white: successful in all passes; black: not successful in any pass).

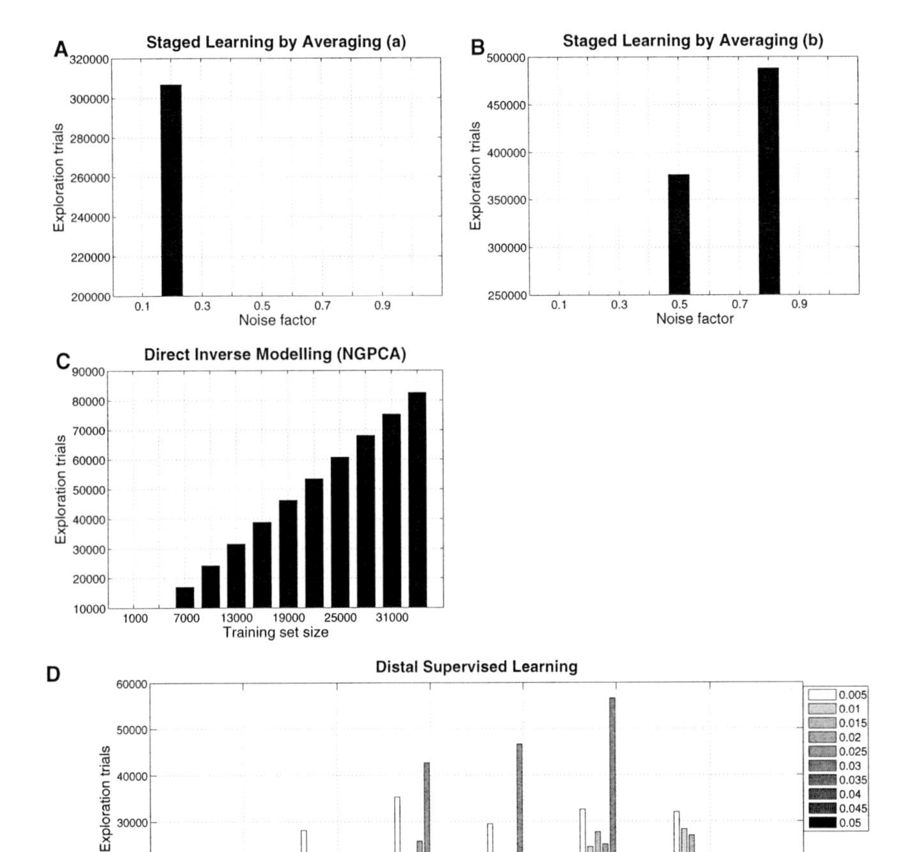

Figure D.28 — Results of the comparison of learning strategies for the **4-link arm with quality function** Q_2 (for further explanation see caption of Fig. D.10).

Appendix E

Notation and Symbols

Often used abbreviations (in alphabetical order)

CLbA	continuous learning by averaging
CNS	central nervous system
DE	differential evolution
DIM	direct inverse modeling
DIM_NGPCA	direct inverse modeling in combination with NGPCA
DSL	distal supervised learning
FC	feedback controller
FDM	forward dynamics model
FEL	feedback error learning
FM	forward model
IDM	inverse dynamics model
IDM_{fb}	feedback IDM
IDM_{ff}	feedforward IDM
IKM	inverse kinematics model
IM	inverse model
LbA	learning by averaging
LbI	learning by input adjustment
MLP	multi-layer perceptron
MM	mapping model
MOSAIC	model of "modular selection and identification for control"
NG	neural gas
NGPCA	abstract recurrent network on the basis of NG and PCA
PCA	principal component analysis
RBFN	radial basis function network
RL	reinforcement learning
SLbA	staged learning by averaging
VM	validator model

Often used symbols in the domain of computational methods (mainly in Chapt. 3)

a^{ik}	activation of node i in layer k (MLP)
$\mathbf{c}$	unit center (RBFN/NG/NGPCA)
E	error (MLP/NG) <u>or</u> energy (DE)
i	unit index
j	unit index <u>or</u> eigenvector index (only NGPCA)
k,l	layer indices (MLP)
K	number of layers (MLP)
m	number of eigenvectors (NGPCA)
n	dimensionality of data space
n^{ik}	node i in layer k (MLP)
N	number of units (NG/NGPCA)
N_{DE}	population size (DE)
N_k	number of units in layer k (MLP/RBFN)
o^{ik}	output of node i in layer k (MLP)
p	pattern index
P	number of patterns
T	number of training steps (NG/NGPCA)
w_j^i	weight of the connection from input unit j to the Gaussian unit i (RBFN)
w_{jl}^{ik}	weight of the connection from unit j in layer l to unit i in layer k (MLP)
$\mathbf{w}$	eigenvector estimate (NGPCA)
$\mathbf{W}$	weight matrix (MLP/RBFN) <u>or</u>
	matrix of estimated eigenvectors (NGPCA)
$\mathbf{x}$	vector drawn from data space (NG/NGPCA/DE)
$\mathbf{x}_{\text{in}}$	vector drawn from input data space (MLP/RBFN)
$\mathbf{x}_{\text{out}}$	vector drawn from output data space (MLP/RBFN)
X	set with training data
ϵ	learning rate (NG/NGPCA)
η	learning rate (MLP amongst others)
λ	eigenvalue estimate (NGPCA)
λ^*	estimated residual variance in a minor eigendirection (NGPCA)
$\mathbf{\Lambda}$	diagonal matrix of eigenvalue estimates (NGPCA)
ρ	neighborhood range (NG/NGPCA)
σ	standard deviation of Gaussian unit (RBFN) <u>or</u>
	general standard deviation
$\{...\}$	set with elements counted by one or several indices

Often used symbols in the domain of motor learning (mainly in Chapts. 2 and 4)

C	controller
F	feedback controller
$\mathbf{G}$	gain matrix
h	sensory output model
$\mathbf{J}$	Jacobi matrix
k	stage index (SLbA)
$\mathbf{K}$	Kalman gain matrix
N_{CON}	number of learning examples for the controller training (DIM/DIM_NGPCA)
N_{EX}	number of exploration trials
$N_{\mathrm{EX}}^{\mathrm{CON}}$	number of exploration trials for the controller (DSL)
$N_{\mathrm{EX}}^{\mathrm{FM}}$	number of exploration trials for the FM (DSL)
N_{FM}	number of learning examples for the FM training (DSL)
P	plant
Q	quality (function)
$\widetilde{Q}$	quality threshold
Q^*	desired controller quality level
Q_C	average controller quality
Q^{DE}	average quality of controller networks on the basis of DE
Q_{PS}	average quality of the learning examples in a pattern set
$\mathbf{u}$	motor command
$\mathbf{x}$	system state (general) or
	sensory context (only for specific kinematic problems)
$\widehat{\mathbf{x}}$	estimated/predicted system state
$\widehat{\mathbf{x}}^-$	uncorrected system state estimate
$\mathbf{x}_d(t)$	desired trajectory in state space
$\mathbf{y}$	plant output
$\widehat{\mathbf{y}}$	estimated/predicted plant output
$\mathbf{y}^*$	desired plant output
$\boldsymbol{\theta}$	joint angles
σ_0	noise factor (CLbA/SLbA)
$\boldsymbol{\tau}$	joint torques

Miscellaneous symbols

$\mathrm{I}_{(x,y)}$	pixel intensity of an image I at pixel coordinates (x, y)
L	number of planar arm links
$\mathbf{m}$	motor command (equivalent to $\mathbf{u}$)
$\mathbf{m}_{\mathrm{FM}}$	motor command as input for visual FM (Chapt. 6)
$\mathbf{m}_{\mathrm{SAC}}$	motor output of saccade controller (Chapt. 6)
N	number of steps in a movement sequence (Chapt. 7)
$\mathbf{p}_{ijkl}$	single point in the grid $\mathbf{P}$ with grid indices i,j,k, and l (Chapt. 5)
$\mathbf{P}$	grid of cumulator units (Chapt. 5)
$\mathbf{s}$	sensory state (equivalent to plant output $\mathbf{y}$) or general vector drawn from arbitrary data space (Sect. 5.2)
$\widehat{\mathbf{s}}$	estimated/predicted sensory state
$\mathbf{s}^*$	desired sensory state
$\mathbf{s}_{\mathrm{KIN}}$	kinesthetic part of $\mathbf{s}_{\mathrm{SAC}}$ (Chapt. 6) or kinesthetic state (Chapt. 7)
$\mathbf{s}_{\mathrm{OR}}$	part of the visual sensory state encoding the block orientation (Chapt. 7)
$\mathbf{s}_{\mathrm{POS}}$	part of the visual sensory state encoding the block position (Chapt. 7)
$\mathbf{s}_{\mathrm{SAC}}$	sensory input for saccade controller (Chapt. 6)
$\mathbf{s}_{\mathrm{VIS}}$	visual part of $\mathbf{s}_{\mathrm{SAC}}$ (Chapt. 6)
$v_{(x_{\mathrm{Out}},y_{\mathrm{Out}})}$	output of the validator model (Chapt. 5)
$\mathbf{w} = (w_x, w_y, w_z)^T$	world coordinates
(x, y)	pixel coordinates in an image
$x_{\mathrm{left}}, y_{\mathrm{left}}, x_{\mathrm{right}}, y_{\mathrm{right}}$	coordinates in left/right camera image
(x, z)	gripper position in the working area (Chapt. 7)
α	gripper orientation (Chapt. 7)
$\mathbf{\Sigma}$	covariance matrix

Bibliography

Abend, W., Bizzi, E., and Morasso, P. Human arm trajectory formation. *Brain*, 105(2):331–348, 1982.

Abrams, R. A., Meyer, D. E., and Kornblum, S. Eye hand coordination — oculomotor control in rapid aimed limb movements. *Journal of Experimental Psychology-Human Perception and Performance*, 16(2):248–267, 1990.

Almassy, N., Edelman, G. M., and Sporns, O. Behavioral constraints in the development of neuronal properties: A cortical model embedded in a real-world device. *Cerebral Cortex*, 8(4):346–361, 1998.

Atchinson, D. A. and Smith, G. *Optics of the human eye*. Butterworth-Heinemann, Oxford, UK, 2000.

Ax, P. *The Phylogenetic System. The Systematization of Organisms on the Basis of their Phylogenesis*. John Wiley & Sons, Chichester, 1987.

Babu, B. V. and Sastry, K. K. N. Estimation of heat transfer parameters in a trickle-bed reactor using differential evolution and orthogonal collocation. *Computers & Chemical Engineering*, 23(3):327–339, 1999.

Bar, M. The proactive brain: Using analogies and associations to generate predictions. *Trends in Cognitive Sciences*, 11(7):280–289, 2007.

Bard, K. A. Neonatal imitation in chimpanzees (Pan troglodytes) tested with two paradigms. *Animal Cognition*, 10(2):233–242, 2007.

Barker, A. L., Brown, D. E., and Martin, W. N. Bayesian estimation and the Kalman filter. Technical Report IPC-TR-94-002, Institute for Parallel Computation, School of Engineering and Applied Science, University of Virginia, USA, 1994.

Battaglia-Mayer, A., Caminiti, R., Lacquaniti, F., and Zago, M. Multiple levels of representation of reaching in the parieto-frontal network. *Cerebral Cortex*, 13(10):1009–1022, 2003.

Baud-Bovy, G. and Viviani, P. Pointing to kinesthetic targets in space. *Journal of Neuroscience*, 18(4):1528–1545, 1998.

Bays, P. M. and Wolpert, D. M. Computational principles of sensorimotor control that minimize uncertainty and variability. *Journal of Physiology-London*, 578(2):387–396, 2007.

Beauchamp, M. S., Petit, L., Ellmore, T. M., Ingeholm, J., and Haxby, J. V. A parametric fMRI study of overt and covert shifts of visuospatial attention. *Neuroimage*, 14(2):310–321, 2001.

Bekkering, H. and Sailer, U. Commentary: Coordination of eye and hand in time and space. *Progress in Brain Research*, 140:365–373, 2002.

Bell, C. C. Memory-based expectations in electrosensory systems. *Current Opinion in Neurobiology*, 11(4):481–487, 2001.

Berthoz, A. *The Brain's Sense of Movement*. Harvard University Press, Cambridge, MA, 2000.

Billard, A. and Schaal, S. Special issue on the brain mechanisms of imitation learning — introduction. *Neural Networks*, 19(3):251–253, 2006.

Blakemore, S. J., Frith, C. D., and Wolpert, D. M. Spatio-temporal prediction modulates the perception of self-produced stimuli. *Journal of Cognitive Neuroscience*, 11(5):551–559, 1999.

Blakemore, S. J., Wolpert, D., and Frith, C. Why can't you tickle yourself? *NeuroReport*, 11(11):R11–R16, August 2000.

Bock, O. Contribution of retinal versus extraretinal signals towards visual localization in goal-directed movements. *Experimental Brain Research*, 64(3): 476–482, 1986.

Bock, O. Localization of objects in the peripheral visual field. *Behavioural Brain Research*, 56(1):77–84, 1993.

Bockisch, C. J. and Miller, J. M. Different motor systems use similar damped extraretinal eye position information. *Vision Research*, 39(5):1025–1038, 1999.

Bongard, J., Zykov, V., and Lipson, H. Resilient machines through continuous self-modeling. *Science*, 314(5802):1118–1121, 2006.

Bortz, J. *Statistik für Sozialwissenschaftler*. Springer-Verlag, Berlin, Heidelberg, New York, fourth edition, 1993.

Bridgeman, B. Failure to detect displacement of the visual world during saccadic eye movements. *Vision Research*, 15(1):719–722, 1975.

Bridgeman, B. A review of the role of efference copy in sensory and oculomotor control systems. *Annals of Biomedical Engineering*, 23(4):409–422, 1995.

Brooks, R. A. A robust layered control system for a mobile robot. *IEEE Journal of Robotics and Automation*, RA-2(1):14–23, 1986.

Brooks, R. A. Intelligence without reason. In *Proceedings of the Twelfth International Joint Conference on Artificial Intelligence (IJCAI-91)*, pages 569–595, San Mateo, CA, 1991a. Morgan Kauffmann.

Brooks, R. A. Intelligence without representation. *Artificial Intelligence Journal*, 47(1-3):139–159, 1991b.

Bruske, J., Hansen, M., Riehn, L., and Sommer, G. Biologically inspired calibration-free adaptive saccade control of a binocular camera-head. *Biological Cybernetics*, 77(6):433–446, 1997.

Buneo, C. A. and Andersen, R. A. The posterior parietal cortex: Sensorimotor interface for the planning and online control of visually guided movements. *Neuropsychologia*, 44(13):2594–2606, 2006.

Buneo, C. A., Jarvis, M. R., Batista, A. P., and Andersen, R. A. Direct visuomotor transformations for reaching. *Nature*, 416(6881):632–636, 2002.

Byrne, R. W. and Russon, A. E. Learning by imitation: A hierarchical approach. *Behavioral and Brain Sciences*, 21(5):667–684, 1998.

Carrozzo, M., McIntyre, J., Zago, M., and Lacquaniti, F. Viewer-centered and body-centered frames of reference in direct visuomotor transformations. *Experimental Brain Research*, 129(2):201–210, 1999.

Clark, A. *Being there. Putting brain, body, and world together again.* MIT Press, Cambridge, MA, 1997.

Cruse, H. The evolution of cognition — a hypothesis. *Cognitive Science*, 27 (1):135–155, 2003.

Cruse, H. and Brüwer, M. The human arm as a redundant manipulator — the control of path and joint angles. *Biological Cybernetics*, 57(1-2):137–144, 1987.

Cruse, H., Brüwer, M., and Dean, J. Control of 3-joint and 4-joint arm movement — strategies for a manipulator with redundant degrees of freedom. *Journal of Motor Behavior*, 25(3):131–139, 1993.

Cruse, H. and Steinkühler, U. Solution of the direct and inverse kinematic problems by a common algorithm based on the mean of multiple computations. *Biological Cybernetics*, 69(4):345–351, 1993.

Cruse, H., Wischmeyer, E., Brüwer, M., Brockfeld, P., and Dress, A. On the cost-functions for the control of the human arm movement. *Biological Cybernetics*, 62(6):519–528, 1990.

Dagher, A., Owen, A. M., Boecker, H., and Brooks, D. J. Mapping the network for planning: A correlational pet activation study with the tower of london task. *Brain*, 122(10):1973–1987, 1999.

Dassonville, P., Schlag, J., and Schlagrey, M. Oculomotor localization relies on a damped representation of saccadic eye displacement in human and nonhuman-primates. *Visual Neuroscience*, 9(3-4):261–269, 1992.

Davidson, P. R. and Wolpert, D. M. Internal models underlying grasp can be additively combined. *Experimental Brain Research*, 155(3):334–340, 2004.

Dean, P., Mayhew, J. E. W., and Langdon, P. Learning and maintaining saccadic accuracy: A model of brainstem-cerebellar interactions. *Journal of Cognitive Neuroscience*, 6(2):117–138, 1994.

Dean, P., Mayhew, J. E. W., Thacker, N., and Langdon, P. M. Saccade control in a simulated robot camera-head system — neural net architectures for efficient learning of inverse kinematics. *Biological Cybernetics*, 66(1):27–36, 1991.

Dean, P. and Porrill, J. Pseudo-inverse control in biological systems: A learning mechanism for fixation stability. *Neural Networks*, 11(7-8):1205–1218, 1998.

Desmurget, M., Prablanc, C., Rossetti, Y., Arzi, M., Paulignan, Y., Urquizar, C., and Mignot, J. C. Postural and synergic control for 3-dimensional movements of reaching and grasping. *Journal of Neurophysiology*, 74(2):905–910, 1995.

Deubel, H. Localization of targets across saccades: Role of landmark objects. *Visual Cognition*, 11(2-3):173–202, 2004.

Deubel, H. and Schneider, W. X. Saccade target selection and object recognition: Evidence for a common attentional mechanism. *Vision Research*, 36 (12):1827–1837, 1996.

Deubel, H., Schneider, W. X., and Bridgeman, B. Postsaccadic target blanking prevents saccadic suppression of image displacement. *Vision Research*, 36 (7):985–996, 1996.

Deubel, H., Schneider, W. X., and Paprotta, I. Selective dorsal and ventral processing: Evidence for a common attentional mechanism in reaching and perception. *Visual Cognition*, 5(1-2):81–107, 1998.

Dorf, R. C. and Bishop, R. H. *Modern Control Systems*. Prentice Hall, tenth edition, 2004.

Doya, K. Complementary roles of basal ganglia and cerebellum in learning and motor control. *Current Opinion in Neurobiology*, 10(6):732–739, 2000a.

Doya, K. Reinforcement learning in continuous time and space. *Neural Computation*, 12(1):219–245, 2000b.

Doya, K., Kimura, H., and Miyamura, A. Motor control: Neural models and systems theory. *International Journal of Applied Mathematics and Computer Science*, 11(1):77–104, 2001.

Duhamel, J. R., Colby, C. L., and Goldberg, M. E. The updating of the representation of visual space in parietal cortex by intended eye-movements. *Science*, 255(5040):90–92, 1992.

Eimer, M., van Velzen, J., Gherri, E., and Press, C. Manual response preparation and saccade programming are linked to attention shifts: ERP evidence for covert attentional orienting and spatially specific modulations of visual processing. *Brain Research*, 1105(1):7–19, 2006.

Fikes, R. and Nilsson, N. STRIPS: A new approach to the application of theorem proving to problem solving. *Artificial Intelligence*, 2(3-4):189–208, 1971.

Fitts, P. M. The information capacity of the human motor system in controlling the amplitude of movement. *Journal of Experimental Psychology*, 47(6): 381–391, 1954.

Flanagan, J. R., Vetter, P., Johansson, R. S., and Wolpert, D. M. Prediction precedes control in motor learning. *Current Biology*, 13(2):146–150, 2003.

Flash, T. The control of hand equilibrium trajectories in multi-joint arm movements. *Biological Cybernetics*, 57(4-5):257–274, 1987.

Flash, T. and Hogan, N. The coordination of arm movements: An experimentally confirmed mathematical model. *The Journal of Neuroscience*, 5 (7):1688–1703, 1985.

Frens, M. A. and Erkelens, C. J. Coordination of hand movements and saccades — evidence for a common and a separate pathway. *Experimental Brain Research*, 85(3):682–690, 1991.

Fritzke, B. *Vektorbasierte Neuronale Netze*. Shaker Verlag, Aachen, 1998.

Gallese, V. and Goldman, A. Mirror neurons and the simulation theory of mind-reading. *Trends in Cognitive Sciences*, 2(12):493–501, 1998.

Gancarz, G. and Grossberg, S. A neural model of saccadic eye movement control explains task-specific adaptation. *Vision Research*, 39(18):3123–3143, 1999.

Gerdes, V. G. J. and Happee, R. The use of an internal representation in fast goal-directed movements: a modelling approach. *Biological Cybernetics*, 70 (6):513–524, 1994.

Gersho, A. and Gray, R. M. *Vector Quantization and Signal Compression*. Springer, Berlin, 1991.

Gerstung, O. Implementierung und Anwendung von Locally Weighted Projection Regression, 2006. Diploma Thesis. Computer Engineering Group, Faculty of Technology, Bielefeld University.

Gibson, J. J. *The Ecological Approach to Visual Perception*. Houghton Mifflin Company, Boston, 1979.

Godwin, L. E. Differential evolution solver class, 1998. http://www.icsi.berkeley.edu/~storn/code.html.

Golub, G. and van Loan, C. *Matrix Computations*. Johns Hopkins University Press, Baltimore, 1996.

Gomi, H. and Kawato, M. Neural network control for a closed-loop system using feedback-error-learning. *Neural Networks*, 6(7):933–946, 1993.

Gomi, H. and Kawato, M. Equilibrium-point control hypothesis examined by measured arm-stiffness during multi-joint movement. *Science*, 272(5258): 117–120, 1996.

Gonzalez, R. C. and Woods, R. E. *Digital Image Processing*. Addison-Wesley Publishing Company, Reading, MA, 1992.

Goodale, M. A. and Milner, A. D. Separate visual pathways for perception and action. *Trends in Neurosciences*, 15(1):20–25, 1992.

Grasso, F. W. Environmental information, animal behavior, and biorobot design. Reflections on locating chemical sources in marine environments. In Webb, B. and Consi, T. R., editors, *Biorobotics*, pages 21–35. MIT Press, Cambridge, MA, 2001.

Graziano, M. S., Taylor, C. S., and Moore, T. Complex movements evoked by microstimulation of precentral cortex. *Neuron*, 34(5):841–851, May 2002.

Grea, H., Desmurget, M., and Prablanc, C. Postural invariance in three-dimensional reaching and grasping movements. *Experimental Brain Research*, 134(2):155–162, 2000.

Greeno, J. G. and Moore, J. L. Situativity and symbols — response to Vera and Simon. *Cognitive Science*, 17(1):49–59, 1993.

Grezes, J. and Decety, J. Does visual perception of object afford action? evidence from a neuroimaging study. *Neuropsychologia*, 40(2):212–222, 2002.

Gross, H.-M., Heinze, A., Seiler, T., and Stephan, V. Generative character of perception: A neural architecture for sensorimotor anticipation. *Neural Networks*, 12(7-8):1101–1129, 1999.

Große, S. Visuelle Vorwärtsmodelle für einen Roboter-Kamera-Kopf, 2005. Diploma Thesis. Computer Engineering Group, Faculty of Technology, Bielefeld University.

Harris, C. M. and Wolpert, D. M. Signal-dependent noise determines motor planning. *Nature*, 394(6695):780–784, August 1998.

Harris, C. M. and Wolpert, D. M. The main sequence of saccades optimizes speed-accuracy trade-off. *Biological Cybernetics*, 95(1):21–29, 2006.

Haruno, M., Wolpert, D. M., and Kawato, M. Multiple paired forward-inverse models for human motor learning and control. In Kearns, M., Solla, S., and

Cohn, D., editors, *Advances in neural information processing systems*, volume 11, pages 31–37, Cambridge, MA, 1999. MIT Press.

Haruno, M., Wolpert, D. M., and Kawato, M. MOSAIC model for sensorimotor learning and control. *Neural Computation*, 13(10):2201–2220, 2001.

Haykin, S. S. *Adaptive filter theory*. Prentice Hall, Upper Saddle River, NJ, fourth edition, 2002.

Held, R. and Freedman, S. J. Plasticity in human sensorimotor control. *Science*, 142(3591):455–462, October 1963.

Held, R. and Hein, A. Movement-produced stimulation in the development of visually guided behaviour. *Journal of Comparative and Physiological Psychology*, 56(5):872–876, 1963.

Hesslow, G. Conscious thought as simulation of behaviour and perception. *Trends in Cognitive Sciences*, 6(6):242–247, 2002.

Hoffmann, H. *Unsupervised Learning of Visuomotor Associations*. MPI Series in Biological Cybernetics. Logos Verlag, Berlin, 2004.

Hoffmann, H. Perception through visuomotor anticipation in a mobile robot. *Neural Networks*, 20(1):22–33, 2007.

Hoffmann, H. and Möller, R. Unsupervised learning of a kinematic arm model. In Kaynak, O., Alpaydin, E., Oja, E., and Xu, L., editors, *Artificial Neural Networks and Neural Information Processing — ICANN/ICONIP 2003, LNCS*, volume 2714, pages 463–470. Springer, Berlin, 2003.

Hoffmann, H. and Möller, R. Action selection and mental transformation based on a chain of forward models. In Schaal, S., Ijspeert, A., Billard, A., Vijayakumar, S., Hallam, J., and Meyer, J.-A., editors, *From Animals to Animats 8, Proceedings of the Eighth International Conference on the Simulation of Adaptive Behavior*, pages 213–222, Los Angeles, CA, 2004. MIT Press.

Hoffmann, H., Schenck, W., and Möller, R. Learning visuomotor transformations for gaze-control and grasping. *bc*, 93(2):119–130, 2005.

Hommel, B. Inverting the Simon effect by intention. *Psychological Research*, 55(4):270–279, 1993.

Hommel, B., Müsseler, J., Aschersleben, G., and Prinz, W. The theory of event coding: A framework for perception and action planning. *Behavioral and Brain Sciences*, 24(5):849–937, 2001.

Hopfield, J. J. Neural networks and physical systems with emergent collective computational abilities. *Proceedings of the National Academy of Sciences of the USA*, 79(8):2554–2558, 1982.

Hopfield, J. J. Neurons with graded response have collective computational properties like those of two-state neurons. *Proceedings of the National Academy of Sciences of the USA*, 81(10):3088–3092, 1984.

Hornik, K., Stinchcombe, M., and White, H. Multilayer feedforward networks are universal approximators. *Neural Networks*, 2(5):359–366, 1989.

Horstmann, A. and Hoffmann, K. P. Target selection in eye-hand coordination: Do we reach to where we look or do we look to where we reach? *Experimental Brain Research*, 167(2):187–195, 2005.

Hubel, D. H. and Wiesel, T. N. Receptive fields, binocular interaction and functional architecture in the cat's visual cortex. *Journal of Physiology*, 160 (1):106–154, 1962.

Iacoboni, M. Neural mechanisms of imitation. *Current Opinion in Neurobiology*, 15(6):632–637, 2005.

Ingber, L. Simulated annealing: Practice versus theory. *Mathematical Computer Modelling*, 18(11):29–57, 1993.

Irwin, D. E. and Gordon, R. D. Eye movements, attention and trans-saccadic memory. *Visual Cognition*, 5(1-2):127–155, 1998.

Ito, M. *The Cerebellum and Neural Control*. Raven Press, New York, 1984.

Jackson, P. L., Lafleur, A. F., Malouin, F., Richards, C., and Doyon, J. Potential role of mental practice using motor imagery in neurologic rehabilitation. *Archives of Physical Medicine and Rehabilitation*, 82(8):1133–1141, 2001.

Jacob, P. and Jeannerod, M. The motor theory of social cognition: A critique. *Trends in Cognitive Sciences*, 9(1):21–25, 2005.

Jain, A. K. *Fundamentals of Digital Image Processing*. Prentice-Hall, New Jersey, 1989.

Jeannerod, M. Mental-imagery in the motor context. *Neuropsychologia*, 33 (11):1419–1432, 1995.

Jeannerod, M. *The Cognitive Neuroscience of Action*. Blackwell, Oxford, UK, 1997.

Jeannerod, M. Neural simulation of action: A unifying mechanism for motor cognition. *NeuroImage*, 14(1):S103–S109, 2001.

Jordan, M. I. Computational aspects of motor control and motor learning. In Heuer, H. and Keele, S. W., editors, *Handbook of Perception and Action, Volume 2: Motor Skills*, pages 87–146. Academic Press, New York, 1996.

Jordan, M. I. and Rumelhart, D. E. Forward models: Supervised learning with a distal teacher. *Cognitive Science*, 16(3):307–354, 1992.

Kalman, R. E. A new approach to linear filtering and prediction problems. *Transactions of the ASME — Journal of Basic Engineering*, 82:35–45, March 1960.

Kawato, M. Feedback-error-learning neural network for supervised motor learning. In Eckmiller, R., editor, *Advanced Neural Computers, Elsevier, North-Holland*, pages 365–372, 1990.

Kawato, M. Internal models for motor control and trajectory planning. *Current Opinion in Neurobiology*, 9(6):718–727, 1999.

Kawato, M., Furukawa, K., and Suzuki, R. A hierarchical neural-network model for control and learning of voluntary movement. *Biological Cybernetics*, 57 (3):169–185, 1987.

Kawato, M. and Gomi, H. The cerebellum and VOR/OKR learning models. *Trends in Neurosciences*, 15(11):445–453, 1992a.

Kawato, M. and Gomi, H. A computational model of four regions of the cerebellum based on feedback-error learning. *Biological Cybernetics*, 68(2):95–103, 1992b.

Keeler, J. D. A dynamical system view of cerebellar function. *Physica D*, 42 (1-3):396–410, 1990.

Klarquist, W. N. and Bovik, A. C. Fovea: A foveated vergent active stereo vision system for dynamic three-dimensional scene recovery. *IEEE Transactions on Robotics and Automation*, 14(5):755–770, 1998.

Knuf, L., Aschersleben, G., and Prinz, W. An analysis of ideomotor action. *Journal of Experimental Psychology: General*, 130(4):779–798, 2001.

Kohonen, T. *Self-Organizing Maps*. Springer, Berlin, 1995.

Kording, K. P. and Wolpert, D. M. Bayesian decision theory in sensorimotor control. *Trends in Cognitive Sciences*, 10(7):319–326, 2006.

Kosslyn, S. M. *Image and Brain*. MIT Press, Cambridge, MA, 1994.

Kosslyn, S. M., Alpert, N. M., Thompson, W. L., Maljkovic, V., Weise, S. B., Chabris, C. F., Hamilton, S. E., Rauch, S. L., and Buonanno, F. S. Visual mental-imagery activates topographically organized visual-cortex — PET investigations. *Journal of Cognitive Neuroscience*, 5(3):263–287, 1993.

Krichmar, J. L. and Edelman, G. M. Machine psychology: Autonomous behavior, perceptual categorization and conditioning in a brain-based device. *Cerebral Cortex*, 12(8):818–830, 2002.

Kröse, B. J. A., Korst, M. J., and Groen, F. C. A. Learning strategies for a vision based neural controller for a robot arm. In Kaynak, O., editor, *Proceedings of the IEEE International Workshop on Intelligent Motion Control*, pages 199–203, Piscataway, NJ, 1990. IEEE.

Kuperstein, M. Adaptive visual-motor coordination in multijoint robots using parallel architecture. In *Proceedings of the IEEE International Conference on Robotics and Automation*, pages 1595–1602, Raleigh, NC, 1987. IEEE.

Kuperstein, M. Neural model of adaptive hand-eye coordination for single postures. *Science*, 239(4845):1308–1311, 1988.

Kuperstein, M. INFANT neural controller for adaptive sensory-motor coordination. *Neural Networks*, 4(2):131–145, 1990.

Leigh, R. J. and Zee, D. S. *The Neurology of Eye Movements*. Oxford University Press, UK, 1999.

Lloyd, S. P. Least squares quantization in PCM. *IEEE Transactions on Information Theory*, 28(2):129–137, 1982.

Luria, S. M. and Kinney, J. A. S. Underwater vision. *Science*, 167(3924): 1454–1461, 1970.

Mallot, H. A. An overall description of retinotopic mapping in the cats visual-cortex area-17, area-18, and area-19. *Biological Cybernetics*, 52(1):45–51, 1985.

Marr, D. *Vision: A Computational Approach*. Freeman & Co, San Francisco, 1982.

Martin, K. A., Moritz, S. E., and Hall, C. R. Imagery use in sport: A literature review and applied model. *Sport Psychologist*, 13(3):245–268, 1999.

Martinetz, T. M., Berkovich, S. G., and Schulten, K. J. "Neural-Gas" network for vector quantization and its application to time-series prediction. *IEEE Transactions on Neural Networks*, 4(4):558–569, July 1993.

Mather, J. A. and Fisk, J. D. Orienting to targets by looking and pointing — parallels and interactions in ocular and manual performance. *Quarterly Journal of Experimental Psychology Section A — Human Experimental Psychology*, 37(3):315–338, 1985.

Matin, L., Matin, E., and Pola, J. Visual perception of direction when voluntary saccades occur. 2. Relation of visual direction of a fixation target extinguished before a saccade to a subsequent test flash presented before saccade. *Perception & Psychophysics*, 8(1):9–14, 1970.

McCarthy, J. and Hayes, P. J. Some philosophical problems from the standpoint of artificial intelligence. In Allen, J., Hendler, J., and Tate, A., editors, *Readings in Planning*, pages 393–435. Kaufmann, San Mateo, CA, 1990.

McIlwain, J. T. *An introduction to the biology of vision*. Cambridge University Press, Cambridge, UK, 1996.

Mehta, B. and Schaal, S. Forward models in visuomotor control. *Journal of Neurophysiology*, 88(2):942–953, 2002.

Mel, B. W. MURPHY: A robot that learns by doing. Technical Report CCSR-88-4, Center for Complex Systems Research, University of Illinois, Urbana-Champaign, USA, 1988.

Melcher, D. Predictive remapping of visual features precedes saccadic eye movements. *Nature Neuroscience*, 10(7):903–907, 2007.

Miall, R. C. Connecting mirror neurons and forward models. *Neuroreport*, 14 (17):2135–2137, 2003.

Miall, R. C., Weir, D. J., Wolpert, D. M., and Stein, J. F. Is the cerebellum a Smith predictor? *Journal of Motor Behavior*, 25(3):203–216, 1993.

Miall, R. C. and Wolpert, D. M. The cerebellum as a predictive model of the motor system: A Smith predictor hypothesis. In Ferrell, W. R. and Proske, U., editors, *Neural Control of Movement*, pages 215–223. Plenum Press, New York, 1995.

Miall, R. C. and Wolpert, D. M. Forward models for physiological motor control. *Neural Networks*, 9(8):1265–1279, 1996.

Miller, W. T., Hewes, R. P., Glanz, F. H., and Kraft, L. G. Rcal-time dynamic control of an industrial manipulator using a neural-network-based learning controller. *IEEE transactions on robotics and automation*, 6(1):1–9, 1990.

Minsky, M. and Papert, S., editors. *Perceptrons*. MIT Press, Cambridge, 1969.

Miyamoto, H., Morimoto, J., Doya, K., and Kawato, M. Reinforcement learning with via-point representation. *Neural Networks*, 17(3):299–305, 2004.

Möller, R. *Wahrnehmung durch Vorhersage — Eine Konzeption der handlungsorientierten Wahrnehmung*. PhD thesis, Faculty of Computer Science and Automation, Ilmenau Technical University, Germany, 1996.

Möller, R. Perception through anticipation — a behavior-based approach to visual perception. In Riegler, A., Peschl, M., and von Stein, A., editors, *Understanding Representation in the Cognitive Sciences*, pages 169–176. Plenum Academic / Kluwer Publishers, New York, 1999.

Möller, R. Interlocking of learning and orthonormalization in RRLSA. *Neurocomputing*, 49(1-4):429–433, 2002.

Möller, R. and Hoffmann, H. An extension of neural gas to local PCA. *Neurocomputing*, 62:305–326, 2004.

Möller, R. and Schenck, W. Bootstrapping cognition from behavior — a computerized thought experiment. *Cognitive Science*, 32(3):504–542, 2008.

Moody, J. and Darken, C. J. Fast learning in networks of locally-tuned processing units. *Neural Computation*, 1(2):281–294, 1989.

Morasso, P. G. and Schieppati, M. Can muscle stiffness alone stabilize upright standing? *Journal of Neurophysiology*, 82(3):1622–1626, 1999.

Muggleton, N. G., Juan, C. H., Cowey, A., and Walsh, V. Human frontal eye fields and visual search. *Journal of Neurophysiology*, 89(6):3340–3343, 2003.

Murata, A., Fadiga, L., Fogassi, L., Gallese, V., Raos, V., and Rizzolatti, G. Object representation in the ventral premotor cortex (area F5) of the monkey. *Journal of Neurophysiology*, 78(4):2226–2230, 1997.

Nabney, I. T. *NETLAB. Algorithms for Pattern Recognition*. Springer, Berlin, Heidelberg, New York, 2002.

Nakanishi, J. and Schaal, S. Feedback error learning and nonlinear adaptive control. *Neural Networks*, 17(10):1453–1465, 2004.

Neggers, S. F. W. and Bekkering, H. Integration of visual and somatosensory target information in goal-directed eye and arm movements. *Experimental Brain Research*, 125(1):97–107, 1999.

Neggers, S. F. W. and Bekkering, H. Ocular gaze is anchored to the target of an ongoing pointing movement. *Journal of Neurophysiology*, 83(2):639–651, 2000.

Newell, A. and Simon, H. Computer science as empirical enquiry: Symbols and search. *Communications of the ACM*, 19(3):113–126, 1976.

Newell, A. and Simon, H. A. GPS: A program that simulates human thought. In Billing, H., editor, *Lernende Automaten*, pages 109–124. Oldenbourg, Munich, 1961.

Nobre, A. C., Gitelman, D. R., Dias, E. C., and Mesulam, M. M. Covert visual spatial orienting and saccades: Overlapping neural systems. *Neuroimage*, 11 (3):210–216, 2000.

Noë, A. *Action in Perception*. MIT Press, Cambridge, MA, 2005.

Norman, J. Two visual systems and two theories of perception: An attempt to reconcile the constructivist and ecological approaches. *Behavioral and Brain Sciences*, 25(1):73–144, 2002.

O'Regan, J. K. and Noë, A. A sensorimotor account of vision and visual consciousness. *Behavioral and Brain Sciences*, 24(5):939–1031, 2001.

Ouyang, S., Bao, Z., and Liao, G.-S. Robust recursive least squares learning algorithm for principal component analysis. *IEEE Transactions on Neural Networks*, 11(1):215–221, January 2000.

Pagel, M., Maël, E., and von der Malsburg, C. Self calibration of the fixation movement of a stereo camera head. *Machine Learning*, 31(1-3):169–186, 1998.

Perry, R. J. and Zeki, S. The neurology of saccades and covert shifts in spatial attention — an event-related fMRI study. *Brain*, 123(11):2273–2288, 2000.

Pfeifer, R. and Scheier, C. *Understanding Intelligence*. MIT Press, Cambridge, MA, 1999.

Piaget, J. *The Origins of Intelligence in Children*. International University Press, New York, 1952.

Prablanc, C., Echallier, J. F., Komilis, E., and Jeannerod, M. Optimal response of eye and hand motor systems in pointing at a visual target. 1. Spatio-temporal characteristics of eye and hand movements and their relationships when varying the amount of visual information. *Biological Cybernetics*, 35 (2):113–124, 1979.

Prinz, W. Perception and action planning. *European Journal of Cognitive Psychology*, 9(2):129–154, 1997.

Pylyshyn, Z. *Computation and Cognition*. MIT Press, Cambridge, MA, 1984.

Ritter, H. J. Parametrized self-organizing maps. In Gielen, S. and Kappen, B., editors, *Proceedings of the International Conference on Artificial Neural Networks*, pages 568–575. Springer, Berlin, 1993.

Rizzolatti, G., Camarda, R., Fogassi, L., Gentilucci, M., Luppino, G., and Matelli, M. Functional organization of inferior area 6 in the macaque monkey. *Experimental Brain Research*, 71(3):491–507, 1988.

Rizzolatti, G. and Fadiga, L. Grasping objects and grasping action meanings: The dual role of monkey rostroventral premotor cortex (area F5). *Novartis Foundation Symposium*, 218:81–103, 1998.

Rizzolatti, G., Fadiga, L., Gallese, V., and Fogassi, L. Premotor cortex and the recognition of motor actions. *Cognitive Brain Research*, 3(2):131–141, 1996.

Rizzolatti, G., Riggio, L., and Sheliga, B. M. Space and selective attention. In Umiltà, C. and Moscovitch, M., editors, *Attention and Performance VI: Conscious and Nonconscious Information Processing*, pages 231–265. MIT Press, Cambridge (MA), 1994.

Robinson, D. A. The use of control system analysis in the neurophysiology of eye movements. *Annual Review of Neuroscience*, 4:463–503, 1981.

Rosenbaum, D. A., Loukopoulos, L. D., Meulenbroek, R. G. J., Vaughan, J., and Engelbrecht, S. E. Planning reaches by evaluating stored postures. *Psychological Review*, 102(1):28–67, 1995.

Rosenbaum, D. A., Meulenbroek, R. J., Vaughan, J., and Jansen, C. Posture-based motion planning: Applications to grasping. *Psychological Review*, 108 (4):709–734, 2001.

Rossetti, Y., Rode, G., Pisella, L., Farné, A., Li, L., Boisson, D., and Perenin, M.-T. Prism adaptation to a rightward optical deviation rehabilitates left hemispatial neglect. *Nature*, 395(6698):166–169, September 1998.

Rumelhart, D. E., Hinton, G., and Williams, R. Learning internal representations by error propagation. In Rumelhart, D. E. and McClelland, J. L., editors, *Parallel distributed processing: Explorations in the microstructure of cognition. Vol. 1: Foundations*, pages 318–362. MIT Press, Cambridge, MA, 1986.

Sailer, U., Flanagan, J. R., and Johansson, R. S. Eye-hand coordination during learning of a novel visuomotor task. *Journal of Neuroscience*, 25(39):8833–8842, 2005.

Schaal, S., Peters, J., Nakanishi, J., and Ijspeert, A. Control, planning, learning, and imitation with dynamic movement primitives. In *Workshop on Bilateral Paradigms on Humans and Humanoids, IEEE International Conference on Intelligent Robots and Systems*, Las Vegas, NV, 2003.

Schaal, S. and Schweighofer, N. Computational motor control in humans and robots. *Current Opinion in Neurobiology*, 15(6):675–682, 2005.

Schenck, W., Hoffmann, H., and Möller, R. Learning internal models for eye-hand coordination in reaching and grasping. In *Proceedings of the European Cognitive Science Conference*, pages 289–294. L. Erlbaum Assoc., Mahwah (NJ), 2003.

Schenck, W., Hoffmann, H., and Möller, R. Grasping to extrafoveal targets: A robotic model. *New Ideas in Psychology*, to appear.

Schenck, W. and Möller, R. Staged learning of saccadic eye movements with a robot camera head. In Bowman, H. and Labiouse, C., editors, *Connectionist Models of Cognition and Perception II*, pages 82–91. World Scientific, New Jersey, London, 2004.

Schenck, W. and Möller, R. Learning strategies for saccade control. *Künstliche Intelligenz*, Iss. 3/06:19–22, 2006.

Schenck, W. and Möller, R. Training and application of a visual forward model for a robot camera head. In Butz, M. V., Sigaud, O., Pezzulo, G., and Baldassarre, G., editors, *Anticipatory Behavior in Adaptive Learning Systems: From Brains to Individual and Social Behavior*, number 4520 in Lecture Notes in Artificial Intelligence, pages 153–169. Springer, Berlin, Heidelberg, New York, 2007.

Schenck, W., Sinder, D., and Möller, R. Combining neural networks and optimization techniques for visuokinesthetic prediction and motor planning. In *ESANN'2008 proceedings — European Symposium on Artificial Neural Networks*, pages 523–528, Bruges (Belgium), 2008. d-side publications.

Schiegg, A., Deubel, H., and Schneider, W. X. Attentional selection during preparation of prehension movements. *Visual Cognition*, 10(4):409–431, 2003.

Schubotz, R. I. Prediction of external events with our motor system: Towards a new framework. *Trends in Cognitive Sciences*, 11(5):211–218, 2007.

Shadmehr, R. and Mussa-Ivaldi, F. Adaptive representation of dynamics during learning of a motor task. *Journal of Neuroscience*, 14(5):3208–3224, 1994.

Shadmehr, R. and Wise, S. P. *The Computational Neurobiology of Reaching and Pointing: A Foundation for Motor Learning*. MIT Press, Cambridge, MA, 2005.

Shimansky, Y. P., Kang, T., and He, J. P. A novel model of motor learning capable of developing an optimal movement control law online from scratch. *Biological Cybernetics*, 90(2):133–145, 2004.

Sinder, D. Roboterarm-Ansteuerung mit Hilfe von visuellen Vorwärtsmodellen, 2006. Diploma Thesis. Computer Engineering Group, Faculty of Technology, Bielefeld University.

Smith, O. J. M. A controller to overcome dead time. *ISA Journal*, 6:28–33, 1959.

Snyder, L. H. Coordinate transformations for eye and arm movements in the brain. *Current Opinion in Neurobiology*, 10(6):747–754, 2000.

Spong, M. W. and Vidyasagar, M. *Robot Dynamics and Control*. John Wiley & Sons, New York, 1989.

Steinkühler, U. and Cruse, H. A holistic model for an internal representation to control the movement of a manipulator with redundant degrees of freedom. *Biological Cybernetics*, 79(6):457–466, 1998.

Storn, R. Designing digital filters with differential evolution. In Corne, D., Dorigo, M., and Glover, F., editors, *New Ideas in Optimisation*, pages 109–126. McGraw-Hill, Maidenhead, UK, 1999.

Storn, R. and Price, K. Differential evolution — a simple and efficient heuristic for global optimization over continuous spaces. *Journal of Global Optimization*, 11(4):341–359, 1997.

Suchman, L. *Plans and Situated Actions. The Problem of Human-machine Communication.* Cambridge University Press, Cambridge, UK, 1987.

Sutton, R. S. and Barto, A. G. *Reinforcement Learning*. MIT Press, Cambridge, MA, 1998.

Szu, H. and Hartley, R. Fast simulated annealing. *Physics Letters A*, 122(3-4): 157–162, 1987.

Tani, J. Model-based learning for mobile robot navigation from the dynamical systems perspective. *IEEE Transactions on Systems, Man, and Cybernetics — Part B*, 26(3):421–436, 1996.

Tao, G. *Adaptive control design and analysis.* John Wiley & Sons, Hoboken, NJ, 2003.

Taylor, J. G. Attention as the control system of the brain. *International Journal of General Systems*, 35(3):361–376, 2006.

Tomasello, M., Savagerumbaugh, S., and Kruger, A. C. Imitative learning of actions on objects by children, chimpanzees, and enculturated chimpanzees. *Child Development*, 64(6):1688–1705, 1993.

Umeno, M. M. and Goldberg, M. E. Spatial processing in the monkey frontal eye field. 1. Predictive visual responses. *Journal of Neurophysiology*, 78(3): 1373–1383, 1997.

Ungerleider, L. G. and Mishkin, M. Two cortical visual systems. In Ingle, D. J., Goodale, M. A., and Mansfield, R. J. W., editors, *Analysis of Visual Behavior*, pages 549–586. MIT Press, Cambridge, MA, 1982.

Uno, Y., Kawato, M., and Suzuki, R. Formation and control of optimal trajectory in human multijoint arm movements. *Biological Cybernetics*, 61(2): 89–101, 1989.

van der Smagt, P. *Visual robot arm guidance using neural networks*. PhD thesis, Faculty of Mathematics and Computer Science, Amsterdam University, Netherlands, 1995.

Varela, F. J., Thompson, E., and Rosch, E. *The Embodied Mind. Cognitive Science and Human Experience*. MIT Press, Cambridge, MA, 1991.

Vaziri, S., Diedrichsen, J., and Shadmehr, R. Why does the brain predict sensory consequences of oculomotor commands? Optimal integration of the predicted and the actual sensory feedback. *Journal of Neuroscience*, 26(16): 4188–4197, 2006.

Vercher, J. L., Magenes, G., Prablanc, C., and Gauthier, G. M. Eye-head-hand coordination in pointing at visual targets - spatial and temporal analysis. *Experimental Brain Research*, 99(3):507–523, 1994.

Vijayakumar, S., D'Souza, A., and Schaal, S. Incremental online learning in high dimensions. *Neural Computation*, 17(12):2602–2634, 2005.

Vijayakumar, S., D'souza, A., Shibata, T., Conradt, J., and Schaal, S. Statistical learning for humanoid robots. *Autonomous Robots*, 12(1):55–69, 2002.

Vijayakumar, S. and Schaal, S. Locally weighted projection regression: An $o(n)$ algorithm for incremental real time learning in high dimensional space. In *Proc. of Seventeenth International Conference on Machine Learning*, pages 1079–1086, 2000.

von Holst, E. and Mittelstaedt, H. Das Reafferenzprinzip. *Die Naturwissenschaften*, 37(20):464–476, 1950.

Wada, Y., Kawabata, Y., Kotosaka, S., Yamamoto, K., Kitazawa, S., and Kawato, M. Acquisition and contextual switching of multiple internal models for different viscous force fields. *Neuroscience Research*, 46(3):319–331, 2003.

Wada, Y. and Kawato, M. A neural network model for arm trajectory formation using forward and inverse dynamics models. *Neural Networks*, 6(7):919–932, 1993.

Wada, Y. and Kawato, M. A via-point time optimization algorithm for complex sequential trajectory formation. *Neural Networks*, 17(3):353–364, 2004.

Walker, M. F., Fitzgibbon, E. J., and Goldberg, M. E. Neurons in the monkey superior colliculus predict the visual result of impending saccadic eye-movements. *Journal of Neurophysiology*, 73(5):1988–2003, 1995.

Walter, J. A., Nölker, C., and Ritter, H. The PSOM algorithm and applications. In *Proceedings of the Symposium on Neural Computation*, pages 758–764, 2000.

Webb, B. What does robotics offer animal behaviour. *Animal behaviour*, 60(5): 545–558, 2000.

Weir, C. R. Proprioception in extraocular muscles. *Journal of Neuro-Ophthalmology*, 26(2):123–127, 2006.

Werbos, P. *Beyond regression: New tools for prediction and analysis in the behavioral sciences*. PhD thesis, Harvard University, Cambridge, MA, 1974.

Widrow, G. and Hoff, M. E. Adaptive switching circuits. In *Institute of Radio Engineers, Western Electronic Show and Convention, Convention Record, part 4*, pages 96–104, 1960.

Wilson, M. Six views of embodied cognition. *Psychonomic Bulletin and Review*, 9(4):625–636, 2002.

Wold, H. Partial least squares. In Kotz, S. and Johnson, N. L., editors, *Encyclopedia of Statistical Sciences*, pages 581–591. Wiley, New York, 1985.

Wolpert, D. M., Doya, K., and Kawato, M. A unifying computational framework for motor control and social interaction. *Philosophical Transactions of the Royal Society of London. Series B*, 358(1431):593–602, 2003.

Wolpert, D. M. and Flanagan, J. R. Sensorimotor learning. In Arbib, M., editor, *The Handbook of Brain Theory and Neural Networks*, pages 1020–1023. MIT Press, Cambridge, MA, second edition, 2003.

Wolpert, D. M., Ghahramani, Z., and Jordan, M. I. Are arm trajectories planned in kinematic or dynamic coordinates? An adaptation study. *Experimental Brain Research*, 103(3):460–470, 1995a.

Wolpert, D. M., Ghahramani, Z., and Jordan, M. I. An internal model for sensorimotor integration. *Science*, 269(5232):1880–1882, 1995b.

Wolpert, D. M. and Kawato, M. Multiple paired forward and inverse models for motor control. *Neural Networks*, 11(7-8):1317–1329, 1998.

Wolpert, D. M., Miall, R. C., and Kawato, M. Internal models in the cerebellum. *Trends in Cognitive Sciences*, 2(9):338–347, 1998.

Yen, E. H. and Vaart, H. R. V. D. On measurable functions, continuous functions and some related concepts. *The American Mathematical Monthly*, 73(9):991–993, 1966.

Zell, A. *Simulation neuronaler Netze*. Oldenbourg Verlag, München, 1997.

Ziemke, T. Rethinking grounding. In Riegler, A., Peschl, M., and von Stein, A., editors, *Understanding Representation in the Cognitive Sciences. Does Representation Need Reality?*, pages 177–190. Plenum Press, New York, 1999.

Ziemke, T., Jirenhed, D.-A., and Hesslow, G. Internal simulation of perception: A minimal neuro-robotic model. *Neurocomputing*, 68:85–104, 2005.

Zimmermann, M. Das somatoviszerale sensorische System. In Schmidt, R. F., Lang, F., and Thews, G., editors, *Physiologie des Menschen*, pages 295–316. Springer, Heidelberg, 29th edition, 2005.